Development and Plasticity in Sensory Thalamus and Cortex

Development and Plasticity in Sensory Thalamus and Cortex

Edited by

Reha Erzurumlu
LSU Health Sciences Center
New Orleans, LA, USA

William Guido
LSU Health Sciences Center
New Orleans, LA, USA

Zoltán Molnár
University of Oxford
Oxford, UK

Reha Erzurumlu
LSU Health Sciences Center
Department of Cell
Biology and Anatomy
New Orleans, LA

William Guido
LSU Health Sciences Center
Department of Cell
Biology and Anatomy
New Orleans, LA

Zoltan Molnar
University of Oxford
Department of Anatomy and
Genetics
Oxford, UK

Library of Congress Control Number: 2006920147

ISBN-10: 0-387-31798-8
ISBN-13: 978-0387-31798-4

Printed on acid-free paper.

Printed in Singapore (KYO)

9 8 7 6 5 4 3 2 1

springer.com

In Memory of Ary S. Ramoa, MD, MSc, Ph.D.

Dr. Ary S. Ramoa, neuroscientist, scholar, teacher, physician, husband, father, and friend to many, passed away on July 1, 2005 from complications of Crohn's disease. Ary was one of the speakers at the satellite symposium. We are all deeply saddened by his loss and this book is dedicated to his memory and contribution to the field.

Ary was born in Rio de Janeiro, Brazil on December 21, 1953. In 1979, he received his M.D. from the State University of Rio de Janeiro. Having already uncovered his passion for research, he continued his training in the laboratory of Dr. Carlos Eduardo Rocha-Miranda at the Federal University of Rio de Janiero. Here, he made some of his first observations of the structure and function of the visual system related to the specific contribution of each eye, a paradigm which later became a particularly successful tool in his own research laboratory. In 1981, Ary received his M.Sc. Degree in Biophysics and moved, with the help of an NIH Fogarty Fellowship, to the United States to earn his Ph.D. His doctoral research at the University of California at Berkeley with Dr. Ralph Freeman continued his examination of the visual system. This was a particularly productive experience for him, in which he not only examined response properties of visual thalamic and cortical neurons, but also was exposed to the tools of developmental manipulation of visual experience and pharmacological manipulation. After receiving his Ph.D. in Physiological Optics in 1986, he pursued postdoctoral training with Dr. Carla Shatz at Stanford University, Dr. Mriganka Sur at the Massachusetts Institute of Technology, and Drs. Pasko Rakic and David McCormick at Yale University. In these positions, he acquired the necessary skills to merge techniques of experiential manipulation, neuropharmacology and electrophysiology to assess the basic mechanisms of visual development. By the time he joined the faculty of the Department of Anatomy and Neurobiology at Virginia Commonwealth University School of Medicine in 1993, Ary had not only studied with some of the most influential visual and developmental neuroscientists in the world, but also had devised a sophisticated, novel and effective approach to his primary research interest.

Supported by multiple NIH and foundation grant awards, Ary's independent work began by addressing the effect of amblyopia on neural development and plasticity. Amblyopia is a neurological disorder characterized by reduced vision in one eye which afflicts millions of people worldwide. Using an animal model of amblyopia, he pioneered the *in vivo* use of viral-mediated gene transfer and antisense techniques in combination with monocular deprivation to discover critical factors involved in loss of visual function. These studies also led to the next logical step: the examination of the factors that regulate recovery of visual cortical function. This work, ongoing at the time of his final illness, showed that different mechanisms control the loss and the recovery phases of visual plasticity. Ary also recently developed another line of research: investigations into the relationship of visual cortical plasticity to Fetal Alcohol Syndrome, a leading cause of mental retardation. Using the animal model of Fetal Alcohol Syndrome which he and his collaborators had devised, early alcohol exposure was documented to reduce neural plasticity necessary for the normal development of visual cortical processing. Collectively, Ary's findings, which have won him national and international attention, have contributed significantly to the fundamental understanding of how sensory experience relates to loss and recovery of brain function. Ultimately, these efforts will provide the conceptual substrate upon which therapeutic strategies can be devised to reduce the impact of sensory deprivation or loss and to enhance the neuroplastic potential for rehabilitation in diverse neurological disorders.

With his devotion not only to research, but also to his friends, family and colleagues, Ary leaves behind a vast number of people whose lives were touched by his presence. He was an outstanding role model for his students and post-docs and provided excellent council, spirit and humor to his fellow researchers and colleagues. Consequently, he received the Virginia Commonwealth University Research Achievement Award in 2004 and in 2005. Ary will also be remembered for his intellect, enthusiasm, perseverance, smile, kind-heartedness, dedication to the advancement of science and love for his family and friends. He was respected worldwide, esteemed by those who knew him and loved dearly by those closest to him. He will truly be missed.

M. Alex Meredith
Virginia Commonwealth University
Richmond, Virginia

Contents

Preface

The idea of this book was conceived during a satellite meeting of the Society for Neuroscience entitled "Development and Plasticity in Sensory Thalamus and Cortex in 2003. The meeting was hosted by the LSU Health Sciences Center, New Orleans and organized by Reha Erzurumlu, William Guido and Zoltán Molnár. That occasion brought together systems and molecular approaches on various key issues on thalamocortical development and plasticity. The sheer volume and concentration of presentations specifically dedicated to these issues distinguished this satellite meeting from other cortical development conferences. The meeting was greeted with great enthusiasm and we all felt that more should follow. When the idea of the book was discussed the majority of the participants decided to contribute with a timely review on their specific field of expertise. Additional leading experts, who were not attending the original New Orleans Meeting, were also invited to contribute with relevant topics.

The 16 chapters each presents topics central to thalamocortical development and plasticity ranging from the establishment of thalamic sensory innervation to imaging cortical function in primates. Hevner and Zecevic start by introducing the concepts of forebrain patterning and axon guidance as well as discussing recent discoveries on the changes in intra- and extracortical circuits during development. Recent progress uncovered some of the mechanisms involved in thalamic axon deployment through the embryonic forebrain to the cortex and the subsequent intracortical guidance mechanisms. The largely transient cell population forming the subplate is central to this process. Understanding of various subplate cell types and their neurochemical properties is driven by excellent transgenic mouse model systems, some of which are discussed here.

Pratt and Price describe the mechanisms which allow thalamic axons to cross several diencephalic and telencaphalic boundaries as they approach the cortex. These compartments are patterned by overlapping gene expression gradients, and as our knowledge on forebrain patterning increases, so does the understanding of the thalamocortical guidance mechanisms. Using in vitro approaches and genetic manipulation it has been possible to dissect out the role that individual transcription factors have on thalamocortical axon guidance and the indirect affects on forebrain patterning.

Shimogori and Grove illustrate the power of transgenic mouse models and in vivo electroporation in the context of understanding subcortical and neocortical guidance mechanisms leading to area-specific thalamocortical innervation. These issues are closely related to the key questions in cortical arealization, as altering the expression of a single gene during early cortical development can not only change the fate of the cortex, but also change the destination of thalamocortical fibres. Surprisingly, the majority of thalamocortical rearrangement in such paradigms occurs at the cortical level. Specifically the early mis-expression of FGF-8 does not change the subcortical trajectories of the thalamocortical projections, but has a drastic effect on the intracortical targeting. This suggests that the pathfinding of the thalamic projections to the subplate relies on different signals, than the subsequent intracortical guidance.

The mechanism of this intracortical rearrangement is not clear, but it has been observed in other model systems (Chapter 11 Rebsam and Gaspar; Chapter 4 Molnár and colleagues). Chapter 4 examines the changes in thalamocortical relations in paradigms when the sensory periphery, the extent of the cortical sheet or the trajectories of the thalamocortical projections are altered. All of these result in large scale area and lamina-specific rearrangements of the thalamocortical projections within the cerebral cortex, which is in contrast to previous suggestions that thalamocortical projections are specified from the time they leave the lower tier of the cortical plate.

Yamamoto and colleagues review the molecular mechanisms involved in lamina-specific thalamocortical targeting. Although thalamic projections innervate every single cortical layer, their major target is layer 4. The challenge is to identify the dominant repulsive and attractive cues in the cortex at the time when cortical lamination and cellular differentiation are incomplete. The tools of modern molecular biology enabled spectacular progress in this field, providing numerous mutants, to test our current ideas on cortical development.

In Chapter 6 the role of Citron K and the effects of its deletion on cortical development are analyzed by Vercelli and colleagues. They show that the barrelfield can develop also in a microencephalic brain, and still responds to peripheral manipulations. Citron K plays a fundamental role in cytokinesis and dendritic development, and its deletion causes neuronal depletion resulting in reduced cortical mantle size which reflects in shrinkage of the barrelfield, and in altered distribution of cortical interneurons.

Juliano and colleagues examine the consequences of a reduction in or absence of layer 4 cells on thalamocortical development and cortical circuit formation. They provide further evidence of a layer 4-specific stop signal.

In chapters 9, 10 and 11 the functional aspects of early thalamocortical interactions are discussed. The first synaptic transmission between thalamic fibres and subplate cells start shortly after the thalamic projections arrive to the cortical subplate. Thalamic projections are able to elicit dynamic patterns of cortical activities which are believed to determine the area and lamina-specific thalamic ingrowth and influence on the cortical circuit formation and plasticity. Due to the pioneering work of Van der Loos and Woolsey, the barrel cortex of the primary somatosensory cortex became a very useful model system in understanding the causal relationships between the periphery related thalamocortical fibre patterning and the formation of the characteristic cytoarchitectonic features, the barrels. Kind, Erzurumlu, Gaspar and their colleagues review the progress emerging from various transgenic mouse lines used to understand the role of pre- and post-synaptic signaling machinery involvement in barrel formation. Disruption of the pre-synaptic release mechanisms prevents thalamocortical axons from forming the periphery related pattern, whereas disruption of the post-synaptic receptor mediated signal transduction mechanisms on cortical cells inhibits the formation of the characteristic cytoarchitectonic barrel pattern. These models increased our understanding of the molecular and cellular mechanisms of thalamocortical axon pathfinding and arbor formation, remodeling and plasticity. Unfortunately our current genetic models do not manipulate these signaling mechanisms selectively and specifically in the thalamocortical synapse; the synapses of the sensory organs, thalamus, and cortex are also affected. The future challenge will be to understand the contributions at each specific level.

The barrel field also provides an excellent system for understanding plasticity. In Chapter 15 Feldman and colleagues describe long term depression mechanisms of sensory map plasticity in the barrel cortex. Both the activity level and activity pattern are instrumental in shaping the thalamocortical network during development. Pallas and colleagues changed the pattern of activity delivered through thalamocortical pathways but not the thalamocortical connectivity itself, in their re-wired brain models. They demonstrated the existence of modality specific activity, which plays an important role in shaping cortical circuitry.

In Chapters 12 to 14 Guido, Hooks and Chen, Huberman and Chapman review recent developments affecting our understanding of the cellular and molecular interactions between retinal projections and the lateral geniculate nucleus of the thalamus. They also propose new rules for the establishment and modification of this pathway. Functional cortical circuits cannot be appreciated without thalamocortical circuits and cortical development cannot be considered without thalamocortical pathway development. But this in turn is influenced by the way that the connectivity between sensory organs and thalamus develops. To this end, the eye-specific termination pattern of retinal ganglion cells is examined in various paradigms. Here the retino-geniculate projection is always examined in the context of the entire sensory pathway, considering the functional implications of alterations in the system. Moore and colleagues further investigate the functional consequences in an fMRI study presented in Chapter 16.

This book is not intended to be exhaustive. We were, for obvious reasons, unable to cover all aspects of thalamocortical development. Since the corticofugal projections dominate the input to the thalamus, the recent progress in the illuminating the mechanisms of corticofugal development should have been reviewed. Transgenic models with reporter gene expression and in vivo electroporation have led to interesting developments in this particular field. Also, perhaps we would have liked to include human imaging studies with possible clinical implications (including the thalamocortical contribution to syndromes such as phantom limb, autism, schizophrenia and synesthesia). Nevertheless, we hope that the book reflects the tremendous enthusiasm and the atmosphere that surrounded the meeting at New Orleans. We are especially grateful to Dr. Nicolas Bazan, Director of the LSU Neuroscience Center. His generosity, hospitality and intellectual support and contributions to the symposium paved the way to this work.

We dedicate this book to Ary Ramoa, who passed away from complications of Crohns disease in July 2005. Ary was a contributor to the symposium. He was a dear friend and colleague of many of the participants and this book is a testament to the contributions he has made. Alex Meredith was kind enough to provide a few words honoring Ary's life and scientific career.

Reha Erzurumlu, William Guido, Zoltán Molnár
New Orleans and Oxford, January 2006,

Contributors

Robert F. Hevner and Nada Zecevic Department of Pathology, University of Washington, Seattle, Washington; and Department of Neuroscience, University of Connecticut Health Center, Farmington, CT, USA.

Thomas Pratt and David J. Price Genes and Development Group, University of Edinburgh, Hugh Robson Building, George Square, Edinburgh EH8 9XD, UK.

Tomomi Shimogori and Elizabeth A. Grove Critical Period Mechanism Research Group, Institute of Physical and Chemical Research (RIKEN), Brain Science Institute, Saitama 351-0198, Japan; and Department of Neurobiology, Pharmacology and Physiology, University of Chicago, Chicago, IL, 60637, USA.

Zoltán Molnár, Guillermina López-Bendito, Daniel Blakey, Alexander Thompson, and Shuji Higashi Department of Human Anatomy and Genetics, University of Oxford, South Park Road, Oxford OX1 3QX, UK, Institute de Neurosciencias, CSIC- Universidad Miguel Hernandez, Campus de San Juan, 03550 San Juan de Alicante, Spain. Division of Neurophysiology, Graduate School, Kyoto Prefectural University of Medicine, Kawaranachi Hirokoji, Kyoto 602-8566 Japan.

Nobuhiko Yamamoto, Makoto Takemoto, Yuki Hattori, and Kenji Hanamura Neuroscience Laboratories, Graduate School of Frontier Biosciences, Osaka University, Suita, Osaka 565-0871, Japan.

Patrizia Muzzi, Paola Camera, Ferdinando Di Cunto, and Alessandro Vercelli Department of Anatomy, Pharmacology and Forensic Medicine, University of Torino, corso M.D'Azeglio 52, 10126 Torino, Italy.

Debra F. McLaughlin, Sylvie Poluch, Beata Jablonska, and Sharon L. Juliano Department of Anatomy, Physiology, and Genetics, and Program in Neuroscience USUHS, 4301 Jones Bridge Rd., Bethesda MD 20814.

Sarah L. Pallas, Mei Xu, and Khaleel A. Razak Department of Biology, Graduate Program in Neurobiology and Behavior, Georgia State University, Atlanta, GA, 30303 USA.

Mark W Barnett, R. F. Watson, and Peter C. Kind Center for Integrative Physiology, University of Edinburgh, EH8 9XD.

Reha S. Erzurumlu and Takuji Iwasato Department of Cell Biology and Anatomy, LSUHSC, New Orleans, LA and PRESTO, Japan Science and Technology Agency and RIKEN brain Science Institute, Saitama, Japan.

Alexandra Rebsam and Patricia Gaspar INSERM U 616, Universite de Paris VI, Hopital Salpetriere, 47, Blvd de 1'Hopital, 75651 Paris Cedex 13.

William Guido Louisiana State Health Sciences Center Department of Cell Biology and Anatomy and the Neuroscience Center of Excellence, New Orleans, LA, 70112.

Bryan M. Hooks and Chinfei Chen Program in Neuroscience, Harvard Medical School, and Neurobiology Program, Children's Hospital, Boston, MA 02115.

Andrew D. Huberman and Barbara Chapman Center for Neuroscience, University of California, Davis, CA 95616, USA.

Kevin J. Bender, Suvarna Deshmukh, and Daniel E. Feldman Division of Biological Sciences, University of California San Diego, La Jolla, CA, 92093-0357.

Aimee J. Nelson, Ph.D, Cheryl A. Cheney, Yin-Ching Iris Chen, PhD, Guangping Dai, PhD, Robert P. Marini, DVM, Graham C. Grindlay, MSc, Yumiko Ishizawa, MD, PhD and Christopher I. Moore, Ph.D McGovern Institute for Brain Research, Massachusetts Institute of Technology, A.A. Martinos Center for Biomedical Imaging, Massachusetts General Hospital, Division of Comparative Medicine, Massachusetts Institute of Technology, Department of Anesthesia and Critical Care, Massachusetts General Hospital, Boston MA, 02115.

1

Pioneer Neurons and Interneurons in the Developing Subplate: Molecular Markers, Cell Birthdays, and Neurotransmitters

Robert F. Hevner and Nada Zecevic

1. Abstract

Subplate neurons are essential for the development of cortical axon pathways, including thalamocortical innervation as well as the formation of some cortico-cortical and descending cortical efferent connections. Previous evidence suggests that the critical subplate neurons are early-born "pioneer" neurons, which extend the first axons out of the cortex to subcortical forebrain regions. However, pioneer neurons are not the only type of neuron in the subplate layer. The subplate contains both glutamatergic and GABAergic neurons, some of which are transitory due to either ongoing cell migration or subsequent cell death. We have studied the cellular composition of the subplate in developing mouse and human cortex by retrograde axon tracing, cell birthdating, and immunohistochemical analysis of specific markers. Our results indicate that pioneer neurons are early-born glutamatergic neurons that express transcription factor Tbr1, transgene *golli-lacZ*, and other markers. In contrast, GABAergic interneurons in the subplate do not make subcortical (pioneer) axon projections, but instead migrate tangentially and radially through the subplate layer, express transcription factor Dlx, and are born both early and late in corticogenesis. Subplate neurons are essential in development of the initial cortical connectivity, and it is thus important to distinguish between the different cell types present in this compartment, using molecular markers. The subplate in humans appears to contain a similar diversity of neuron types as in mice, but is markedly thicker than in mice, as confirmed by the broad band of Tbr1 expression extending below the cortical plate in humans.

Keywords: Cerebral cortex, Dlx, GABA, *golli-lacZ*, *reeler*, Tbr1

2. Introduction

The subplate is a neuronal layer in the developing cerebral cortex, located deep to the cortical plate and superficial to the intermediate zone (reviewed by Allendoerfer and Shatz, 1994). The subplate is present in all mammals, though its thickness, distinctness as a morphological layer, and persistence in adulthood vary among species. In rodents, the subplate is relatively thin and morphologically distinct, and many subplate neurons persist into adulthood as layer VIb of the mature cortex. At the opposite end of the spectrum, in primates, the subplate is relatively thick (up to four times thicker than the cortical plate in human somatosensory cortex at 22 gestational weeks) and indistinct, blurring with the cortical plate above and the intermediate zone below (Kostovic and Rakic, 1990). Also, relatively few subplate neurons persist to adulthood in primates, mainly as "interstitial neurons" in the white matter (Kostovic and Rakic, 1980). Even in simpler mammalian species where the subplate forms a distinct layer, "subplate neurons" are also recognized in the embryonic intermediate zone, and are thought to persist in small numbers into adulthood as interstitial neurons.

Subplate neurons are important because they play an essential role in the development of cortical axon connections. They give rise to the first ("pioneer") efferent axons out of the cortex, and provide the first postsynaptic targets for afferent thalamocortical axons (McConnell et al., 1989; Molnár et al., 1998a). Ablation of subplate neurons causes localized absence of thalamocortical innervation, as well as defects of some cortical efferent pathways (Ghosh et al., 1990; Ghosh et al., 1993; McConnell et al., 1994). The mechanisms by which subplate neurons mediate development of these several axon pathways are not all clear. In the case of reciprocal thalamocortical and corticothalamic connections, early interactions between cortical pioneer axons and thalamic axons in the internal capsule may be critical for guiding both sets of axons. This mechanism is known as the "handshake hypothesis" (Molnár and Blakemore, 1995).

Efforts to characterize subplate neurons have been difficult because the subplate and intermediate zone contain heterogeneous types of neurons. In addition to pioneer neurons, which are glutamatergic and are produced from progenitors in the cortical neuroepithelium, the subplate also contains numerous interneurons, which are produced mainly from progenitors in the regions of the basal forebrain known as the medial ganglionic eminence (MGE), caudal ganglionic eminence (CGE), and lateral ganglionic eminence (LGE) (Anderson et al., 1997; Wichterle et al., 2001; Marín and Rubenstein, 2001; Nery et al., 2002; Xu et al., 2004). Many immature interneurons migrate tangentially and radially through the developing subplate at the same time as pioneer axons project to the internal capsule. Also, some interneurons cease migration in the subplate, and account for a proportion of neurons in adult layer 6b and white matter (Fairén et al., 1986).

Proposed markers of subplate pioneer neurons include p75 neurotrophin receptor (p75NTR), kynurenine aminotransferase (KAT), *golli-lacZ* transgene, transcription factor Tbr1, SP-1 antibody to cytosolic peptide, and others (Antonini and Shatz, 1990; Allendoerfer et al., 1990; Allendoerfer and Shatz, 1994; Bicknese et al., 1994; Dunn et al., 1995; Landry et al., 1998; Hevner et al., 2001; Csillik et al., 2002; Heuer et al., 2003). With regard to cell birthdays, many subplate neurons, especially the pioneer neurons, are produced at the earliest stages of neurogenesis (Allendoerfer and Shatz, 1994). However, our own studies in mice suggest that a significant minority of subplate interneurons are produced later in neurogenesis (Hevner et al., 2004). In primates, early neurons in the preplate contain both GABAergic and non-GABAergic (presumably glutamatergic) neurons (Zecevic and Milosevic, 1997; Zecevic et al., 1999). After the primate cortical plate is formed, the subplate grows in size by late generated subplate neurons that probably originate in the cortical subventricular zone (Smart et al., 2002; Zecevic et al., 1999, 2005). Previous work by Antonini and Shatz (1990) showed that interneurons and pioneer projection neurons are distinct sets of neurons. In this chapter, we report further studies of the developing mouse and human cortex, in which we show that subplate pioneer neurons and interneurons can be distinguished on the basis of axonal projections, cell birthdays, and—most conveniently—molecular markers. In addition, we show that the same neuron populations can be distinguished in the malformed (overall inverted) cortex of *reeler* mice, in which subplate pioneer neurons and interneurons occupy abnormal positions in superficial cortex ("superplate").

3. Identifying Subplate Neurons Using Markers and Cell Birthdays

3.1. Transcription Factors Tbr1 and Dlx in the Subplate

The subplate is the main location of cortical "pioneer" neurons, defined as neurons that send the earliest axon projections out of the cortex to the internal capsule (McConnell et al., 1989). Our goal was to determine molecular characteristics that distinguish subplate pioneer neurons from cortical interneurons. We hypothesized that the pioneer neurons are glutamatergic projection neurons, which form distinct subsets from GABAergic interneurons and their precursors. To identify pioneer neurons in preparation for immunohistochemistry with glutamatergic and GABAergic markers, we placed crystals of DiI (a fluorescent retrograde and anterograde axon tracer) in the internal capsule of fixed E16.5 mouse forebrains (Fig. 1*A* and *B*). As predicted from previous studies, many subplate neurons were labeled retrogradely, as well as scattered neurons in the intermediate zone and cortical plate (Fig. 1*B*). The DiI also labeled axon bundles in the subplate and intermediate zone of the cortex, and several neurons in the dorsal thalamus (Fig. 1*A*). Sections from the DiI-labeled tissues were used for immunohistochemistry to detect Tbr1, a marker of the glutamatergic lineage, and Dlx, a marker of the GABAergic lineage (Hevner et al., 2001; Anderson et al, 1997;

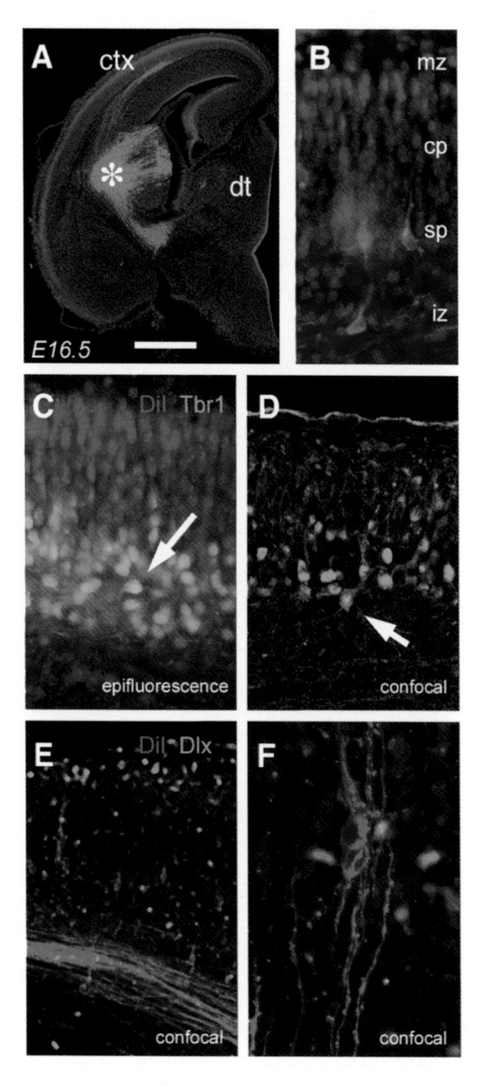

Figure 1 Pioneer neurons, labeled retrogradely with DiI, express Tbr1 but not Dlx in E16.5 mouse cortex. *A*, Pioneer neurons in the subplate of the cortex (ctx) were labeled retrogradely with DiI placed in the internal capsule (asterisk) of fixed E16.5 mouse forebrain. Cells in the dorsal thalamus (dt) were also labeled retrogradely. *B*, Within the cortex, retrogradely labeled cells were located mainly in the subplate (sp), as well as the deep cortical plate (cp) and intermediate zone (iz). Rarely, the marginal zone (mz) contained retrogradely labeled cells (not shown). *C–D*, DiI-labeled cells (red) expressed Tbr1 (green). *Arrows* indicate double labeled cells. (*C*) shows standard epifluorescence microscopy and (*D*) shows a confocal image. *E–F*, DiI-labeled cells (red) did not express Dlx (green). Both (*E*) and (*F*) are confocal images. Scale bar: 1 mm in *A*; 40 μm in *B–D*; 80 μm in *E*; 20 μm in *F*.

Stühmer et al., 2002). We found that the DiI-labeled subplate pioneer neurons expressed Tbr1 (Fig. 1*C* and *D*) but not Dlx (Fig. 1*E* and *F*). These results supported our hypothesis and revealed that within the subplate, Tbr1$^+$ pioneer neurons and Dlx$^+$ interneurons are different neuron types, mixed together in close apposition.

3.2. Transgene *golli-lacZ* in Mouse Subplate Neurons

The *golli-lacZ* transgene was previously identified as a specific marker for early-born pioneer neurons in the subplate (Landry et al., 1998). Expression of the transgene was markedly reduced in the malformed cortex of *Tbr1* knockout mice, suggesting that the pioneer neurons required Tbr1 for differentiation (Hevner et al., 2001). We hypothesized that all *golli-lacZ*$^+$ pioneer neurons express Tbr1, but not Dlx. We studied the parietal cortex of E16.5 *golli-lacZ* transgenic mouse embryos, using immunohistochemistry to detect β-galactosidase, the *lacZ* gene product. The β-galactosidase$^+$ neurons were located mainly in the subplate, though some were also located in the cortical plate (Fig. 2*A*). Two-color immunofluorescence and confocal microscopy confirmed that 100% of the *golli-lacZ*$^+$ cells expressed Tbr1 (Fig. 2A–E). Conversely, most, but not

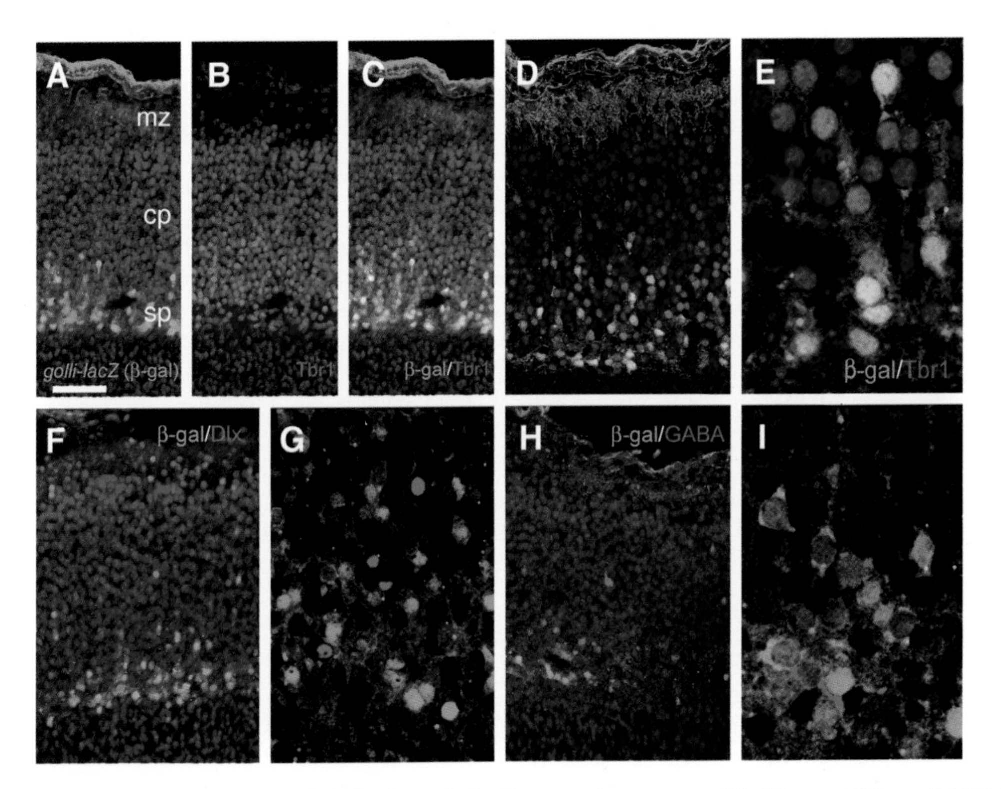

Figure 2 Pioneer neurons, labeled by *golli-lacZ* expression, express Tbr1 but not Dlx or GABA in E16.5 mouse cortex. *A*, Immunofluorescence to detect β-galactosidase (green), expressed from the *golli-lacZ* transgene, labeled neurons in the subplate (sp), deep cortical plate (cp), and (rarely) marginal zone (mz). *B–C*, Tbr1 immunofluorescence in the same section as (*A*) showed that 100% of β-galactosidase$^+$ cells (green) expressed Tbr1 (red). In the merged image (*C*), double-labeled cells appeared yellow. *D–E*, Confocal imaging confirmed Tbr1 expression in β-galactosidase$^+$ cells. *F–G*, β-galactosidase$^+$ cells (green) did not express Dlx (red). (*F*) shows epifluorescence microscopy, and (*G*) shows a confocal image. *H–I*, β-galactosidase$^+$ cells (green) did not express GABA (red). (*H*) shows epifluorescence microscopy, and (*I*) shows a confocal image. Note the close approximation of β-galactosidase$^+$/Tbr1$^+$ projection neurons and Dlx$^+$/GABA$^+$ interneurons in the subplate. Scale bar: 100 μm in *A–C*, *F*, and *H*; 80 μm in *D*; 30 μm in *G*; 20 μm in *E* and *I*.

all Tbr1$^+$ cells expressed β-galactosidase, suggesting that glutamatergic neurons in the subplate are heterogeneous.

Double labeling immunofluorescence for β-galactosidase and interneuron markers demonstrated that β-galactosidase$^+$ pioneer neurons did not express Dlx (Fig. 2*F* and *G*) or GABA (Fig. 2*H* and *I*). Overall, β-galactosidase$^+$ pioneer neurons appeared to be the predominant population in E16.5 cortex, outnumbering Dlx$^+$ or GABA$^+$ interneurons by approximately 3 or 4 to 1. Similar to the results from DiI labeling (above), the β-galactosidase labeling indicated that interneurons and pioneer neurons were mixed together in the subplate. In some examples, the GABAergic interneurons were completely surrounded by β-galactosidase$^+$ pioneer neurons, and by other, unidentified cells that expressed neither GABA nor β-galactosidase (Fig. 2*I*). The apposition of these different cell types illustrates the diversity of cells in the subplate, and highlights the importance of distinguishing among them with criteria other than just location in the subplate.

3.3. Cell Birthdays of Different Neuron Types in the Subplate

Subplate neurons include some of the earliest-born cells in the cerebral cortex, as shown in many birthdating studies (reviewed by Allendoerfer and Shatz, 1994). However, the subplate also contains significant numbers of late-born neurons (Hevner et al., 2004). We hypothesized that interneurons accounted for most of the late-born neurons in the subplate, and that pioneer neurons were born mainly or exclusively during early neurogenesis. To test this hypothesis, we studied the birthdays of molecularly defined cell types (Tbr1$^+$ and GABA$^+$) by double labeling for BrdU and Tbr1 or Dlx (Hevner et al., 2003a; Hevner et al., 2004). We restricted our analysis to the subplate, defined for purposes of this study (in perinatal mice) as the deepest 10% of the cortical thickness. We used a "binning" method to outline the subplate layer for cell counting in digital images of immunofluorescence (Hevner et al., 2004).

Analysis of cell birthdays in the neonatal (P0.5) cortex indicated that Tbr1$^+$ glutamatergic neurons in the subplate were born early in corticogenesis, from E10.5 to E13.5 (Fig. 3*A*), but GABA$^+$ neurons in the subplate were produced early and late in corticogenesis, from E10.5 to E16.5 (Fig. 3*B*). The Tbr1$^+$ cells correspond to pioneer neurons and some other projection neuron types (see above), while the GABA$^+$ cells represent interneurons. These results support the hypothesis that pioneer neurons are early-born cells, while subplate interneurons have a broad range of cell birthdays. Since interneurons continue to migrate radially after P0.5 (Hevner et al., 2004), some of the late-born GABA$^+$ cells may move to different positions in the mature cortex. These results illustrate that late-born cells in the subplate are generally interneurons, but early-born cells in the subplate may be pioneer neurons, other glutamatergic neurons, or interneurons.

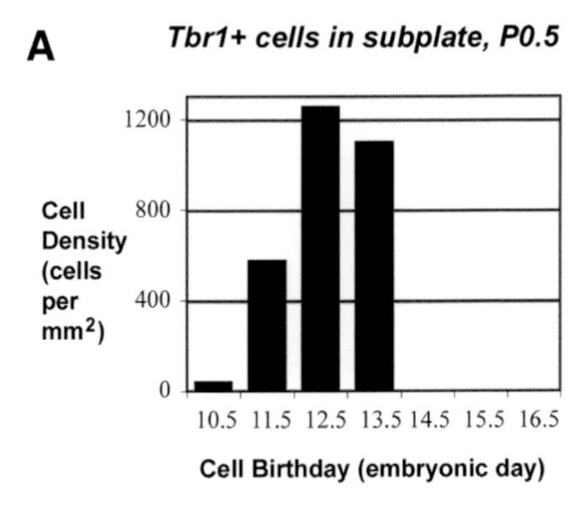

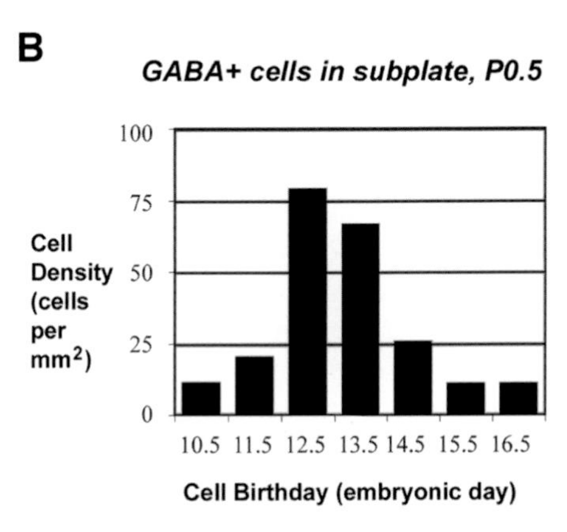

Figure 3 Subplate pioneer neurons are born early in neurogenesis, but subplate interneurons are born both early and late. *A*, $Tbr1^+$ cells in the deepest 10% of the cortical thickness (roughly approximating the subplate, but also including part of layer 6) were born from E10.5 to E13.5. *B*, $GABA^+$ cells in the same zone were born from E10.5 to E16.5. Cell birthdays were determined by double labeling for BrdU and Tbr1 or Dlx in P0.5 parietal cortex as described previously (Hevner et al., 2003b; Hevner et al., 2004).

3.4. Markers and Cell Birthdays in Developing *reeler* Cortex

To determine if markers of subplate pioneer neurons and interneurons can be used to characterize abnormal cortical development, we studied the expression of markers in *reeler* embryonic cortex. The cortex of *reeler* mice is malformed due to mutation of the *reelin* gene, which encodes a large secreted protein that regulates the migration of cortical neurons (reviewed by Rice and Curran, 2001). The *reeler* cortex has been described as roughly "inverted" since early-born neurons migrate to superficial positions within the cortical plate instead of their normal deep positions, and late-born neurons migrate to deep rather than superficial positions (Caviness, 1982). Despite overall inversion, the layer-specific phenotypes of neurons born at each embryonic age appear to be unchanged, as indicated by expression of several molecular markers (Ferland et al., 2003; Hevner et al., 2003b; Inoue et al., 2004). Subplate pioneer neurons, which are early-born cells, settle in superficial rather than deep positions in the *reeler* cortex, thus forming a "superplate" (reviewed by Rice and Curran, 2001). Nevertheless, the early-born neurons still send pioneer axon projections to the internal capsule and retain the ability to guide afferent thalamocortical axons, as shown by studies using DiI and other fluorescent tracers (Yuasa et al., 1994; Molnar et al., 1998b).

To determine if subplate pioneer neurons retain specific molecular properties and can be distinguished from interneurons in *reeler* cortex, we bred the *golli-lacZ* transgene into *reeler* mice and studied the expression of β-galactosidase, Tbr1, Dlx, and GABA by double labeling immunofluorescence. Immunoreactivity for β-galactosidase was located mainly in superficial, subpial positions, consistent with the formation of a "superplate" (Fig. 4*A*). In addition, some β-galactosidase$^+$ neurons were scattered, or formed isolated groups in the upper (superficial) half of the cortical plate (Fig. 4*A*). The distribution of β-galactosidase$^+$ cells was similar to that of chondroitin sulfate proteoglycan, another proposed subplate marker (Bicknese et al., 1994; Sheppard and Pearlman, 1997). Double immunofluorescence for β-galactosidase and Tbr1 showed that all of the β-galactosidase$^+$ pioneer neurons expressed Tbr1 in *reeler* (Fig. 4*B* and *C*), as in normal cortex (Fig. 2*A–E*). Also, the *reeler* superplate contained some Tbr1$^+$/β-galactosidase$^-$ cells, showing the same diversity of molecular phenotypes as in normal cortex. Double labeling for β-galactosidase and Dlx showed that many Dlx$^+$ interneurons mingled near the β-galactosidase$^+$ cells in *reeler*, although the β-galactosidase$^+$ cells themselves did not express Dlx (Fig. 4*D* and *E*). Double immunofluorescence for β-galactosidase and GABA confirmed that β-galactosidase$^+$ pioneer neurons and GABA$^+$ interneurons were also distinct, but mingled with each other in the *reeler* cortex (data not shown). These results suggest that the superplate in *reeler*, like the subplate in normal cortex, contains a mixture of interneurons and pioneer neurons, which can be distinguished by molecular expression. Moreover, these results also show the value of molecular markers for assessing cortical malformations.

3.5. Markers in Human Fetal Subplate

The subplate has undergone massive expansion in mammalian evolution, and attains a much greater thickness in primates than in other species (Kostovic and Rakic, 1990; Hevner, 2000; Smart et al., 2002). We demonstrated the thick subplate in the human occipital (striate) cortex by DiI labeling from the optic radiations in mid-gestational fixed fetal tissue (Fig. 5*A–C*). Analysis of Tbr1 expression in frozen sections of the human fetal cortex showed that Tbr1 was strongly expressed in deep layers of the cortical plate, and in neurons in a broad zone corresponding to subplate (Fig. 5*D*). Similarly, immunohistochemistry for Dlx showed many Dlx$^+$ cells in the cortical plate as well as the subplate (Fig. 5*E–G*). Interestingly, many of the Dlx$^+$ cells in the subplate

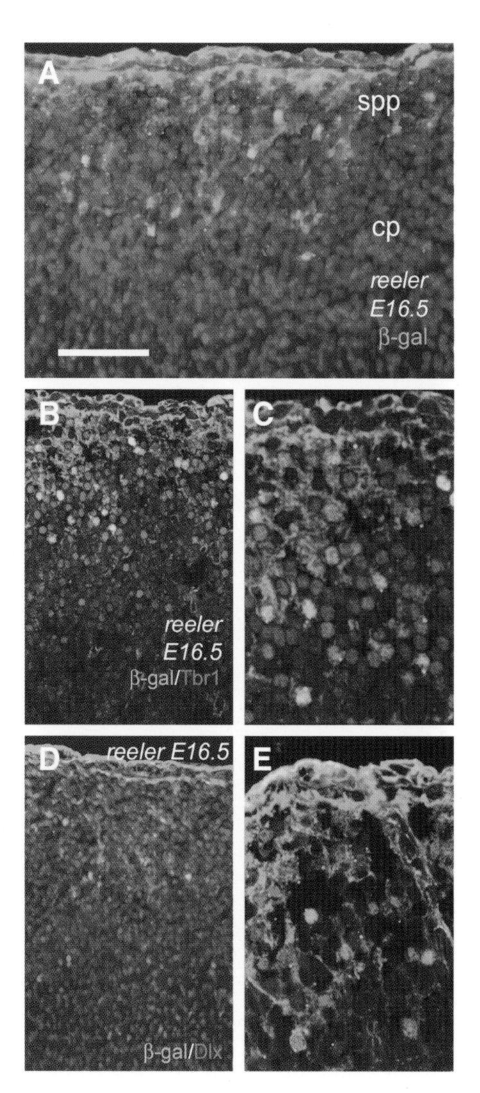

Figure 4 In E16.5 *reeler* mouse cortex, pioneer neurons are ectopic in the "superplate," but the distinctions between pioneer neurons and interneurons are maintained. *A*, Immunofluorescence to detect β-galactosidase (green) labeled neurons in the superplate (spp) and superficial portions of the disorganized cortical plate (cp). *B–C*, Double labeling for β-galactosidase (green) and Tbr1 (red) showed that Tbr1 was expressed in 100% of β-galactosidase$^+$ cells. *D–E*, β-galactosidase$^+$ cells (green) did not express Dlx (red). (*D*) shows epifluorescence microscopy, and (*E*) shows a confocal image. Scale bar: 100 μm in *A*, *B*, and *D*; 50 μm in *C* and *E*.

had an elongated nuclear morphology, suggestive of active migration (Fig. 5*F* and *G*). In addition, both GABAergic and calretinin$^+$ cells were crossing tangentially and radially through the subplate (Fig. 5*H–J*). These results suggest that the subplate in humans, though expanded relative to mice, probably contains a similar heterogeneity of neuron

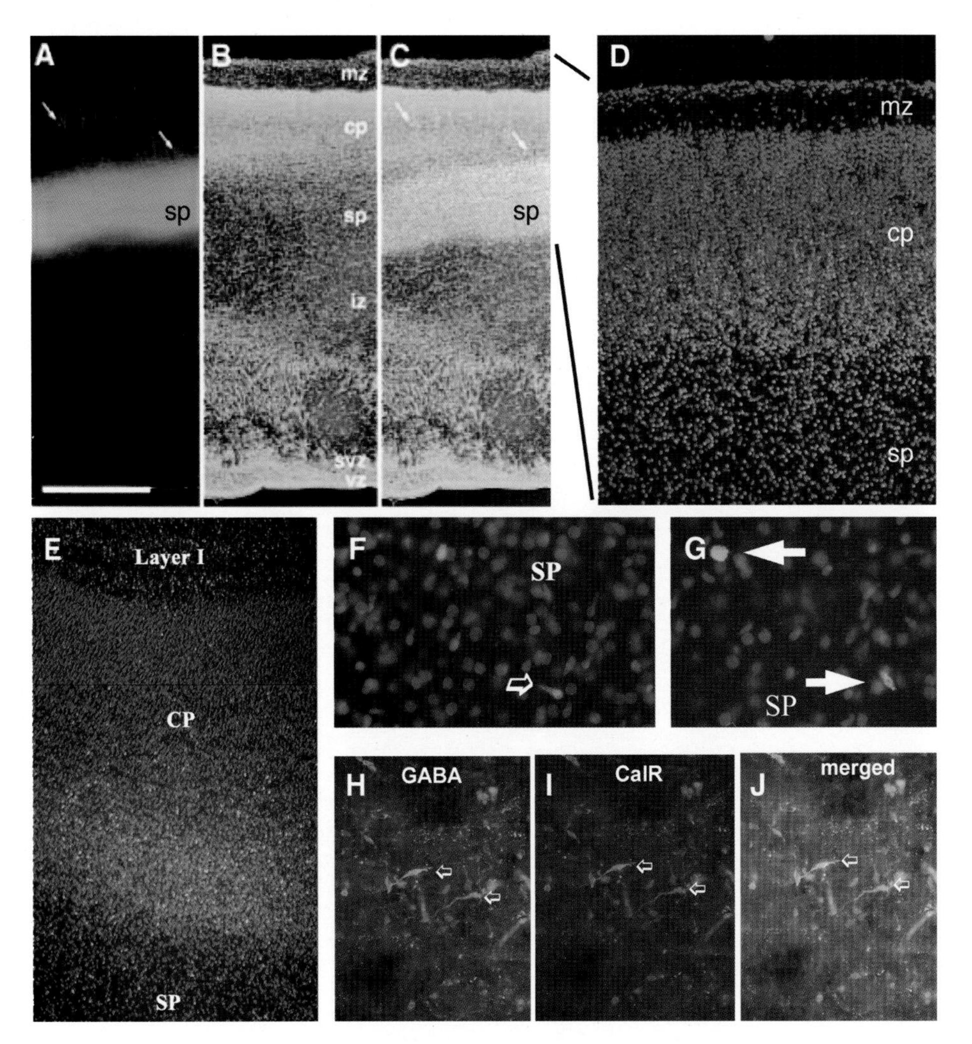

Figure 5 The subplate zone in humans. *A–C*, The subplate in primary visual cortex was labeled by DiI transport from the optic radiations in a fetal brain (22 gestational weeks). DiI labeling (red) is shown in (*A*), DAPI labeling of cells in the same section is shown in (*B*), and the merged image is shown in (*C*). The subplate (sp) contained dense fiber labeling and numerous retrogradely labeled cells. Scattered retrogradely labeled cells were also present in the cortical plate (cp). The marginal zone (mz), intermediate zone (iz), subventricular zone (svz), and ventricular zone (vz) are also indicated. *D*, Tbr1 immunofluorescence of a cortex from another 22 gestational week fetus shows abundant Tbr1^{+} cells (red) in the deep (lower) cortical plate and subplate (blue is DAPI counterstain). *E–G*, Immunofluorescence for Dlx (green) in mid-gestation cortex (blue is DAPI counterstain). Dlx^{+} cells (*arrows*) were scattered throughout the cortical plate (CP) and subplate (SP). Some of the Dlx^{+} cells had elongated nuclei, consistent with migration (arrows). *H–J*, Interneurons in the subplate layer of the 17 gestational weeks human fetus: *H*, tangentially oriented GABAergic (green) neurons (arrows); *I*, calretinin (CalR) immunoreactivity (red) in the same section; *J*, merged image shows the colocalization (yellow) of these two markers in the same cells. *A–C* adapted from Hevner (2000). Scale bar: 1 mm in *A–C*; 400 μm in *D–E*; 100 μm in *F–G*; 120 μm in *H–J*.

types including pioneer neurons, other glutamatergic (Tbr1^{+}) neurons, mature interneurons, and migrating (immature) interneurons.

4. Mouse Subplate Neuron Types Defined by Markers and Cell Birthdays

Previous work by Antonini and Shatz (1990) suggested that the subplate contains at least two distinct types of cells: peptidergic or GABAergic interneurons, and glutamatergic pioneer neurons. Our results expand on that idea, to suggest that the subplate contains a heterogeneous population of multiple neuron types, including pioneer neurons, other glutamatergic projection neurons, mature interneurons, and migrating interneurons. The overall conclusion is that individual subplate neurons cannot be identified precisely on the basis of location alone, but should also be characterized by other means, especially molecular markers. This conclusion also applies to layers of the mature cortical plate, each of which contains multiple types of projection neurons and interneurons (Hevner et al., 2003b; Xu et al., 2004).

4.1. Pioneer Neurons

Pioneer neurons are often referred to simply as "subplate neurons" (Allendoerfer and Shatz, 1994), since they probably account for the largest proportion of neurons in the subplate. Functionally, pioneer neurons are critically important for thalamocortical innervation, organization of cortical columns, and formation of descending projections to the thalamus and other targets (Ghosh et al., 1990; Ghosh et al., 1993; McConnell et al., 1994; Kanold et al., 2003). Some of these functions may be accomplished by direct interactions between pioneer axons and thalamic axons in the internal capsule, as proposed by the "handshake hypothesis" (Molnár and Blakemore, 1995) and supported by studies of many mutant mouse strains (Hevner et al., 2002; López-Bendito and Molnár, 2003).

Strictly speaking, pioneer neurons are defined on the basis of very early projections into the internal capsule. As shown in the present study and other studies, pioneer neurons are located mainly but not exclusively in the subplate. Pioneer neurons also occur, in fewer numbers, in the cortical plate and the marginal zone (Molnár et al., 1998a; Landry et al., 1998). Expression of the *golli-lacZ* transgene appears to identify all of the pioneer neurons, in the cortical plate and marginal zone as well as the subplate (Landry et al., 1998), although technical limitations have precluded a direct demonstration of this. Our results in the present study were consistent with this idea, since all pioneer neurons (traced retrogradely with DiI) expressed Tbr1 and all *golli-lacZ*-expressing neurons expressed Tbr1, but it remains possible that some *golli-lacZ*-expressing neurons are not pioneer neurons. Likewise, it is possible that some pioneer neurons do not express *golli-lacZ*, since we

Table 1 Comparisons of subplate pioneer neurons and interneurons.

	Projection pioneer neurons	Interneurons
Neurotransmitter:	Glutamate	GABA
Origins:	Cortex	MGE, LGE, CGE
Cell birthdays:	E10.5–13.5	E10.5–16.5
Axon projections:	Internal capsule	Local or migrating
Molecular markers:	Tbr1, Emx1, p75NTR, golli-lacZ, KAT, CTGF	Dlx, Lhx6

Abbreviations: CGE, caudal ganglionic eminence; CTGF, connective tissue growth factor; KAT, kynurenine aminotransferase; LGE, lateral ganglionic eminence; MGE, medial ganglionic eminence; p75NTR, p75 neurotrophin receptor.

observed some Tbr1$^+$/β-galactosidase$^-$ cells in the subplate (Fig. 2*E*). Pioneer neurons are thought to express several other molecular markers in addition to Tbr1 and *golli-lacZ* transgene, which allow them to be distinguished from interneurons (Table 1). These include kynurenine aminotransferase, p75 neurotrophin receptor, connective tissue growth factor, and Emx1, among others (Allendoerfer et al., 1990; Allendoerfer and Shatz, 1994; Chan et al., 2001; Csillik et al., 2002; McQuillen et al., 2002; Heuer et al., 2003).

Early neurogenesis is thought to be a characteristic property of pioneer neurons in the subplate (reviewed by Allendoerfer and Shatz, 1994), and our current results from BrdU birthdating of Tbr1$^+$ cells in the subplate supported this conclusion (Fig. 3*A*). Many pioneer neurons are thought to undergo apoptosis during postnatal life, while others are thought to persist as layer VIb or VII of the cortex, or as "interstitial cells" in the subcortical white matter (reviewed by Allendoerfer and Shatz, 1994). In short, pioneer neurons are a dynamic set of cells throughout development, and studies of their properties at different ages may produce different results.

4.2. Non-Pioneer Glutamatergic Neurons

Our results from double labeling with β-galactosidase and Tbr1 suggested that the glutamatergic cell population in the subplate is heterogeneous, since we found both Tbr1$^+$/β-galactosidase$^+$ and Tbr1$^+$/β-galactosidase$^-$ cells within the subplate (Fig. 2*A–E*). We interpret the Tbr1$^+$/β-galactosidase$^-$ cells as non-pioneer glutamatergic neurons, possibly corticothalamic or cortico-cortical projection neurons. Previous studies by Antonini and Shatz (1990) indicated that the subplate sent glutamatergic projections to the thalamus and to the contralateral cerebral cortex. Since pioneer neurons are defined by their early axon projections to the internal capsule, we suspect that they are a different population from the contralateral projection neurons. Also, as noted above, it is possible that some pioneer neurons are β-galactosidase$^-$. Clearly, further studies will be necessary to better characterize the glutamatergic neuronal components in the subplate.

4.3. Mature Interneurons

The subplate expresses several markers associated with GABAergic interneurons (e.g., neuropeptides, calbindin, and glutamic acid decarboxylase) during development, and in subplate remnants during adult life (Antonini and Shatz, 1990; Allendoerfer and Shatz, 1994; Arias et al., 2002). GABAergic interneurons are present in all layers of the cerebral cortex, including the subplate (or layer VIb/VII in mature animals), and display a wider range of cell birthdays than glutamatergic neurons in the same layers (Fairén et al., 1986, Peduzzi, 1988; Hevner et al., 2004). The GABAergic interneurons in the subplate do not send pioneer axons outside of the cortex (Antonini and Shatz, 1990). The function of GABAergic interneurons in the subplate is likely to be similar as in other layers of the cortex, i.e., modulation of local functional activity.

4.4. Migrating Interneurons

Studies in the past 10 years have shown that the majority of interneurons in rodents are produced outside the cerebral cortex, in the MGE, LGE, and CGE, and must migrate tangentially as well as radially to reach their final positions in the cortex (Anderson et al., 1997; Wichterle et al., 2001; Marín and Rubenstein, 2001; Nery et al., 2002; Hevner et al., 2004; Xu et al., 2004). Interestingly, in humans, a larger proportion of interneurons are produced in the cortex (Letinic et al., 2002; Rakic and Zecevic, 2003). The migrating interneurons express specific transcription factors such as Dlx and Lhx6 (Anderson et al., 1997; Flames et al., 2004), as well as GABA and other markers that allow them to be distinguished from pioneer neurons and other types of projection neurons (Table 1). The radial phase of interneuron migration occurs over a more prolonged period than radial migration of glutamatergic projection neurons, and interneuron radial migration through the subplate is still highly dynamic during the first postnatal week in mice (Hevner et al., 2004). In the present study, we observed many interneurons in the E16.5 mouse subplate with an elongated morphology, consistent with active migration (Fig. 2*F–I*). In sum, interneurons in the embryonic subplate are a potentially migratory and transient subset of subplate neurons.

5. Subplate/"Superplate" Neuron Types in *reeler* Mouse Cortex

The malformed cortex in *reeler* mice shows a surprising preservation of molecular and connectional properties. The abnormal relation between cell birthday and laminar position does not appear to alter layer-related molecular expression (Hevner et al., 2003b). Connections between the cortex and subcortical structures are overall intact (reviewed by Rice et al., 2001), and even somatosensory "barrels" form in the reeler cortex (Polleux et al., 1998). Moreover, early interactions between pioneer neurons and afferent thalamocortical axons are maintained (Molnár et al.,

1998b). Previous birthdating studies suggest that the neonatal *reeler* superplate contains projection neurons and interneurons with typical cell birthdays of subplate neurons (Hevner et al., 2004). In this context, it is not surprising that the molecular properties of subplate neurons appeared to be normal. Even though the pioneer neurons were located in the superplate rather than the subplate, they nevertheless expressed *golli-lacZ* transgene and Tbr1, and were mixed with other neuron types, including $Tbr1^{+}/\beta$-galactosidase^{-} cells, and Dlx^{+} and $GABA^{+}$ interneurons (Fig. 4). Thus, the normal components of the subplate were present and expressed typical molecular properties in the *reeler* cortex, despite ectopic positions. The present studies add subplate neurons to the list of cortical neuron types with preserved molecular expression in the *reeler* cortex, and further demonstrate the potential value of molecular markers for characterization of cortical malformations. In future studies, it is hoped that panels of markers will be developed to analyze malformations of the human cortex, such as polymicrogyria, lissencephaly, and focal cortical dysplasia.

6. Subplate Neuron Types in Human Cortex Compared to Mouse

To understand abnormalities of human cortical development, it will be important to investigate normal development of the human cortex. The subplate is relatively expanded in humans, and is thought to play an even bigger developmental role in humans than in rodents and other species, reflecting the complexity of human cortical connections (Kostovic and Rakic, 1990; Kostovic and Judas, 2002). From the early embryonic stages, GABAergic, calretinin^{+} and Dlx^{+} neurons were present in the human preplate (Zecevic and Milosevic, 1997; Zecevic et al., 1999; Rakic and Zecevic, 2003). Later expansion of the subplate layer in primates is probably due to late generated neurons in the cortical subventricular zone (Smart et al., 2002; Zecevic et al., 2005). In the present study, we found that the human cortex contained a similar mixture of $Tbr1^{+}$ glutamatergic neurons and Dlx^{+} GABAergic interneurons as in mice. The retrograde labeling of many neurons in the subplate and cortical plate from DiI injection of the optic radiations was consistent with the presence of pioneer neurons in the subplate, with fewer scattered in the cortical plate (Fig. 5*A–C*). The broad band of Tbr1 expression extending far below the cortical plate reinforced this interpretation (Fig. 5*D*). The abundance of Dlx^{+} and calretinin^{+} GABAergic interneurons in the cortical plate and subplate, many of which had an elongated morphology consistent with ongoing migration, suggested that the subplate of humans, like that of rodents, contains migrating as well as stationary interneurons. Our results suggest that the subplate in humans, though relatively thick, contains the same heterogeneity of neuron types as in other species.

7. Conclusions

The present study has documented the existence of multiple neuron types (glutamatergic and GABAergic) in the subplate of normal and *reeler* mice, as well as humans. However, the present study has also highlighted the inadequacy of current molecular markers for distinguishing precisely among all the neuron types in the subplate. Such markers would be valuable for studying normal development, as well as cortical malformations in animals and humans. Additional markers could help resolve issues about the transience of specific subplate neuron types due to cell death or migration (Woods et al., 1992; Robertson et al., 2000; Arias et al., 2002), and could answer questions about the evolution of the subplate in humans. What is the significance of the thick subplate in humans? Have specific types of subplate neurons been selectively amplified in the human brain? Are species differences in cortical connections and organization related to particular neuron types? When do different subplate neuron types complete phases of migration or apoptosis? The answers to these questions will have important implications for understanding cortical development and plasticity.

8. Acknowledgments

This work was supported by National Institutes of Health grants K02 NS045018 (RFH) and R01 NS041489 (NZ), and by the Edward Mallinckrodt, Jr. Foundation (RFH). R.F.H. is recipient of the Shaw Professorship in Neuropathology. We thank Jhumku Kohtz (Northwestern University) for anti-Dlx antibodies.

9. References

Allendoerfer, K. L., Shelton, D. L., Shooter, E. M., and Shatz, C. J. (1990). Nerve growth factor receptor immunoreactivity is transiently associated with the subplate neurons of the mammalian cerebral cortex. Proc. Natl. Acad. Sci. USA 87:187–190.

Allendoerfer, K. L., and Shatz, C. J. (1994). The subplate, a transient neocortical structure: its role in the development of connections between thalamus and cortex. Annu. Rev. Neurosci. 17:185–218.

Anderson, S. A., Eisenstat, D. D., Shi, L., and Rubenstein, J. L. R. (1997). Interneuron migration from basal forebrain to neocortex: dependence on Dlx genes. Science 278:474–476.

Antonini, A., and Shatz, C. J. (1990). Relation between putative transmitter phenotypes and connectivity of subplate neurons during cerebral cortical development. Eur. J. Neurosci. 2:744–761.

Arias, M. S., Baratta, J., Yu, J., and Robertson, R. T. (2002). Absence of selectivity in the loss of neurons from the developing cortical subplate of the rat. Brain Res. Dev. Brain Res. 139:331–335.

Bicknese, A. R., Sheppard, A. M., O'Leary, D. D. M., and Pearlman, A. L. (1994). Thalamocortical axons extend along a chondroitin sulfate proteoglycan-enriched pathway coincident with the neocortical subplate and distinct from the efferent path. J. Neurosci. 14:3500–3510.

Caviness, V. S., Jr. (1982). Neocortical histogenesis in normal and reeler mice: a developmental study based upon [^{3}H]thymidine autoradiography. Brain Res. Dev. Brain Res. 4:293–302.

Chan, C.-H., Godinho, L. N., Thomaidou, D., Tan, S.-S., Gulisano, M., and Parnavelas, J. G. (2001). *Emx1* is a marker for pyramidal neurons of the cerebral cortex. Cereb. Cortex 11:1191–1198.

Csillik, A. E., Okuno, E., Csillik, B., Knyihár, E., and Vécsei, L. (2002). Expression of kynurenine aminotransferase in the subplate of the rat and its possible role in the regulation of programmed cell death. Cereb. Cortex 12:1193–3211.

Dunn, J. A., Kirsch, J. D., and Naegel, J. R. (1995). Transient immunoglobulin-like molecules are present in the subplate zone and cerebral cortex during postnatal development. Cereb. Cortex 5:494–505.

Fairén, A., Cobas, A., and Fonseca, M. (1986). Times of generation of glutamic acid decarboxylase immunoreactive neurons in mouse somatosensory cortex. J. Comp. Neurol. 251:67–83.

Ferland, R. J., Cherry, T. J., Preware, P. O., Morrissey, E. E., and Walsh, C. A. (2003). Characterization of Foxp2 and Foxp1 mRNA and protein in the developing and mature brain. J. Comp. Neurol. 460:266–279.

Flames, N., Long, J. E., Garratt, A. N., Fischer, T. M., Gassmann, M., Birchmeier, C., Lai, C., Rubenstein, J. L. R., and Marín, O. (2004). Short- and long-range attraction of cortical GABAergic interneurons by neuregulin-1. Neuron 44:251–261.

Ghosh, A., Antonini, A., McConnell, S. K., and Shatz, C. J. (1993). Requirement for subplate neurons in the formation of thalamocortical connections. Nature 347:179–181.

Ghosh, A., and Shatz, C. J. (1993). A role for subplate neurons in the patterning of connections from thalamus to neocortex. Development 117:1031–1047.

Heuer, H., Christ, S., Friedrichsen, S., Brauer, D., Winckler, M., Bauer, K., and Raivich, G. (2003). Connective tissue growth factor: a novel marker of layer VII neurons in the rat cerebral cortex. Neuroscience 119:43–52.

Hevner, R. F. (2000). Development of connections in the human visual system during fetal mid-gestation: a DiI-tracing study. J. Neuropathol. Exp. Neurol. 59:385–392.

Hevner, R. F., Shi, L., Justice, N., Hsueh, Y.-P., Sheng, M., Smiga, S., Bulfone, A., Goffinet, A. M., Campagnoni, A. T., and Rubenstein, J. L. R. (2001). Tbr1 regulates differentiation of the preplate and layer 6. Neuron 29:353–366.

Hevner, R. F., Miyashita-Lin, E., and Rubenstein, J. L. R. (2002). Cortical and thalamic axon pathfinding defects in *Tbr1*, *Gbx2*, and *Pax6* mutant mice: evidence that cortical and thalamic axons interact and guide each other. J. Comp. Neurol. 447:8–17.

Hevner, R. F., Neogi, T., Englund, C., Daza, R. A. M., and Fink, A. (2003a). Cajal-Retzius cells in the mouse: transcription factors, neurotransmitters, and birthdays suggest a pallial origin. Brain Res. Dev. Brain Res. 141:39–53.

Hevner, R. F., Daza, R. A. M., Rubenstein, J. L. R., Stunnenberg, H., Olavarria, J., and Englund, C. (2003b). Beyond laminar fate: toward a molecular classification of cortical projection/pyramidal neurons. Dev. Neurosci. 25:139–151.

Hevner, R. F., Daza, R. A. M., Englund, C., Kohtz, J., and Fink, A. (2004). Postnatal shifts of interneuron position in the neocortex of normal and *reeler* mice: evidence for inward radial migration. Neuroscience 124:605–618.

Inoue, K., Terashima, T., Nishikawa, T., and Takumi, T. (2004). Fez1 is layer-specifically expressed in the adult mouse neocortex. Eur. J. Neurosci. 20:2909–2916.

Kanold, P. O., Kara, P., Reid, R. C., and Shatz, C. J. (2003). Role of subplate neurons in functional maturation of visual cortical columns. Science 301:521–525.

Kostovic, I., and Judas, M. (2002). The role of the subplate zone in the structural plasticity of the developing human cerebral cortex. Neuroembryology 1:145–153.

Kostovic, I., and Rakic, P. (1980). Cytology and time of origin of interstitial neurons in the white matter in infant and adult human and monkey telencephalon. J. Neurocytol. 9:219–242.

Kostovic, I., and Rakic, P. (1990). Developmental history of the transient subplate zone in the visual and somatosensory cortex of the macaque monkey and human brain. J. Comp. Neurol. 297:441–470.

Landry, C. F., Pribyl, T. M., Ellison, J. A., Givogri, M. I., Kampf, K., Campagnoni, C. W., and Campagnoni, A. T. (1998). Embryonic expression of the myelin basic protein gene: identification of a promoter region that targets transgene expression to pioneer neurons. J. Neurosci. 18:7315–7327.

Letinic, K., Zoncu, R., and Rakic, P. (2002). Origin of GABAergic neurons in the human neocortex. Nature 417:645–649.

López-Bendito, G., and Molnár, Z. (2003). Thalamocortical development: how are we going to get there? Nat. Rev. Neurosci. 4:276–289.

Marín, O., and Rubenstein, J. L. R. (2001). A long, remarkable journey: tangential migration in the telencephalon. Nat. Rev. Neurosci. 2:780–790.

McConnell, S. K., Ghosh, A., and Shatz, C. J. (1989). Subplate neurons pioneer the first axon pathway from the cerebral cortex. Science 245:978–982.

McConnell, S. K., Ghosh, A., and Shatz, C. J. (1994). Subplate pioneers and the formation of descending connections from cerebral cortex. J. Neurosci. 14:1892–1907.

McQuillen, P. S., DeFreitas, M. F., Zada, G., and Shatz, C. J. (2002). A novel role for p75NTR in subplate growth cone complexity and visual thalamocortical innervation. J. Neurosci. 22:3580–3593.

Molnár, Z., and Blakemore, C. (1995). How do thalamic axons find their way to the cortex? Trends Neurosci. 18:389–397.

Molnár, Z., Adams, R., and Blakemore, C. (1998a). Mechanisms underlying the early establishment of thalamocortical connections in the rat. J. Neurosci. 18:5723–5745.

Molnár, Z., Adams, R., Goffinet, A. M., and Blakemore, C. (1998b). The role of the first postmitotic cortical cells in the development of thalamocortical innervation in the *reeler* mouse. J. Neurosci. 18:5746–5765.

Nery, S., Fishell, G., and Corbin, J. G. (2002). The caudal ganglionic eminence is a source of distinct cortical and subcortical cell populations. Nat. Neurosci. 5:1279–1287.

Peduzzi, J. D. (1988). Genesis of GABA-immunoreactive neurons in the ferret visual cortex. J. Neurosci. 8:920–931.

Polleux, F., Dehay, C., and Kennedy, H. (1998). Neurogenesis and commitment of corticospinal neurons in *reeler*. J. Neurosci. 18:9910–9923.

Rakic, S., and Zecevic, N. (2003). Emerging complexity of cortical layer I in humans. Cereb. Cortex 13:1072–1083.
Rice, D. S., and Curran, T. (2001). Role of the reelin signaling pathway in central nervous system development. Annu. Rev. Neurosci. 24:1005–1039.
Robertson, R. T., Annis, C. M., Baratta, J., Haraldson, S., Ingeman, J., Kageyama, G. H., Kimm, E., and Yu, J. (2000). Do subplate neurons comprise a transient population of cells in developing neocortex of rats? J. Comp. Neurol. 426:632–650.
Sheppard, A. M., and Pearlman, A. L. (1997). Abnormal reorganization of preplate neurons and their associated extracellular matrix: an early manifestation of altered neocortical development in the reeler mutant mouse. J. Comp. Neurol. 378:173–179.
Smart, I. H. M., Dehay, C., Giroud, P., Berland, M., and Kennedy, H. (2002). Unique morphological features of the proliferative zones and postmitotic compartments of the neural epithelium giving rise to striate and extrastriate cortex in the monkey. Cereb. Cortex 12:37–53.
Stühmer, T., Anderson, S. A., Ekker, M., and Rubenstein, J. L. R. (2002). Expression from a *Dlx* gene enhancer marks adult mouse cortical GABAergic neurons. Cereb. Cortex 12:75–85.
Wichterle, H., Turnbull, D. H., Nery, S., Fishell, G., and Alvarez-Buylla, A. (2001). In utero fate mapping reveals distinct migratory pathways and fates of neurons born in the mammalian basal forebrain. Development 128:3759–3771.
Woods, J. G., Martin, S., and Price, D. J. (1992). Evidence that the earliest generated cells of the murine cerebral cortex form a transient population in the subplate and marginal zone. Brain Res. Dev. Brain Res. 66:137–140.
Xu, Q., Cobos, I., De La Cruz, E., Rubenstein, J. L., and Anderson, S. A. (2004). Origins of cortical interneuron subtypes. J. Neurosci. 24:2612–2622.
Yuasa, S., Kitoh, J., and Kawamura, K. (1994). Interactions between growing thalamocortical afferent axons and the neocortical primordium in normal and reeler mutant mice. Anat. Embryol. 190:137–154.
Zecevic, N., and Milosevic, A. (1997). The initial development of the GABA-immunoreactivity in the human cerebral cortex. J. Comp. Neurol. 380:495–506.
Zecevic, N., Milosevic, A., Rakic, S., and Marin-Padilla, M. (1999). Early development and composition of the human primordial plexiform layer: an immunohistochemical study. J. Comp. Neurol. 412:241–254.
Zecevic, N., Chen, Y., and Filipovic, R. (2005). Contributions of cortical subventricular zone to the development of the human cerebral cortex. J. Comp. Neurol., 491:109–122.

2

Dual Roles of Transcription Factors in Forebrain Morphogenesis and Development of Axonal Pathways

Thomas Pratt and David J. Price

Introduction

During its development the brain must generate a variety of neural structures and organise the correct axonal connections between and within them. In this Chapter we concentrate on how transcription factors specify both these processes in the developing eye and forebrain. It is now well-established that regionally expressed transcription factors regulate the morphogenesis of each region of the brain. More recently, many of these same transcription factors have been implicated in regulating the development of axonal pathways including those providing sensory inputs to the cerebral cortex. In some cases there is evidence that the effects of transcription factors on axonal development involve direct, cell autonomous actions.

The recent sequencing of the mouse and human genomes has allowed estimates of the number of protein coding genes required to generate a mouse and a human. It appears that about 30,000 proteins are sufficient to generate a mammal. Given the enormous complexity of the finished product, the construction of the animal during development would seem to demand that the available genes are used efficiently. One way of doing this would be to allow a particular gene to participate in several developmental processes. The use of the same transcription factors for both tissue morphogenesis and axonal growth and guidance may be an example of the efficient use of available genetic resources.

In this Chapter we consider three possible mechanisms of gene action. The first regulates morphogenesis, the second and third regulate axon guidance. (1) A gene may coordinate the proliferation, differentiation, migration and death of cells required to generate tissue shape or cell type composition, for example the cup-shaped retina with its six cell types organised in their characteristic laminated pattern. (2) A gene may control the properties of a cell projecting an axon, for example by regulating the expression of proteins on the navigating growth cone of a thalamocortical axon. (3) A gene may influence axon navigation by regulating the properties of the environment through which the growth cone must navigate, for example by regulating the proteins expressed at the optic chiasm where retinal axons are sorted into the optic tract.

Transcription factors are proteins that bind to DNA and regulate the transcription of genes into messenger RNA (mRNA) and control the amount available to translate into protein. A given transcription factor may regulate the expression of many target genes. Mouse genetics have allowed the importance of transcription factors in eye and forebrain development to be tested by examining the consequences of perturbing their expression. An emerging theme is that many transcription factors have dual roles in forebrain morphogenesis and development of axonal pathways and the next section examines the roles of the transcription factors Foxd1, Foxg1, Islet2, Pax2, Pax6, Vax1, Vax2 and Zic2 in these processes. We examine the behaviour of RGC axons at the optic chiasm in particular detail. The final section examines the several roles of Pax6 in specifying the morphology and connectivity of the forebrain.

Untangling the Roles of Transcription Factors in Regulating Both Tissue Morphogenesis and Axonal Development

In some ways, examining a mutant phenotype can be likened to a crash investigation where the aim is to identify the cause of the crash from a mangled pile of wreckage. Tissue morphogenesis generally precedes axon navigation and so disrupting a gene with a role in both morphogenesis and axon navigation may produce a mutant animal with an abnormally shaped brain and with axon pathfinding errors. It is not always obvious whether the axon pathfinding errors are a mechanical consequence of a change in brain shape, or whether they reflect a subsequent direct [and in this context more interesting] alteration in the adhesive or other properties of the navigating growth cone and the cells through which it navigates. As in the case of the crash investigation, identifying the primary cause of observed defects is a vital concern.

There are several experimental approaches available to dissect the causality of axon guidance mistakes in mutant mice where the (1) the gene is expressed in both the cells projecting axons and in the tissues through which they navigate or (2) in which disrupted brain shape precedes axon navigation and can complicate the analysis of axon guidance phenotypes. Mouse mosaics comprising mixtures of wild-type and mutant cells are powerful tools for determining the site of action of a particular gene. These can be in the form of chimeras produced by the fusion of a wild-type and a mutant embryo or conditional gene knockouts in which the gene of interest is mutated in a genetically defined subset of cells at a specific time point in their differentiation. Because they contain wild-type cells, mosaics also have the potential to minimise any alterations in brain shape that might complicate the analysis of unconditional mutants. Another approach is to combine wild-type and mutant tissues in culture. Both *in vivo* and *in vitro* approaches provide the opportunity to observe the behaviour of axons projected by mutant cells into a wild-type environment and *vice versa*. If axons projected by mutant cells make navigation errors when navigating a wild-type environment, or wild-type axons are able to navigate correctly through a mutant environment, this shows that the gene is required to program the responsiveness of the growth cone to its environment. Finding mutant axons navigate a wild-type environment correctly shows that the gene is required outwith the growth cone to supply it with guidance cues. Another possibility is that both wild-type and mutant axons navigate correctly through both wild-type and mutant environments, in which case the navigation errors seen in the unconditional mutant are in fact secondary to other factors such as aberrant morphogenesis.

Transcription Factors and the Development of the Visual Pathway

Normal Development of the Eye and Visual Pathway

During normal development, at around embryonic day 9 (E9) in the mouse, the retina, retinal pigment epithelium, and optic stalk are formed from an out-bulging of the ventral diencephalic neuroepithelium that undergoes a series of folding manoeuvres in concert with ectodermal tissue that will form the cornea and lens (reviewed by Smith et al., 2002). The retina and retinal pigment epithelium form distally. The retina is initially open at its ventral surface (the choroid fissure) but this soon closes to complete the familiar eye ball shape. The optic stalk is formed from more proximal diencephalic tissue. The retina then differentiates to generate several cell types including retinal ganglion cells (RGCs) that project axons to the brain (Cepko et al., 1996). The first RGC axons exit the retina at the optic nerve head at E12 and navigate along the optic stalk to form the optic nerve, which connects to the ventral surface of the brain at the optic chiasm. In mice the vast majority of retinal axons cross the ventral midline at the optic chiasm and join the contralateral optic tract whereas a minority do not cross and join the ipsilateral tract (Fig. 1A). The optic tract then grows over the surface of the thalamus and onto the superior colliculus.

The following sections examine the consequences of mutating transcription factors in transgenic mice for the formation of the structures of the eye and chiasm and the navigation of RGC axons along the optic nerves, through the chiasm, and into the optic tract. The transcription factors are dealt with in pairs reflecting functional relationships revealed by complementary expression domains (Fig. 1) and defective axon navigation phenotypes in mutants. These examples serve to illustrate the dual roles of transcription factors in tissue morphogenesis and axon guidance, the experimental approaches used to dissect these processes, and the challenges posed by these types of experiment.

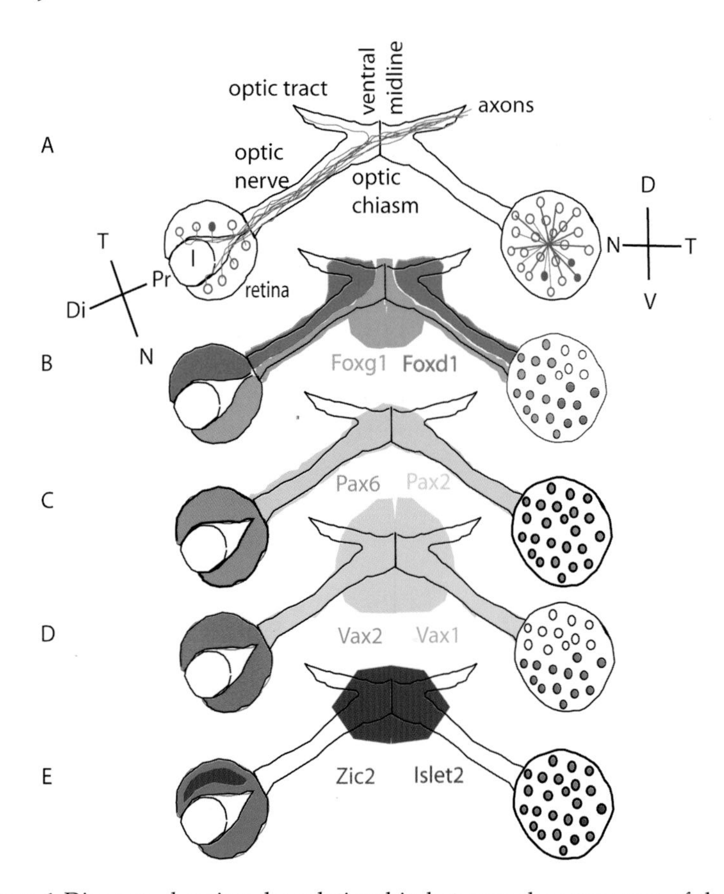

Figure 1 Diagram showing the relationship between the structures of the developing visual system in the eye and ventral forebrain, the trajectory of its axons, and the expression of transcription factors regulating its formation. (A) Retinal ganglion cells (RGCs) project axons along the inner surface of the retina to the optic nerve head where they exit the eye to form the optic nerve. The optic nerve contacts the ventral surface of the hypothalamus at the optic chiasm where axons are sorted into the optic tracts. The retina on the left is viewed in horizontal section, the retina on the right is viewed head on, parallel to the optic nerve. Ipsilaterally and contalaterally projecting RGC bodies are represented by filled and open ovals respectively. (B–E) The expression of transcription factors is mapped onto the RGCs and the structures of the developing visual system through which their axons navigate: (B) Foxg1 and Foxd1; (C) Pax2 and Pax6; (D) Vax1 and Vax2; (E) Zic2 and Islet2. Abbreviations: D, dorsal; Di, distal; N, nasal; l, lens; Pr, proximal. Literature on which this diagram is based is cited in the text.

Foxg1 and Foxd1

Foxg1 (formerly called BF1) and Foxd1 (formerly called BF2) are forkhead box winged helix transcription factors expressed throughout the development of the eye and optic chiasm. The expression of these genes is strikingly complementary with Foxg1 expressed in the nasal retina and anterior optic chiasm and Foxd1 being restricted to the temporal retina and posterior chiasm (Fig. 1B; Hatini et al., 1994; Xuan et al.,

1995; Huh et al., 1999; Marcus et al., 1999). Experiments in the chick have shown that forced expression of Foxd1 and Foxg1 in the retina directly controls the retinotectal mapping of RGC axons (Yuasa et al., 1996; Takahashi et al., 2003), indicating that these genes are capable of directly influencing the properties of the navigating RGC growth cone. Mice lacking these genes exhibit defects in several aspects of eye and forebrain morphogenesis and retinal axon guidance. Careful examination of their mutant phenotypes reveals that these genes may well be involved in simultaneously regulating the properties of the navigating retinal growth cone and in defining the properties of the environment through which it navigates.

The most obvious consequences of depriving the embryo of Foxg1 are the abnormal shape of the eyes and forebrain (Xuan et al., 1996). The abnormal shape of the forebrain is due mainly to an extremely hypoplastic telencephalon. The eye develops an elongated retina which fails to close properly, resulting in coloboma, and the lens is small (Huh et al., 1999). These morphological defects are not restricted to nasal territory which normally expresses Foxg1, suggesting a non-autonomous role for Foxg1 in morphogenesis of temporal eye structures. The eye lacks an optic stalk and the retina connects directly to the base of the brain at the optic chiasm. Loss of Foxg1 does not dramatically affect the dorso-ventral patterning of the eye, as evidenced by the fact that the reciprocal gradients of the receptor tyrosine kinase EphB2 and its ligand ephrinB2 are maintained in the mutant. Naso-temporal polarity is not abolished in the mutant: Foxg1 gene activation remains predominantly nasal and ipsilaterally projecting RGCs are located predominantly in temporal retina, as in wild-types (Pratt et al., 2004). The expression domain of Foxd1 does, however, encroach upon nasal territory, which normally expresses Foxg1 but not Foxd1 (Huh et al., 1999). In spite of this the mutant generates retinal ganglion cells (RGCs) which project axons along the inner surface of the retina, where they fasciculate and enter the optic tract via the optic chiasm. Although the overall trajectory of retinal axons in the mutant strongly resembles that seen during normal development (Pratt et al., 2002), Foxg1 is required for at least one important aspect of axon pathfinding. In the *Foxg1*$^{-/-}$ mutant the ipsilateral projection is massively increased and matches the size of the contralateral projection. Foxg1 therefore normally suppresses the ipsilateral projection of RGC axons. In the nasal retina RGCs normally express Foxg1 and so might repress the expression of proteins required for ipsilateral projection or might activate the expression of proteins required for contralateral projection. In the temporal retina it is more likely that Foxg1 assists the contralateral projection of RGCs, which never express Foxg1, by regulating the expression of navigational instructions supplied to RGC growth cones by cells at the optic chiasm and other points along their journey (Pratt et al., 2004). It remains an open question as to whether the expression of Foxg1 by nasal RGCs is directly involved in the midline crossing behaviour of these axons.

Foxd1 is normally expressed in temporal retina and optic stalk and in the posterior chiasm. Its complementary expression to Foxg1 might suggest that these related genes perform similar functions in their respective domains of the developing visual pathway, but comparison of the Foxg1 and Foxd1 mutant phenotypes shows it is not that simple. The morphology of the Foxd1 mutant eye is not greatly disturbed, but there are alterations to the expression of genes whose expression normally coincides with Foxd1. These include a loss of the ipsilateral determinants Zic2 (a transcription factor, see below) and EphB1 (Williams et al., 2003) from the ventral-temporal retina and an invasion of Foxg1 expression into temporal territory normally occupied by Foxd1. Perhaps surprisingly, in light of the loss of ipsilateral determinants from the ventro-temporal retina, the Foxd1 mutant exhibits an increased ipsilateral projection. Closer examination shows that the ipsilateral projection from the ventro-temporal retina is indeed reduced consistent with a cell autonomous role for Foxd1 in these RGCs. The increased ipsilateral projection arises mostly from RGCs located outside the normal domain of Foxd1 expression in the ventro-temporal retina. RGCs located outside the ventro-temporal retina would not normally express Foxd1 and would normally cross the midline at the optic chiasm to join the contralateral optic tract. This increased ipsilateral projection is attributed to alterations of the molecular properties of the $Foxd1^{-/-}$ chiasm including a reduction in expression of Zic2 and Islet1 (both transcription factors) and an expansion of the expression domain of Slit2 (Herrera et al., 2004). Slit family members Slit1 and Slit2 are expressed around the optic chiasm as it develops (Erskine et al., 2000) and their mutant phenotypes indicate a repulsive role for these proteins in preventing RGC axons from wandering from their normal path (Plump et al., 2002).

Foxg1 and Foxd1 mutually repress each other's expression, either directly or indirectly, but it is at present unknown whether Foxd1 and Foxg1 each regulate the expression of the same target genes in the retina and optic chiasm or whether the presence of different cofactors in these two structures allows participation in distinct molecular programs. It is also unknown whether their target genes involved in regulating morphogenesis are the same as those engaged in axon navigation.

Pax6 and Pax2

Pax2 and Pax6 are dynamically expressed during the early development of the eye. As morphogenesis proceeds Pax6 becomes restricted to more distal structures including the lens, retinal pigment epithelium, and retina. Pax2 is expressed in the optic fissure as it closes, in the optic stalk, and in the preoptic area of the ventral diencephalon, where contralaterally projecting RGC axons will cross the midline at the optic chiasm (Fig. 1C). A combination of elegant transgenic and *in vitro* experiments demonstrated that Pax2 and Pax6 bind to regulatory elements in each other's promoters to mutually repress transcription (Schwarz et al., 2000). Pax2 and Pax6 are required for the formation of optic stalk and optic cup respectively, as shown by the lack of optic cup in Pax6 mutant embryos (Hill et al., 1991) and optic stalk in Pax2 mutant embryos (Torres et al., 1996).

The *Pax6* gene has retained its ability to specify the formation of an eye in species as diverse as *Drosophila*, *Xenopus*, mouse and humans. Loss of Pax6 results in a failure of the eye to form. Although Pax6 has not yet been shown to have a role in the navigation of retinal axons, *Pax6* is expressed by projecting RGCs (Baumer et al., 2002) and so is poised to fulfil this function. Certainly, in other parts of the developing brain Pax6 has functions in axon guidance as well as in tissue morphogenesis and regulates genes implicated in axon guidance (see below).

Pax6 is expressed in both surface ectoderm and optic vesicle tissues, which integrate to generate the structures of the eye. These fail to progress past their very early development in embryos completely lacking Pax6. This complicates the examination of the functions of Pax6 in subsequent events in eye formation, including its roles in morphogenesis and axon guidance. This problem has recently been addressed by the use of Cre-lox technology to selectively disrupt Pax6 in discrete parts of the developing eye. The studies have shown that removing Pax6 from the developing surface ectoderm produces an eye lacking a lens but possessing a retina with RGCs able to project axons (Ashery-Padan et al., 2000). Removing Pax6 function after the retina forms results in a retina comprising mainly amacrine cells at the expense of other retinal cell types including RGCs (Marquardt et al., 2001). Examination of $Pax6^{+/+} \leftrightarrow Pax6^{-/-}$ mouse chimeras has shown that Pax6 is required in the optic vesicle for maintenance of contact with the overlying lens epithelium, a necessary event in eye formation, providing a clue that Pax6 may be involved in defining the adhesive properties of these cells. Pax6 appears to act in a cell autonomous manner in these aspects of eye development (Collinson et al., 2000; Quinn et al., 1996). The dosage of Pax6 is important as increasing (Schedl et al., 1996) or decreasing (Hill et al., 1991) *Pax6* gene dosage in the eye both result in aberrant eye development.

As discussed above, an important aspect of the developing retina with consequences for the trajectory of its axons is the establishment of naso-temporal and dorso-ventral polarity defined by the expression of proteins including the transcription factors Foxg1 and Foxd1 and the EphB2 receptor tyrosine kinase. In embryos where Pax6 has been conditionally ablated from the retina, expression of both Foxg1 and Foxd1 is lost indicating that Pax6 may be required in the generation of nasal-temporal polarity (Baumer et al., 2002). In the chick retina Pax6 is expressed in a ventral^{High} to dorsal^{Low} gradient coincident with the gradient of EphB2 expression (Ziman et al., 2003). Although no such retinal Pax6 gradient has been reported in the mouse, Pax6 may be involved in specifying the dorso-ventral polarity as in the absence of Pax6 the optic vesicle loses its dorsal expression of the transcription factor Tbx5 while the ventral expression domain of Vax1 is expanded (Baumer et al., 2002). Genetic dissection of the *Pax6* locus has revealed that Pax6 expression is controlled independently in different parts of the developing eye. For example, although Pax6 is expressed throughout the developing retina, expression in its distal regions is specifically driven by an 'α element'. Furthermore this element continues to direct Pax6 expression in a subset of RGCs as they project axons into the brain (Baumer et al., 2002).

The expression of Pax2 is complementary to that of Pax6. Whereas Pax6 expression is restricted to the structures of the developing eyeball (lens, retina, retinal pigmented epithelium), Pax2 is expressed in the developing optic stalk and optic chiasm. Mice lacking Pax2 produce elongated retinas, probably at the expense of optic stalk tissue, reminiscent of those seen in Foxg1 mutant embryos described above. The Pax2 mutant retinas are able to project RGC axons which form an optic nerve. The optic nerves from the two eyes do not converge to form the optic chiasm as in wild-types but instead project ipsilaterally to their targets in the brain. Pax2 mutants are therefore classed as achiasmatic (Torres et al., 1996). As Pax2 is not expressed by RGCs but is expressed at the location where the chiasm normally forms it is likely that Pax2 is needed to specify the formation of the preoptic area, whose cells normally support the contralateral projection of RGC axons (Torres et al., 1996).

Vax1 and Vax2

Vax1 and Vax2 are homeodomain containing transcription factors that exhibit complementary expression patterns in the developing visual system. Vax2 is restricted to the developing retina and Vax1 is expressed by cells at the midline where RGC axons form the optic chiasm (Fig. 1D; Hallonet et al., 1998; Bertuzzi et al., 1999; Hallonet et al., 1999). Vax1 is required for morphogenesis of the eye as the optic cup fails to close properly resulting in coloboma in embryos lacking Vax1. The boundary between mutant optic cup and optic stalk is poorly defined with regions normally occupied by optic stalk exhibiting retinal features including retinal pigment epithelium. Although RGCs form in these mutants, their axons navigate abnormally and, instead of approaching the midline to form the optic chiasm, become stalled shortly after leaving the eye (Bertuzzi et al., 1999; Hallonet et al., 1999). As Vax1 is not expressed by RGCs this defect is most likely to reflect a requirement for Vax1 in producing the correct environment for navigating axons. Indeed, Netrin-1, that is normally expressed at the optic nerve head and at the point where the optic nerve connects to the brain and is believed to guide axons along their path (Deiner et al., 1997; Deiner et al., 1999), is missing in Vax1 null-mutants. This provides a plausible molecular mechanism for the inability of RGC axons to reach the chiasm (Bertuzzi et al., 1999).

Vax2 expression is restricted to the ventral region of the prospective neural retina. In embryos lacking Vax2 the optic cup fails to close resulting in coloboma. Vax2 appears to specify ventral character. Its absence causes loss of the expression of EphB2, which is normally present in ventral retina, and expansion of ephrinB2 expression, which is normally restricted to dorsal retina, throughout the mutant retina. Vax2 mutant mice generate RGCs which, unlike those in Vax1 mutants, are able to navigate to the optic chiasm and into the brain. As ipsilaterally projecting RGCs are present in ventral retina and in Vax2 mutants the ventral retina acquires a dorsal character, it might be predicted that the ipsilateral projection would be lost in these mutants. This was reported to be the case in one line of Vax2 null-mutant mice (Barbieri et al, 2002) but in a different line of Vax2 null-mutant mice produced by another group (Mui et al., 2002) the dorsalisation of retina produced an increased ipsailateral projection. This discrepancy may reflect differences in the mutant Vax2 alleles or in their genetic backgrounds.

Zic2 and Islet2
Zic2 is a zinc finger protein homologous to the *Drosophila* gene odd-paired that is widely expressed in neural and non-neural tissues in the mouse. In the developing visual system Zic2 is restricted to ventrotemporal retina and cells around the chiasm (Fig. 1E). In the E15 retina at the time RGC axons are sorted into ipsilateral and contralateral optic tracts, Zic2 expression is restricted to the ventro-temporal quadrant of the retina from which the ipsilateral projection arises. Zic2 is also expressed at the optic chiasm. Targeted disruption of the *Zic2* gene produced a $Zic2^{kd}$ allele (*kd* indicates a 'knockdown' allele in which *Zic2* function is reduced rather than completely abolished as in a 'knockout' allele). $Zic2^{kd/kd}$ embryos have profound morphological brain defects including hypoplasia of the dorsal telencephalon, disruption to midline structures, and eye defects. In contrast $Zic2^{kd/+}$embryos have morphologically normal eyes and brains (Nagai et al., 2000). In addition to this early role in specifying the morphology of brain structures associated with the optic tract, Zic2 also appears to directly control the trajectory of retinal axons. The size of the ipsilateral projection is reduced in $Zic2^{kd/+}$ embryos and *in vitro* experiments showed that RGCs forced to express Zic2 produce axons that are repelled by the optic chiasm. A comparison of Zic2 expression across species with different degrees of binocular vision shows a positive correlation between the number of RGCs expressing Zic2 and the size of the ipsilateral projection (Herrera et al., 2003). Although these experiments are consistent with Zic2 regulating the navigation properties of RGC growth cones, Zic2 is also expressed at the optic chiasm so it is conceivable that Zic2 also influences the navigation of RGC axons by regulating the expression of guidance cues at the optic chiasm. In fact, in the $Foxd1^{-/-}$ mutant described above (Herrera et al., 2004) reduced expression of Zic2 at the chiasm is associated with an increased ipsialteral projection.

Islet2 is a LIM homeodomain containing transcription factor. Islet2 expressing RGCs are located throughout the retina and project contralaterally (Fig. 1E). In embryos lacking Islet2 the ipsilateral projection is increased with the increased projection mapping exclusively to the ventrotemporal retina, coincident with an increase in the number of Zic2 expressing RGCs. This suggests that in the ventrotemporal quadrant,

Islet2 represses Zic2 expression by RGCs and therefore prevents them from projecting ipsilaterally (Pak et al., 2004). Ipsilaterally projecting RGCs express the receptor tyrosine kinase EphB1 which causes their axons to be repelled by its ligand ephrinB2 expressed on cells at the optic chaism (Nakagawa et al., 2000, Williams et al. 2003). It remains to be determined whether Zic2 specifies ipsilateral projections by directly positively regulating the transcription of EphB1 and whether Zic2 transcription is in turn negatively regulated by Islet2.

One feature of the above genes is that they are needed to regulate the structures of the eye and forebrain and the degree of ipsilateral and contralateral projection by RGCs. This is intriguing since, whereas the physical structure of the eye and the developing visual pathway is highly conserved between vertebrates, the fine details of axon organisation within the ubiquitous X-shape formed by the optic nerves, chiasm, and tract varies considerably. For example, there is considerable variation between species in the proportion of axons sorted into the ipsilateral and contralateral optic tracts. It might seem a risky strategy to employ the same gene to regulate the shape of the eye, that is relatively fixed in evolution, and the fine tuning of its RGC projections, that is far more plastic. Perhaps these different aspects come under the control of distinct regulatory genetic elements that can evolve independently. Further diversity can be achieved by the production of several functionally distinct isoforms with distinct transcriptional properties from a single gene, for example by differential splicing.

Transcription Factors that Regulate the Development of the Thalamocortical Tract

The thalamus can be thought of as a 'relay station' for sensory information from the periphery (sight, touch, taste, and hearing) passing through the thalamus *en route* to the cerebral cortex for processing and interpretation. In the mouse, axons exit the dorsal thalamus at E12.5 and grow through the ventral thalamus. They make a sharp lateral turn at the hypothalamus and enter the ventral telencephalon through the internal capsule (Braisted et al., 1999, Tuttle et al., 1999, Auladell et al., 2000). The thalamic axons then grow into the cerebral cortex where they form synapses with layer 4 neurons. The basic thalamocortical circuitry is complete at this point. The navigation of the thalamocortical actions has complex spatial (as the tract describes a three dimensional geometry) and temporal (as all thalamic axons do not navigate synchronously) dimensions. The section below concentrates on how the complex spatial and temporal expression of the transcription factor Pax6 contribute to several aspects of the formation of the structures of the thalamocortical tract and the navigation of its axons.

Several transcription factors have been implicated in the control of thalamocortical development on the basis of defects in this pathway in mice with null mutations in the corresponding genes (reviewed recently in Lopez-Bendito and Molnar, 2003). These factors include Emx2, Tbr1, Gbx2, Mash1, Ebf1, Foxg1 and Pax6. Loss of Gbx2, Mash1, Foxg1 or Pax6 results in failure of thalamic axons to innervate the cortex (Miyashita-Lin et al., 1999; Tuttle et al., 1999; Pratt et al., 2000 & 2002); loss of other transcription factors cause more subtle targeting defects. Loss of these factors also cause morphological defects of the thalamus and/or the tissues through which thalamocortical axons normally grow. Expression of Gbx2 is normally restricted to the thalamus and loss of this factor causes defects of thalamic differentiation (Miyashita-Lin et al., 1999); it is likely, therefore, that thalamic cells have an intrinsic requirement for Gbx2 to allow their innervation of the cortex. Foxg1, on the other hand, is not expressed by thalamic cells but is expressed by ventral telencephalic territory through which thalamic axons normally grow. Failure of thalamic axons to enter the telencephalon in *Foxg1*$^{-/-}$ mouse embryos is, therefore, most likely secondary to defects in the ventral telencephalon (Pratt et al., 2002). For other factors, the likely mechanisms are less clear since, in many cases, they are expressed in the thalamus and at other sites along the route taken by thalamocortical axons. In the case of Pax6, experiments outlined in the next sections have been carried out to test whether there might be a thalamic requirement for it to allow axons to navigate correctly.

How Pax6 Regulates the Morphogenesis of Thalamus and Cortex

Pax6 is expressed in the developing diencephalon. Up until about E12 in the mouse, Pax6 is expressed in diencephalic regions that will become both the major elements of the thalamus. These elements are known traditionally as the dorsal and ventral thalamus, although they are probably better renamed as thalamus and prethalamus respectively. The thalamus is the major recipient of afferents from the sensory periphery and sends its thalamocortical efferents to the cerebral cortex. After E12, Pax6 expression in the diencephalon becomes more restricted, mainly to the prethalamus , that lies rostral to the zona limitans intrathalamica (zli), although expression persists in the proliferating ventricular zone of the thalamus. In mice lacking Pax6, there are major defects in the development of these regions of the diencephalon. Their structure appears abnormal, with a reduction in the size and distortion in the shape of particularly the prethalamus. This is most likely due to a reduction of cell proliferation throughout the diencephalon in the absence of Pax6 (Warren and Price, 1997). The major components of the diencephalon are present in mutants, but there are changes in the patterns of gene expression. These include changes in the expression of other regionally-expressed transcription factors (Grindley et al., 1997; Stoykova et al., 1996; Warren and Price, 1997; Pratt et al., 2000). For example, the expression domains of Nkx2.2 and Lim1 (also known as Lhx1) are expanded throughout the diencephalon, suggesting that a primary action of Pax6 is to generate correct patterning in this region of the brain (Pratt et al.,

2000). *Pax6*$^{-/-}$ cells do not intermingle freely with their wild-type counterparts in the thalamus of *Pax6*$^{+/+}$ $\leftrightarrow$ *Pax6*$^{-/-}$ mouse chimeras indicating that Pax6 defines the adhesive properties of thalamic cells (Pratt et al., 2002). Thalamocortical axons start to grow at about E13-4 in both wild-type mice and in mice lacking Pax6 but, in mutants, they fail to navigate correctly through the ventral telencephalon and, even by the time of birth, when these mutants die, there is no cortical innervation from the thalamus (Auladell et al., 2000; Kawano et al., 1999; Pratt et al., 2000).

Pax6 is also expressed in the developing telencephalon. It is expressed dorsally in the developing cortex and hippocampus and also in some ventral regions, mainly in the region of the amygdala, through which thalamocortical axons normally grow. In the developing cortex and hippocampus Pax6 is expressed in proliferating progenitor cells but is downregulated in differentiating neurons. It is expressed from before the folding of the neural plate throughout neurogenesis. Recent work has shown that radial glial cells, which have been known for decades to guide the migration of neuronal precursor cells, are in fact neuronal progenitor cells and that they express Pax6 (Heins et al., 2002). Loss of Pax6 causes numerous defects in the morphology of the developing cerebral cortex. The cortex is smaller than normal, and cells become densely packed into numerous dense clusters in the intermediate zone (Schmahl et al., 1993; Caric et al., 1997). This has been ascribed to changes in the cell-surface properties of the mutant cells (Warren et al., 1999; Talamillo et al., 2003; Tyas et al., 2003). There is a failure of late-born cells to migrate into the cortical plate. This defect can be corrected by transplanting late-born cells into wild-type cortex, indicating that it is not a cell-autonomous defect but more likely secondary to defects of other cells (Caric et al., 1997). There are two main contenders for the primary source of this migration defect. First, the radial glial cells, which produce and provide guidance for migrating neuronal precursors, show defective morphology in the absence of Pax6 (Gotz et al., 1998). Second, thalamocortical axons can stimulate migration of cortical precursors and so loss of these inputs might impair migration in mutants (Edgar and Price, 2001).

How Forebrain Axon Pathways are Altered in Mutants Lacking Pax6

The early brain contains a primitive network of axonal tracts and there have been many studies of the development of these pathways in a variety of species. The first major longitudinal (i.e. coursing rostrocaudally) tract to form is the tract of the postoptic commissure (TPOC) which runs along the ventrolateral diencephalic surface and continues into the midbrain as the ventral longitudinal tract. Mouse embryos lacking Pax6 show pathfinding defects in the developing TPOC (Mastick et al., 1997; Andrews and Mastick, 2003; Nural and Mastick, 2004). Whereas in wild-type embryos TPOC axons spread out when they contact Pax6-expressing diencephalic neurons, in mutants they make errors indicating that Pax6 is required for local cues guiding the navigational behaviour of TPOC axons as they enter its expression domain.

It has been shown that the cell adhesion molecule R-cadherin (Cdh4) is lost from the region in which TPOC navigational errors occur in mice lacking Pax6 and that axonal growth through this region can be restored by replacing R-cadherin. This indicates that the action of Pax6 in regulating early TPOC tract formation is mediated by the regulation of a cell adhesion molecule in the region through which the axons would grow. Expression of R-cadherin is also lost in the embryonic cerebral cortex of mice lacking Pax6 (Stoykova et al, 1997). In the cortex, this loss is thought to explain changes in the tangential and possibly radial migratory properties of neuronal precursors and hence the cellular constitution and morphology of this tissue. It seems, therefore, that the regulation of cell adhesion molecules by Pax6 is not only necessary for the correct development of tissues but also the subsequent navigation of axons through those structures. In the case of the TPOC, the transcription factor Pax6 is not expressed by the projecting neurons (Mastick et al., 1997; Andrews and Mastick, 2003) so its regulation of axonal navigation appears to be secondary to actions on regional expression of cell adhesion molecules by cells encountered by navigating axons.

Similarly, there is a cell non-autonomous role for Pax6 in regulating the guidance of the catecholaminergic neurons of the substantia nigra (SN) and the ventral tegmental area (VTA) (Vitalis et al., 2000). This is known to be cell non-autonomous since SN-VTA neurons do not express Pax6. Mice lacking Pax6 show defective pathfinding by SN-VTA projections as they cross regions that do express Pax6. It has been suggested that this can be attributed to an expansion of the expression domain of the axon guidance molecule Netrin-1. Jones et al. (2002) suggested that Pax6 is required for the normal development of thalamocortical axonal connections by regulating expression of surface molecules including Sema5A and Sema3C in the regions through which the axons grow.

There is also evidence that PAX6 is essential for the development of axon tracts in the human brain. It is well-known that humans heterozygous for mutations in PAX6 suffer from congenital aniridia but more recent work using magnetic resonance imaging (MRI) has revealed either the absence or hypoplasia of the anterior commissure of the brain in a large proportion of aniridia cases (Sisodiya et al., 2001).

The thalamus and cortex form at similar stages of gestation. Thalamic axons grow through the diencephalon, turn sharply laterally to enter the ventral telencephalon, cross the medial and lateral ganglionic eminences and then turn dorsally to penetrate the cortex. The mechanisms thought to direct thalamocortical axons to the cortex include guidance from (i) pioneering axons growing from cortex towards thalamus and (ii) a transient set of axons growing from the ventral telencephalon to the thalamus (Metin and Godement 1996; Molnar et al., 1998; Molnar 1998; Braisted et al 1999). In $Pax6^{-/-}$ mutants, neither of these form correctly (Kawano et al., 1999; Hevner et al., 2002; Jones et al., 2002; Pratt et al., 2002) and so it is possible that defects of thalamocortical axons are secondary to the absence of normal descending projections. Jones et al. (2002) examined the corticofugal projections in mice lacking Pax6 and described abnormalities of these axons at the corticostriatal junction. Jones et al. (2002) and Pratt et al. (2002) showed defects of ventral telencephalic cells within the internal capsule associated with altered early thalamic growth.

Is there any evidence that Pax6 plays a primary role in the projecting thalamic cells themselves, allowing them to navigate to their cortical targets? Evidence that this is the case has come from co-culture studies (Pratt et al., 2000). Explants from either wild-type or *Pax6*$^{-/-}$mutant embryonic thalamus were co-cultured with wild-type ventral telencephalon and it was found that while axons from wild-type thalamus navigated correctly through wild-type ventral telencephalon, axons from mutant thalamus did not (Fig. 2). This indicates that the

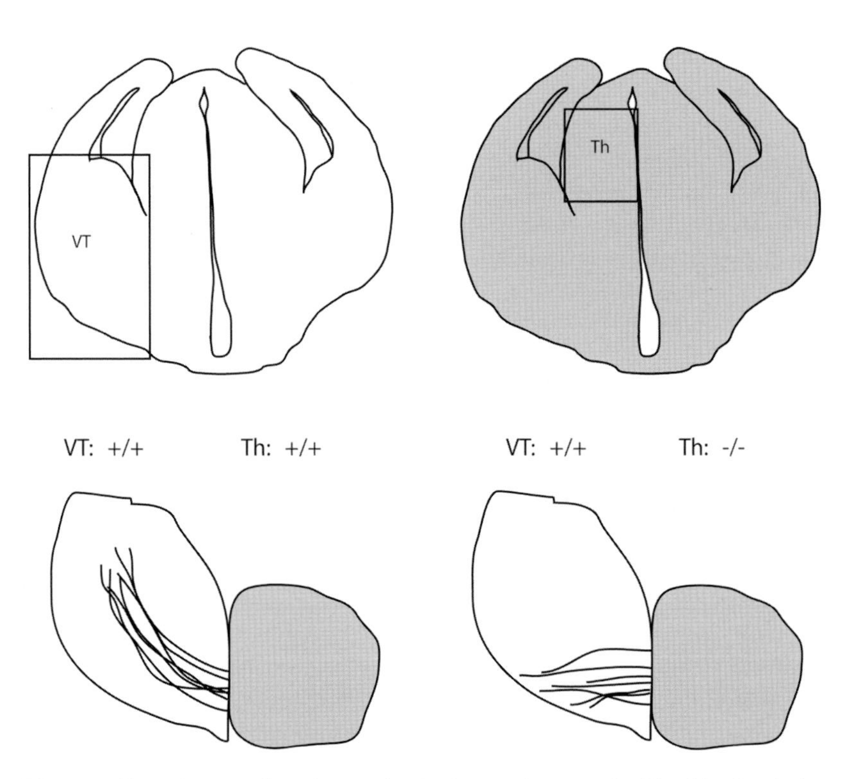

Figure 2 Experiments showing an intrinsic requirement for Pax6 in the thalamus for thalamocortical development (Pratt et al., 2000). Explants of ventral telencephalon (VT) were taken from wild-type mice. Thalamic explants were taken from mice in which all cells express green fluorescent protein linked to tau (tauGFP); these mice were either wild-type or *Pax6*$^{-/-}$. Explants of wild-type ventral telencephalon were placed with explants of tauGFP-expressing wild-type or mutant thalamus and axons labelled with tauGFP could be seen growing into the ventral telencephalon. If the thalamus was wild-type, then these axons navigated through the ventral telencephalon in the direction of the cortex. If the thalamus was *Pax6*$^{-/-}$, then these axons failed to navigate correctly. Since the ventral telencephalon is wild-type in both cases, there must be a defect in the *Pax6*$^{-/-}$dorsal thalamus.

navigational defects of *Pax6*$^{-/-}$ thalamic cells are not corrected if they are confronted with a normal environment through which to grow—the gene must be needed by the thalamus itself for normal development of its cortical projections.

Does Pax6 Regulate Separate Sets of Genes in Morphogenesis and Guidance?

To regulate morphogenetic processes of cell proliferation, migration and fate determination, Pax6 controls the expression of a wide range of molecules, including transcription factors, cell adhesion and cell-cell signalling molecules, hormones and structural proteins (Simpson and Price, 2002). At present, too little is known about the targets of Pax6 to know whether or not Pax6 might regulate the same, overlapping or distinct sets of target genes during early morphogenesis and later axon guidance. As discussed above, there is strong evidence that Pax6 regulates cell-cell adhesion during brain morphogenesis and this control is likely to be equally important during axon pathfinding. Further work is needed to discover what the targets of Pax6 are and whether they change during development.

Regulation of Genes that Might be Involved in Guidance

It is most likely that this involves regulating the transcription of members of the molecular network that connects guidance cues with the cytoskeleton to control growth cone behaviour. It is possible that a lack of Pax6 alters the expression of a number of members of the network and that the combined effect causes a failure of thalamic responsiveness. A simplified list of many known members of the network is given in Fig. 3; the top rows include guidance cues shown to play or likely to play important roles in thalamocortical development. There is evidence from other systems that the expression of some of these molecules is affected by Pax6.

(i) Semaphorins are a large family of secreted and membrane-associated proteins that are chemorepellant or chemoattractant. They are grouped into 8 classes; vertebrate semaphorins are in classes 3–7 (Semaphorin Nomenclature Committee, 1999). Semaphorins are expressed in and around the developing thalamocortical pathway (Skaliora et al., 1998). *In vitro*, thalamocortical axons are responsive to at least one of these, secreted Sema3A (Bagnard et al., 2001). Mice lacking the transmembrane Sema6A, which is proposed to act in thalamocortical axons as a guidance receptor, have thalamocortical defects similar to (although less severe than) those in *Pax6*$^{-/-}$embryos (Leighton et al., 2001). Thus, Sema6A is a good candidate as one potential direct or indirect target of Pax6 in dorsal thalamus. In addition, expression of Sema5A and Sema3C are altered in the telencephalon of mice lacking Pax6 and this has been suggested to contribute to thalamocortical defects in these mutants (Jones et al., 2002). Neuropilins are receptors

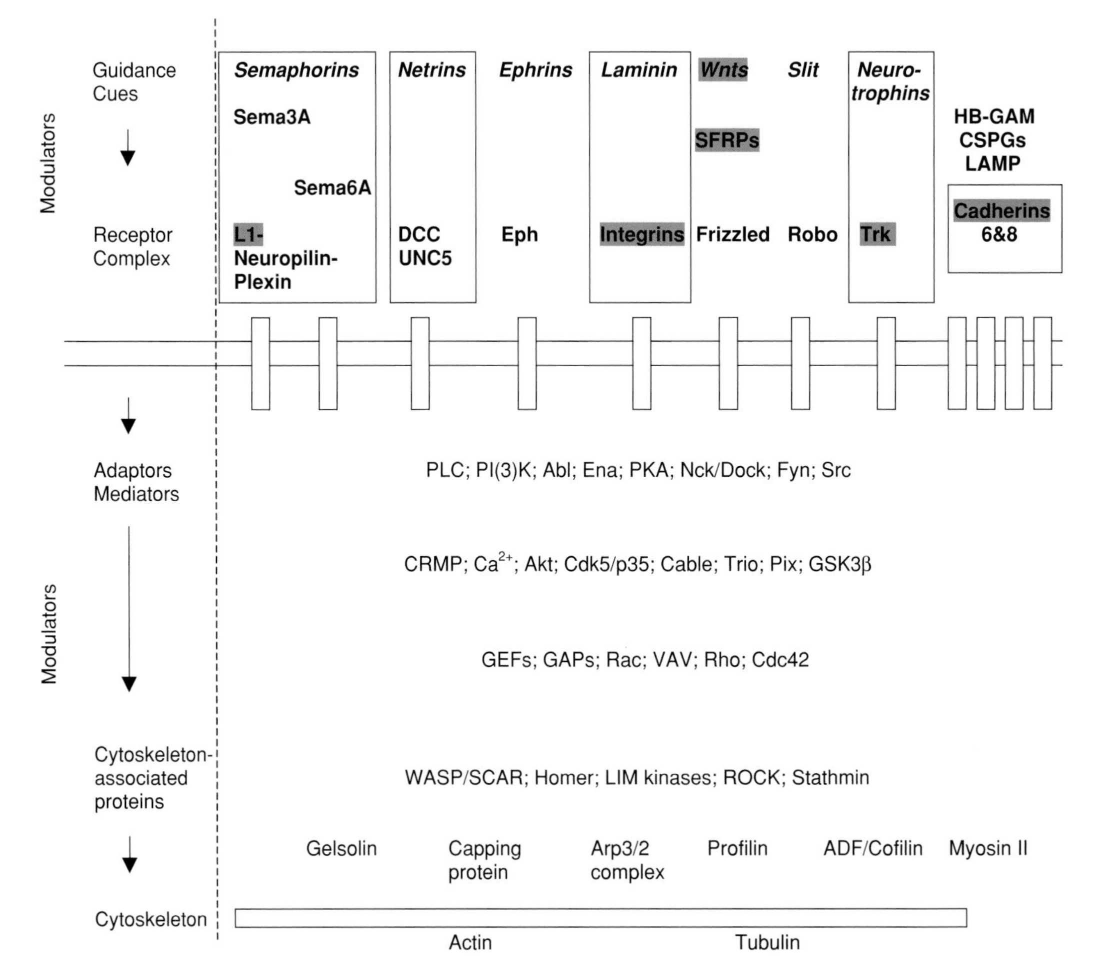

Figure 3 This diagram shows many of the molecules or classes of molecule that are likely to be involved in directing axonal growth, from extracellular cues to cytoskeletal rearrangement. Types of molecule are listed to the left of the broken vertical line; individual molecules or families of molecule are listed to the right. Molecules that are outside the cell or are components of the cell membrane are in bold at the top of the diagram. In many cases the cues interact with receptors, which are lined up below the corresponding ligands. Highlighting is used to identify molecules whose expression is regulated by Pax6. Families of molecule enclosed by boxes are those which include members that are prime candidates for being regulated in the thalamus, directly or indirectly, by Pax6. Many molecules inside the cell might have their expression affected by Pax6, but evidence is lacking at present: for simplicity, the pathways that may link these molecules are not drawn. Many of the families of molecule indicated are very large. Literature on which this diagram is based is cited in Song and Poo (2001) and in the text.

for class 3 secreted semaphorins; they complex with plexin and neural cell adhesion molecule L1 to form Sema3A receptors (Rohm et al., 2000; Castellani et al., 2000). *In vitro* experiments have indicated that *Pax6* can bind to specific sequences in the L1 promoter (Chalepakis et al., 1994), although L1 is still expressed at high level in *Pax6*$^{-/-}$embryos (Vitalis et al., 2000). Nevertheless, defects of

L1-neuropilin-plexin receptors may also contribute to thalamocortical defects in *Pax6*$^{-/-}$embryos.

(ii) Netrins include (a) diffusible proteins, whose attractive effects are mediated *via* receptors of the DCC (Deleted in colorectal cancer) family (DCC and neogenin) and whose repulsive effects require members of the UNC5 family, and (b) a membrane-linked member expressed at sites that include embryonic thalamus (Nakashiba et al., 2000). Diffusible Netrin1 is present in ventral telencephalon and, acting via DCC and neogenin receptors on dorsal thalamic axons, it may play a role in guiding thalamocortical axons through the ventral telencephalon (Braisted et al., 2000). Defects of thalamocortical axons are much less severe in loss-of-function mutation of Netrin1 than in *Pax6*$^{-/-}$embryos. One intriguing possibility is that *Pax6*$^{-/-}$dorsal thalamic neurons upregulate UNC5 receptors thereby converting a normally chemoattractive effect of Netrin1 into a chemorepulsive effect and so preventing thalamocortical development.

(iii) Ephrins and Eph receptor tyrosine kinases are involved in processes including growth cone guidance (Wilkinson, 2001) and Eph receptors and ephrins are expressed in the developing thalamocortical system. In particular, a role for ephrin-A5 in thalamocortical development has been suggested on the basis of expression and *in vitro* data (Gao et al., 1998; Mackarehtschian et al., 1999), although thalamocortical axons do form in mice lacking ephrin-A5 these do exhibit subtle mapping errors in their synaptic connections with the cerebral cortex (Vanderhaeghen et al., 2000). Interestingly, work in other systems has shown that the actions of Eph receptors and ephrin-A5 involve activation of integrins including β1-integrin, which may be directly regulated by Pax6 at least in the lens (Duncan et al., 2000; Davy and Robbins, 2000). Integrins have been shown to play an important role in growth cone motility (Condic and Letourneau, 1997).

(iv) Neurotrophins, which act *via* Trk receptors, have been implicated as chemoattractants in the developing nervous system (Gallo and Letourneau, 2000) and thalamic axons do respond to members of this family (Lotto and Price, 1995). Furthermore, there is evidence that Pax6 directly or indirectly regulates the expression of Trk receptors in the developing cortex, although the thalamus was not investigated (Warren et al., 1999).

(v) Other diffusible molecules that need to be considered include members of the Wnt family, which signal through Frizzled receptors and whose actions are modulated by secreted frizzled related proteins (SFRPs). Their possible involvement is suggested by findings that Wnt7a regulates axonal development in cerebellum (Lucas and Salinas, 1997) and that Pax6 regulates forebrain expression of Wnt7b and SFRP-2 (Kim et al., 2001). It is possible that expression of Frizzled receptors or SFRPs in thalamus may be disrupted in *Pax6*$^{-/-}$ embryos. Robo receptors, highly conserved molecules that mediate the chemorepulsive activity of secreted Slits (Erskine et al., 2000), are expressed in dorsal thalamus and so may be involved in thalamocortical axon guidance (Nakagawa and O'Leary, 2001).

(vi) Cell adhesion and extracellular matrix molecules (ECMs) are essential in axon guidance. Pax6 is required for normal cortical expression of members of the cadherin family of calcium-dependent cell adhesion glycoproteins (Stoykova et al., 1997; Bishop et al., 2000). Cadherins present in the developing thalamocortical system include cadherin-6 and cadherin-8 (Rubenstein et al., 1999); their expression in the thalamus of *Pax6*$^{-/-}$embryos remains to be investigated. Chondroitin sulphate proteoglycans (CSPGs), heparin-binding growth-associated molecule (HB-GAM) and limbic system-associated membrane protein (LAMP) are suggested to play roles in thalamocortical development (Mann et al., 1998; Kinnunen et al., 1999); the possibility that their expression is regulated by Pax6 has yet to be tested.

Conclusion

Mutant mice show that the transcription factors discussed here are required for the morphogenesis of forebrain structures projecting and receiving axons and for axon navigation in the forebrain. This efficient use of genes may explain the massive biological diversity delivered by a relatively small number of genes. We predict that the list of transcription factors playing multiple roles in tissue morphogenesis and axon guidance will increase. The next challenge will be the comprehensive identification of their transcriptional targets and a molecular biological dissection of their various functions.

References

Andrews GL, Mastick GS. (2003) R-cadherin is a Pax6-regulated, growth-promoting cue for pioneer axons. J Neurosci. 23:9873–80

Ashery-Padan R, Marquardt T, Zhou X, Gruss P. (2000) Pax6 activity in the lens primordium is required for lens formation and for correct placement of a single retina in the eye. Genes Dev. 14:2701–11.

Auladell, C. Perez-Sust P, Super H, Soriano E. (2000) The early development of thalamocortical and corticothalamic projections in the mouse. Anat Embryol. 201, 169–179.

Bagnard, D., Chounlamountri N, Puschel AW, Bolz J. (2001) Axonal surface molecules act in combination with semaphorin 3A during the establishment of corticothalamic projections. Cer. Cortex 11: 278–285.

Barbieri, A. M., Broccoli, V., Bovolenta, P., Alfano, G., Marchitiello, A., Mocchetti, C., Crippa, L., Bulfone, A., Marigo, V., Ballabio, A. and Banfi, S. (2002). Vax2 inactivation in mouse determines alteration of the eye dorsal-ventral axis, misrouting of the optic fibres and eye coloboma. Development 129, 805–813

Baumer N, Marquardt T, Stoykova A, Ashery-Padan R, Chowdhury K, Gruss P. (2002). Pax6 is required for establishing naso-temporal and dorsal characteristics of the optic vesicle.Development. 129:4535–45.

Bertuzzi S, Hindges R, Mui SH, O'Leary DD, Lemke G. (1999). The homeodomain protein vax1 is required for axon guidance and major tract formation in the developing forebrain. Genes Dev. 13:3092–105.

Bishop, K.M. Goudreau G, O'Leary DD. (2000) Regulation of area identity in the mammalian neocortex by Emx2 and Pax6. Science, 288: 344–349

Braisted, J.E. Catalano SM, Stimac R, Kennedy TE, Tessier-Lavigne M, Shatz CJ, O'Leary DD. (2000) Netrin-1 promotes thalamic axon growth and is required for proper development of the thalamocortical projection. J. Neurosci. 20: 5792–5801.

Braisted, J.E., Tuttle R, O'leary DD. (1999). Thalamocortical axons are influenced by chemorepellent and chemoattractant activities localized to decision points along their path. Dev. Biol. 208, 430–440.

Caric D, Gooday D, Hill RE, McConnell SK, Price DJ. (1997) Determination of the migratory capacity of embryonic cortical cells lacking the transcription factor Pax-6. Development 124:5087–96.

Castellani, V., Chedotal A, Schachner M, Faivre-Sarrailh C, Rougon G. (2000) Analysis of the L1-deficient mouse phenotype reveals cross-talk between Sema3A and L1 signalling pathways in axonal guidance. Neuron 27: 237–249.

Cepko CL, Austin CP, Yang X, Alexiades M, Ezzeddine D. (1996). Cell fate determination in the vertebrate retina.Proc Natl Acad Sci U S A 93:589–95.

Chalepakis, G. Wijnholds J, Giese P, Schachner M, Gruss P. (1994) Characterization of Pax6 and Hoxa-1 binding to the promoter region of the neural cell adhesion molecule, L1. DNA Cell Biol. 13: 891–900.

Collinson, M., Hill, R.E. and West, J.D. (2000) Different roles for Pax6 in the optic vesicle and facial epithelium mediate early morphogenesis of the murine eye. Development 127: 945–956.

Condic, M.L. and Letourneau, P.C. (1997) Ligand-induced changes in integrin expression regulate neuronal adhesion and neurite outgrowth. Nature 389: 852–856.

Davy, A. and Robbins, S.M. (2000) Ephrin-A5 modulates cell adhesion and morphology in an integrin-dependent manner. EMBO J. 19: 5396–5405.

Deiner MS, Sretavan DW. (1999) Altered midline axon pathways and ectopic neurons in the developing hypothalamus of netrin-1- and DCC-deficient mice. J Neurosci. 19:9900–12.

Deiner MS, Kennedy TE, Fazeli A, Serafini T, Tessier-Lavigne M, Sretavan DW. (1997) Netrin-1 and DCC mediate axon guidance locally at the optic disc: loss of function leads to optic nerve hypoplasia. Neuron. 19:575–89

Duncan, M.K. Kozmik Z, Cveklova K, Piatigorsky J, Cvekl A. (2000) Overexpression of PAX6(5a) in lens fiber cells results in cataract and upregulation of $\alpha 5\beta 1$ integrin expression. J. Cell Sci. 113: 3173–3185.

Edgar, J.M. and Price, D.J. (2001) Radial migration in the cerebral cortex is enhanced by signals from thalamus. European Journal of Neuroscience 13: 1745–1754.

Erskine, L. et al (2000) Retinal ganglion cell axon guidance in the mouse optic chiasm: expression and function of Robos and Slits. J. Neurosci. 20: 4975–4982.

Gallo, G. and Letourneau, P.C. (2000) Neurotrophins and the dynamic regulation of the neuronal cytoskeleton. J. Neurobiol., 44: 159–173.

Gao, P-P. Yue Y, Zhang JH, Cerretti DP, Levitt P, Zhou R. (1998) Regulation of thalamic neurite outgrowth by the Eph ligand ephrin-A5: implications in the development of thalamocortical projections. Proc. Natl. Acad. Sci. USA 95: 5329–5334.

Gotz M, Stoykova A, Gruss P. (1998) Pax6 controls radial glia differentiation in the cerebral cortex. Neuron. 21:1031–44.

Grindley, JC, Hargett, LK, Hill, RE, Ross, A., and Hogan BLM (1997). Disruption of Pax6 function in mice homozygous for the $Pax6^{Sey\text{-}Neu}$ mutation produces abnormalities in the early development and regionalization of the diencephalon. Mech. Devel., 64, 111–126

Hallonet M, Hollemann T, Pieler T, Gruss P. (1999). Vax1, a novel homeobox-containing gene, directs development of the basal forebrain and visual system. Genes Dev. 23:3106–14.

Hallonet M, Hollemann T, Wehr R, Jenkins NA, Copeland NG, Pieler T, Gruss P. (1998). Vax1 is a novel homeobox-containing gene expressed in the developing anterior ventral forebrain. Development. 14:2599–610.

Heins N, Malatesta P, Cecconi F, Nakafuku M, Tucker KL, Hack MA, Chapouton P, Barde YA, Gotz M. (2002) Glial cells generate neurons: the role of the transcription factor Pax6. Nat Neurosci. 5:308–15

Herrera, E., Brown, L., Aruga, J., Rachel, R. A., Dolen, G., Mikoshiba, K., Brown, S. and Mason, C. A. (2003). Zic2 patterns binocular vision by specifying the uncrossed retinal projection. Cell 114, 545–57

Herrera E, Marcus R, Li S, Williams SE, Erskine L, Lai E, Mason C. (2004) Foxd1 is required for proper formation of the optic chiasm. Development. 131:5727–39.

Hevner RF, Miyashita-Lin E, Rubenstein JL. (2002) Cortical and thalamic axon pathfinding defects in Tbr1, Gbx2, and Pax6 mutant mice: evidence that cortical and thalamic axons interact and guide each other. J Comp Neurol. 447: 8–17.

Hill RE, Favor J, Hogan BL, Ton CC, Saunders GF, Hanson IM, Prosser J, Jordan T, Hastie ND, van Heyningen V. (1991) Mouse small eye results from mutations in a paired-like homeobox-containing gene. Nature 354:522–5.

Huh, S., Hatini, V., Marcus, R. C., Li, S. C. and Lai, E. (1999). Dorsal-ventral patterning defects in the eye of BF-1-deficient mice associated with a restricted loss of shh expression. Developmental Biology 211, 53–63.

Jones L, Lopez-Bendito G, Gruss P, Stoykova A, Molnar Z. (2002) Pax6 is required for the normal development of the forebrain axonal connections. Development 129:5041–52.

Kawano, H., Fukuda T, Kubo K, Horie M, Uyemura K, Takeuchi K, Osumi N, Eto K, Kawamura K. (1999) Pax-6 is required for thalamocortical pathway formation in fetal rats. J. Comp. Neurol. 408, 147–160.

Kim, A.S. Anderson SA, Rubenstein JL, Lowenstein DH, Pleasure SJ. (2001) Pax-6 regulates expression of SFRP-2 and Wnt-7b in the developing CNS. J. Neurosci. 21, RC132 (1–5)

Kinnunen, A. Niemi M, Kinnunen T, Kaksonen M, Nolo R, Rauvala H. (1999) Heparan sulphate and HB-GAM (heparin-binding growth-associated molecule) in the development of the thalamocortical pathway of rat brain. Eur. J. Neurosci. 11: 491–502.

Leighton, P.A., Mitchell KJ, Goodrich LV, Lu X, Pinson K, Scherz P, Skarnes WC, Tessier-Lavigne M. (2001) Defining brain wiring patterns and mechanisms through gene trapping in mice. Nature 410: 174–179.

Lopez-Bendito G, Molnar Z. (2003) Thalamocortical development: how are we going to get there? Nat Rev Neurosci. 4:276–89.

Lotto, R.B. and Price, D.J. (1995) The stimulation of thalamic neurite outgrowth by cortical derived growth factors in vitro; the influence of cortical age and activity. Eur. J. Neurosci., 7: 318–328.

Lucas, F.R. and Salinas, P.C. (1997) WNT-7a induces axonal remodeling and increases synapsin I levels in cerebellar neurons. Dev. Biol. 192: 31–44.

Mackarehtschian, K. Lau CK, Caras I, McConnell SK. (1999) Regional differences in the developing cerebral cortex revealed by ephrin-A5 expression. Cer. Cortex 9: 601–610.

Mann, F., Zhukareva V, Pimenta A, Levitt P, Bolz J. (1998) Membrane-associated molecules guide limbic and nonlimbic thalamocortical projections. J. Neurosci. 18: 9409–9419.

Marcus RC, Shimamura K, Sretavan D, Lai E, Rubenstein JL, Mason CA. (1999) Domains of regulatory gene expression and the developing optic chiasm: correspondence with retinal axon paths and candidate signaling cells. J Comp Neurol.403:346–58.

Marquardt, T., Ashery-Padan R, Andrejewski N, Scardigli R, Guillemot F, Gruss P. (2001) Pax6 is required for the multipotent state of retinal progenitor cells. Cell 105: 43–55.

Mastick, G. S., Davis, N. M., Andrews, G. L., Easter, S. S. Jr. (1997). Pax6 functions in boundary formation and axon guidance in the embryonic mouse forebrain. Development 124, 1985–1997.

Metin, C. and Godement, P. (1996) The ganglionic eminence may be an intermediate target for corticofugal and thalamocortical axons. J. Neurosci. 16: 3219–3235.

Miyashita-Lin EM, Hevner R, Wassarman KM, Martinez S, Rubenstein JL. (1999) Early neocortical regionalization in the absence of thalamic innervation. Science 285:906–9

Molnar, Z. (1998) Development of thalamocortical connections. Springer-Verlag.

Molnar Z, Adams R, Blakemore C. (1998) Mechanisms underlying the early establishment of thalamocortical connections in the rat. J Neurosci. 18:5723–45

Nagai T, Aruga J, Minowa O, Sugimoto T, Ohno Y, Noda T, Mikoshiba K. (2000). Zic2 regulates the kinetics of neurulation. Proc Natl Acad Sci U S A. 15, 1618–23.

Mui, S. H., Hindges, R., O'Leary, D. D., Lemke, G. and Bertuzzi S. (2002). The homeodomain protein Vax2 patterns the dorsoventral and nasotemporal axes of the eye. Development 129, 797–804.

Nakagawa, Y. and O'Leary, D.D.M. (2001) Combinatorial expression patterns of LIM-homeodomain and other regulatory genes parcellate developing thalamus. J. Neurosci. 21: 2711–2725.

Nakashiba, T. Ikeda T, Nishimura S, Tashiro K, Honjo T, Culotti JG, Itohara S. (2000) Netrin-G1: a novel glycosyl phosphatidylinositol-linked mammalian netrin that is functionally divergent from classical netrins. J. Neurosci. 20: 6540–6550.

Nural HF, Mastick GS. (2004) Pax6 guides a relay of pioneer longitudinal axons in the embryonic mouse forebrain. J Comp Neurol. 479:399–409

Pak W, Hindges R, Lim YS, Pfaff SL, O'Leary DD. (2004). Magnitude of binocular vision controlled by islet-2 repression of a genetic program that specifies laterality of retinal axon pathfinding. Cell. 119:567–78.

Plump AS, Erskine L, Sabatier C, Brose K, Epstein CJ, Goodman CS, Mason CA, Tessier-Lavigne M. (2002). Slit1 and Slit2 cooperate to prevent premature midline crossing of retinal axons in the mouse visual system. Neuron. 33:219–32.

Pratt, T., Vitalis, T., Warren, N., Edgar, J.M., Mason, J.O. and Price, D.J. (2000) A role for Pax6 in the normal development of dorsal thalamus and its cortical connections. Development, 127: 5167–5178.

Pratt, T., J.C. Quinn, T.I. Simpson, J.D. West, J.O. Mason, D.J. Price (2002) Disruption of early events in thalamocortical tract formation in mice lacking the transcription factors Pax6 or Foxg1. Journal of Neuroscience, 22: 8523–8531.

Pratt, T., N.M.M.-L. Tian, T.I. Simpson, J.O. Mason and D.J. Price (2004) The winged helix transcription factor Foxg1 facilitates retinal ganglion cell axon crossing of the ventral midline in the mouse. Development 2004 131: 3773–3784.

Quinn, J.C., West. J.D. and Hill, R.E. (1996) Multiple functions for Pax6 in mouse eye and nasal development. Genes and Development 10: 435–446.

Rohm, B., Ottemeyer A, Lohrum M, Puschel AW. (2000) Plexin/neuropilin complexes mediate repulsion by the axonal guidance signal semaphorin 3A. Mech. Dev. 93: 95–104.

Rubenstein, J.L.R. Anderson S, Shi L, Miyashita-Lin E, Bulfone A, Hevner R. (1999) Genetic control of cortical regionalization and connectivity. Cer. Cortex 9: 524–532.

Schedl A, Ross A, Lee M, Engelkamp D, Rashbass P, van Heyningen V, Hastie ND. (1996) Influence of PAX6 gene dosage on development: overexpression causes severe eye abnormalities. Cell. 86:71–82.

Schmahl W, Knoedlseder M, Favor J, Davidson D. (1993) Defects of neuronal migration and the pathogenesis of cortical malformations are associated with Small eye (Sey) in the mouse, a point mutation at the Pax-6-locus. Acta Neuropathol (Berl). 86:126–35.

Schwarz M, Cecconi F, Bernier G, Andrejewski N, Kammandel B, Wagner M, and Gruss P. (2000) Spatial specification of mammalian eye territories by reciprocal transcriptional repression of Pax2 and Pax6. Development 127: 4325-34.

Semaphorin Nomenclature Committee (1999) Unified Nomenclature for the Semaphorins/Collapsins. Cell 97: 551–552

Sisodiya SM, Free SL, Williamson KA, Mitchell TN, Willis C, Stevens JM, Kendall BE, Shorvon SD, Hanson IM, Moore AT, van Heyningen V. (2001) PAX6 haploinsufficiency causes cerebral malformation and olfactory dysfunction in humans. Nat. Genet. 28:214–6.

Simpson, T.I and Price, D.J. (2002) Pax6; a pleiotropic player in development. Bioessays, 24: 1041–1051.

Skaliora, I., Singer W, Betz H, Puschel AW. (1998) Differential patterns of semaphorin expression in the developing rat brain. Eur. J. Neurosci. 10: 1215–1229.

Song, H-j. and Poo, M-m. (2001) The cell biology of neuronal navigation. Nature Cell Biol. 3: E81–E88.

Smith R.S., Kao, W.W.-Y, John, S.W.M (2002) Ocular Development in Systematic Evaluation of the Mouse Eye Eds Smith R.S. CRC Press.

Stoykova, A. Fritsch R, Walther C, Gruss P. (1996) Forebrain patterning defects in *Small eye* mutant mice. Development 122: 3453–3465.

Stoykova, A., Gotz M, Gruss P, Price J. (1997) *Pax6*-dependent regulation of adhesive patterning, *R-cadherin* expression and boundary formation in developing forebrain. Development 124: 3765–3777

Takahashi H, Shintani T, Sakuta H, Noda M. (2003) CBF1 controls the retinotectal topographical map along the anteroposterior axis through multiple mechanisms. Development. 130:5203–15.

Talamillo A, Quinn JC, Collinson JM, Caric D, Price DJ, West JD, Hill RE. (2003) Pax6 regulates regional development and neuronal migration in the cerebral cortex. Dev Biol. 255:151–63.

Torres, M., Gomez-Pardo, E. and Gruss, P. (1996). Pax2 contributes to inner ear patterning and optic nerve trajectory. Development 122, 3381–91.

Tuttle, R., Nakagawa Y, Johnson JE, O'Leary DD. (1999). Defects in thalamocortical axon pathfinding correlate with altered cell domains in Mash-1-deficient mice. Development 126, 1903–1916.

Tyas DA, Pearson H, Rashbass P, Price DJ. (2003) Pax6 regulates cell adhesion during cortical development. Cereb Cortex. 13:612–9.

Vanderhaeghen, P. Lu Q, Prakash N, Frisen J, Walsh CA, Frostig RD, Flanagan JG. (2000) A mapping label required for normal scale of body representation in the cortex. Nature Neurosci. 3: 358–365.

Vitalis, T., Cases, O., Engelkamp, D., Verney, C. and Price, D.J. (2000) Defects of tyrosine hydroxylase-immunoreactive neurons in the brain of mice lacking the transcription factor Pax6. J. Neurosci., 20: 6501–6516.

Warren, N. and Price, D.J. (1997) Roles of Pax-6 in murine diencephalic development. Development, 124: 1573–1582.

Warren, N., Caric, D., Pratt, T., Clausen, J.A., Asavaritikrai, P., Mason, J.O., Hill, R.E. and Price, D.J. (1999) The transcription factor, *Pax6*, is required for cell proliferation and differentiation in the developing cerebral cortex. Cer. Cortex, 9, 627–635.

Wilkinson, D.G. (2001) Multiple roles of Eph receptors and ephrins in neural development. Nature Rev. Neurosci. 2: 155–164.

Williams, S. E., Mann, F., Erskine, L., Sakurai, T., Wei, S., Rossi, D.J., Gale, N. W., Holt, C. E., Mason, C. A. and Henkemeyer, M. (2003). Ephrin-B2 and EphB1 mediate retinal axon divergence at the optic chiasm. Neuron 39, 919–35.

Xuan, S., Baptista, C. A., Balas, G., Tao, W., Soares, V. C. and Lai, E. (1995). Winged helix transcription factor BF-1 is essential for the development of the cerebral hemispheres. Neuron 14, 1141–52.

Yuasa J, Hirano S, Yamagata M, Noda M. (1996) Visual projection map specified by topographic expression of transcription factors in the retina. Nature. 382:632–5.

Ziman M, Rodger J, Lukehurst S, Hancock D, Dunlop S, Beazley L. (2003). A dorso-ventral gradient of Pax6 in the developing retina suggests a role in topographic map formation.Brain Res Dev Brain Res. 140:299–302.

3

Subcortical and Neocortical Guidance of Area-specific Thalamic Innervation

Tomomi Shimogori and Elizabeth A. Grove

Abstract

The specialized functions of neocortical areas depend on patterned innervation from the thalamus. Recent evidence suggests several key molecules for correct connections of thalamocortical axons (TCAs) between specific thalamic nuclei and cortical areas. First, the correct area map has to be generated in cortex, and thalamic nuclei need to be specified in correct patterns. Second, axon guidance molecules that are expressed in the TCA subcortical pathway need to be established. Finally, guidance within the cortex is needed for TCAs to navigate to their target area and prevent them from being misrouted. Here, we discuss thalamocortical axon guidance mechanisms, especially somatosensory axon guidance, during the development of the rodent brain.

Introduction

Virtually all the information we receive from the outside world, such as visual, auditory and somatosensory sensation, passes through the thalamus to the cerebral cortex. Functioning of the nervous system relies on the precision of its wiring pattern. To set up the correct functional pattern of projections in the nervous system, as a one of a model, first, peripheral neurons need to be born with correct positional information. Next, axonal growth cones must find their targets and migrating along pathways that may be long and complex. Finally they reach the correct region and pass the information to next neuron for further processing. One of the most studied axon path finding procedure is the projection from the retina to the optic tectum (superior colliculus), which is a complexed pathay (Dingwell et al., 2000). To help this complicated journey, intermediate targets divide the long pathway into several steps. Retinal axons grow into the optic nerve, separate ipsi- and contralateral axonal projection at the optic chiasm, grow caudally along the diencephalon

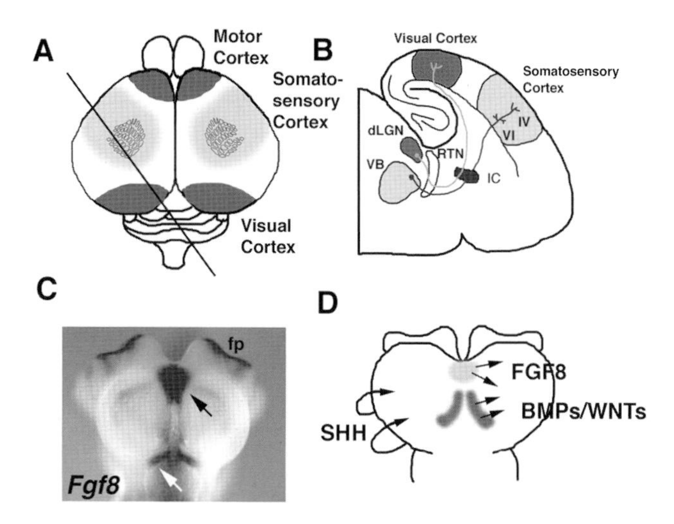

Figure 1 Scheme of cortical area map, TCAs projection pathway in postnatal brain and expression pattern of signaling molecules in embryonic brain. **A,** Major sensory cortical areas are shown in gray. Anterior is to the top. Partial barrelfield (whisker pad region) is shown in gray somatosensory region. **B,** Plane of thalamocortical slice, angle and approximate position of slice is shown in A; line on the left hemisphere. Thalamic nuclei, such as dLGN and VB send axons to RTN and IC. After they reach cortex, they project to visual cortex and somatosensory cortex individually. VB axons arborize in layer 6 and layer 4 in which barrel field clusters form. **C,** In situ hybridization expression pattern of Fgf8 in embryonic day (E) 10.5 mouse brain. Anterior is to the top. Black arrow shows the expression in telencephalon and white arrow shows the expression in telencephalon-diencephalon boundary. **D,** Illustration of expression of other signaling molecules at E10.5 mouse brain. dLGN: dorsal lateral geniculate nucleus, VB: ventro basal nucleus, fp: face primordium.

and finally enter the tectum. At each step, attractive cues orient and guide the growth cone towards its next target, or expression of repulsive cues prevent misrouting of the axons (O'Leary and Wilkinson, 1999; Williams et al., 2004).

Within the mammalian forebrain, a major ascending pathway to cortex is made up of thalamocortical axons (TCAs), which, in part, relay sensory information into organized cortical areas (Fig. 1A) to process the information further. During development, TCAs follow a complex path (simplified in Fig. 1B): neurons born in sensory nuclei such as somatosensory (VB: ventrobasal nuclei) and visual nuclei (dLGN: dorsal lateral geniculate nuclei), project axons towards the reticular thalamic nuclei (RTN), making sharp turn to enter the striatum, and another turn at the cortical–subcortical telencephalic boundary to enter the internal capsule (IC) and finally the cerebral cortex (Miller et al., 1993; Auladell et al, 2000; López-Bendito and Molnár, 2003; Molnár et al., 2003).

Somatosensory thalamus and cortex in rodents contain topological representations of the facial whisker pad (Woolsey and Van der Loos, 1970). The thalamic representation of a single whisker ("barreloid") is presumed to project exclusively to the cortical representation ("barrel").

During a brief period of postnatal development, thalamocortical axons establish two tiers of terminations in layer 6a and in layer 4, and form whisker-specific clusters within layer 4 (Senift and Woolsey 1991., Erzurumlu and Jhaveri 1990., Rebsam et al, 2002., Lee et al., 2005). For the correct projection of single barreloid in VB to S1 in cortex, which is again long and complex, there is a need to have very well organized mechanism. This includes the initial patterning of the cortical area map, expression of guidance molecules in each part of the subcortical path, and, finally, functional activity for at least some aspects of the innervation and for its maintenance (Garel and Rubenstein 2004, Erzurumlu and Kind 2001). The aim of this review will be to place these insights about the somatosensory fields in context with the literature that already exists.

The Cortical Area Map is Set up Prior to the Arrival of TCAs

Gene targeted knock out studies in mice indicate that area-specific molecular features are specified independent of thalamic input. Gene expression patterns in the embryonic brain that prefigure area boundaries still develop in mice, lacking the transcription factor Gbx2, in which thalamocortical innervation is severely disrupted (Fig. 2B) (Miyashita-Lin et al. 1999). Similar observations are reported in mice deficient in the transcription factor Mash-1 (Tuttle et al. 1999). This correlation of thalamic innervation and the cortical map seems to fail in the opposite direction too. The topography of TCA projections is shifted in mice deficient in the transcription factors Ebf1 and Dlx1/2, yet, in this case, cortical regional gene expression is unaltered (Garel et al. 2003) (Fig. 2C). Thus, molecular regionalization that anticipates the cytoarchitectonic area map arises in spite of disturbed or even absent thalamic input.

Patterning of the head and forebrain, which incorporate cerebral cortex, depends on sources of signaling molecules including Bone Morphogenetic Proteins (BMPs), WNTs, and their antagonist proteins (Bachiller et al. 2000, Kiecker and Niehrs 2001, Nordstrom et al. 2002). Signaling centers determine dorsal/ventral or anterior/posterior (A/P) axes in the spinal cord and at the midbrain-hindbrain boundary (Fig. 1C and D) (Briscoe and Ericson, 2001, Crossley et al. 1996, Lee et al. 2000, Shamim et al. 1999, Wurst and Bally-Cuif, 2001). We previously found that the neocortical area map can be rearranged by manipulating FGF8 levels in the cortical primordium (Fukuchi-Shimogori and Grove, 2001) (Fig. 3). Using in utero microelectroporation to augment or diminish an anterior telencephalic FGF8 source, or to create a new source, areas can be shifted posteriorly or anteriorly in the map, or even partially duplicated (Fukuchi-Shimogori and Grove, 2001). Duplication of the somatosensory barrel fields, whose formation and maintenance reflects innervation from the VB in thalamus (Jeanmonod et al, 1981; Van der Loos and Woolsey, 1973.), suggested, first, that the new, induced barrels were innervated, and, second, that our molecular manipulations of

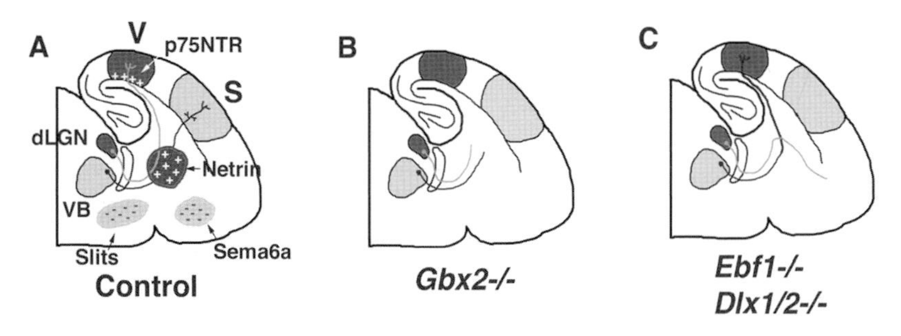

Figure 2 Expression of guidance molecules and axon trajectory and projection defects in knock out mice. **A,** Thalamocortical slice shows the expression of guidance molecules along the thalamocortical trajectory. Expression of Slits in hypothalamus and Sema6a in basal ganglia prevents thalamic axon projection into those areas and netrin expression in IC attracts axons. In cortex, p75NTR shows area identity and is suggested to work as an attractant. **B,** Gbx2 deficient mouse show no TCA arrival in cortex without any patterning defect in thalamus and cortex. **C,** TCAs in Ebfl and Dlx1/2 knock out mouse are misrouted and send axons to the wrong area in cortex.

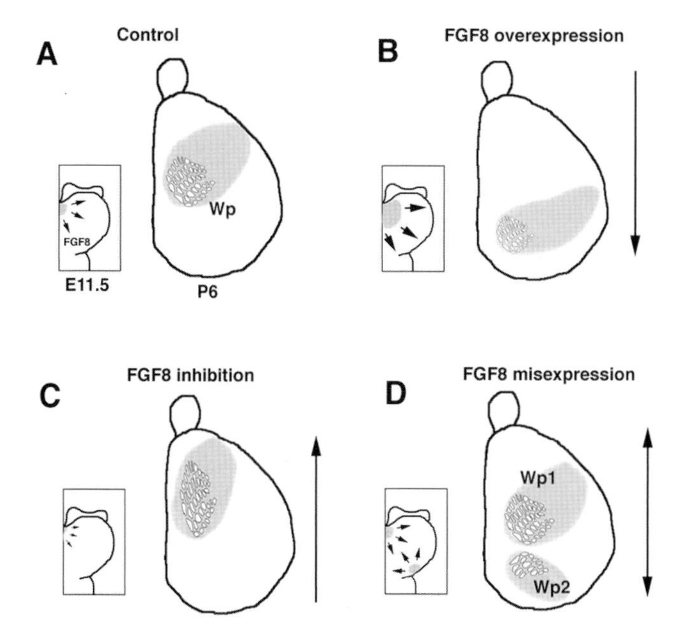

Figure 3 Scheme of cortical area pattern shifts in brains in which FGF8 is augmented, inhibited or misexpressed posteriorly. **A,** Position of somatosensory area, and whisker pad (Wp) barrel subfield in control brain. The rough size of endogenous FGF8 source at the time when FGF8 is modified by in utero electroporation is shown on left. **B,** In the FGF8 augmented brain, primary somatosensory cortex is shifted posteriorly and the shape and size of Wp is shrunken. **C,** An FGF8 depleted brain shows anterior shifted somatosensory cortex. The Wp is elongated toward anterior pole of cortex. **D,** When FGF8 is ectopically misexpressed at posterior pole, second somatosensory area is formed just posterior to the native field and its Wp pattern is a mirror image of native Wp.

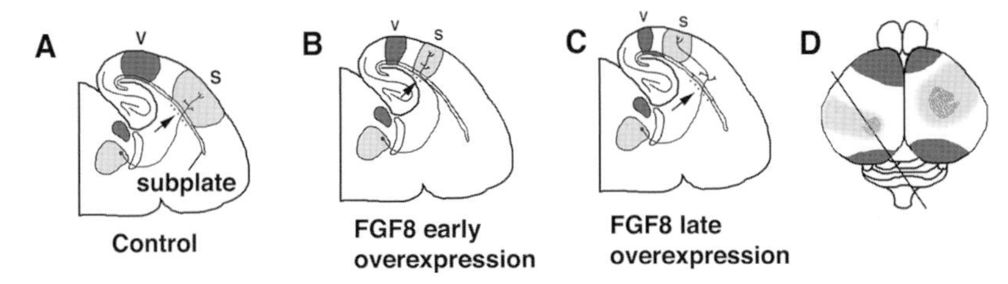

Figure 4 VB axon trajectory is shown in FGF8 early and late augmented thalamocortical section. **A,** VB axons follow the area identity in subplate to penetrate into cortex (arrow) and arborize in layer 6a and layer 4 in register. **B,** In the early FGF8 overexpressed brain, area identity in subplate and cortex are both shifted. Hence, VB axon entry into cortex and arbors in layer 6a and layer 4 are all in line. **C,** In the late FGF8 overexpressed brain, area identity of subplate is remain as same as control brain. In spite of non-shift of subplate area identity, neocortical layer area identity is shifted, which causes a "mis-match" between the VB axon entry point and arbor point in cortex. **D,** Approximate position of thalamocortical section in FGF8 augmented brain. Both in early and late FGF8 overexpressed brain show similar area shift.

the cortical area map could be used to investigate the cellular cues that direct thalamic input to specific areas of neocortex (Rakic, 2001).

TCAs CAN Find Final Targets Even They are Misguided on the Way

We followed the development of VB axons at P5, when functional areas are detectable, by placing a deposit of the fluorescent carbocyanide dye DiI into the VB (Shimogori and Grove, 2005). In control brains, in primary somatosensory cortex (S1), VB axons formed two sets of arbors in different layers, layers 6a and 4 (Fig. 4A). The cortical layers are known to form an in inside-out manner: early borne neurons migrate and form deeper layers and late born neurons migrate further to make upper layers (Bayer and Altman, 1990). From previous findings it is suggested that cells in cortex receive the information for their area identity when they are born in ventricular zone (VZ) (Bishop et al, 2000; Muzio et al, 2000; Zhou et al, 2001). Thus, maybe it is possible to change the area identity in layer specific manner by changing the timing of expression of a signaling molecule. Because electroporation-mediated gene transfer can be performed at several embryonic ages, FGF8 was overexpressed in the anterior telencephalon at two different embryonic stages; embryonic day (E) 10.5 and E11.5 (Shimogori and Grove, 2005). Despite the different timing of FGF8 misexpression, brains showed the same patterns of clusters of barrel field in shifted position. However, in the different conditions, TCAs entered into cortex from different positions along the anterior-posterior axis and arborized in layer 6a in different positions (Fig 4B and C, arrow). In spite of this different arbor position in layer 6a, genes that show area and layer-specific gene expression patterns, such as Tbr1 and CoupTFI (Hevner et al, 2001, Zhou et al., 2001) are shifted identically in both electroporation conditions (Shimogori and Grove, 2005).

Axon Guidance Cues in the Subplate

These findings provide new support for the subplate guidance role, previously reported (Ghosh et al., 1990; Ghosh and Shatz, 1992; Ghosh and Shatz 1993). In previous experiments, the subplate was locally ablated beneath visual cortex, with the result that axons from the lateral geniculate nucleus did not grow into the cortical plate (Ghosh and Shatz 1993). These observations suggest that the subplate is important for penetration of TCAs into cortex. Different groups have reported candidate genes for axon guidance molecules in the subplate (Yun et al., 2003; Mackarehtschian et al., 1999; McQuillen et al., 2002). One of the candidate genes, the Eph receptor tyrosine kinases: the Eph ligand ephrinA5 is expressed in the ventricular zone (VZ) in an anterior strong to posterior weak gradient in the rat telencephalon (Mackarehtschian et al., 1999). Closer to birth, ephrinA5 is expressed in the cortical plate and subplate. We therefore explored positional information represented by ephrinA5 in the subplate and cortical plate in FGF8 electroporated brains. In the early FGF8 augmented brain, close to the initiation of subplate formation, electroporation successfully generated a posterior shift in regional identity in the subplate, as well as in layer 4 of presumptive S1 (Shimogori and Grove, 2005). In contrast, after FGF8 electroporation at E11.5 or E12.5, regional markers of neocortical layers were shifted posteriorly, as previously reported (Fukuchi-Shimogori and Grove, 2001), but, dramatically, the subplate was not shifted (Shimogori and Grove, 2005). Analysis of ephrinA5 mutants has revealed a distortion of the shape and size of the S1 map but did not show any disruption of the topographic precision of the projection, suggesting that genetic redundancy may be obscuring the full extent of the role of ephrin/Eph genes in this system (Vanderhaeghen et al., 2000).

The second possible candidate molecule expressed in subplate is the low-affinity NGF receptor p75, which has a gradient expression pattern that is stronger caudally, in the presumptive visual area (McQuillen et al., 2002). Mice lacking p75NTR have diminished innervation of visual cortex from dLGN, but projections from the thalamus to somatosensory and auditory cortex are normal. These observations strongly support an important role for the subplate as TCAs invade their target region of cortex.

Existence of Axon Guidance Cues IN Cortex

In the later FGF8 augmented brain, VB axons exit from the internal capsule (IC) in a normal position, but after they pass the subplate, they project within the cortex to the shifted S1 (Fig. 4C). This suggests the existence of guidance cues in cortex independent from subplate cues. To test the hypothesis, VB axons were traced in brains, which have duplicate barrel fields (Shimogori and Grove, 2005). A second, ectopic, barrel field was induced by a new source of FGF8 in the posterior cortical primordium (Fig. 3D). Both the native and ectopic fields were innervated

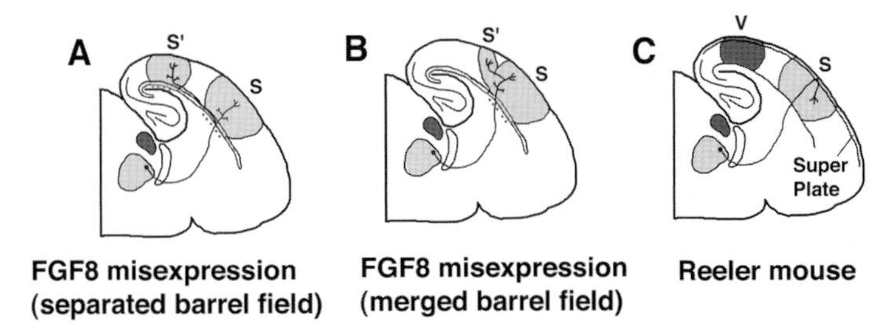

Figure 5 VB axon trajectories in brains with two barrel fields. **A,** A brain with separated barrel fields and matched area identities in subplate and in cortex. Two branches from single VB axons in subplate are heading toward their two target areas, which are in opposite directions. **B,** In a brain which two barrel fields are merged because of their close duplication, the subplate does not appear to be separated in two distinct areas. Hence, VB axons enter from one point and after they cross the subplate, they split in two in the cortex. **C,** Reeler mutant mice show inverted layers, including an inverted superplate, which becomes a superplate (Caviness, 1982). VB axons track to the superplate and only then return to the cortical plate.

by the VB (Fig. 5). In brains, with two barrel fields, VB axons could be followed and seen to bifurcate in the subplate (Fig. 5A), sending branches towards the two separate fields. In brains with merged barrel fields, VB axons also crossed between the two fields but in the cell layers of neocortex instead of in subplate (Fig. 5B). We presume that the induction of double barrel fields by two sources of FGF8 reflects the duplication of positional information within the neocortical primordium (Grove and Fukuchi-Shimogori, 2003). The present findings indicate that this doubling of positional value translates into a duplication of robust, regional axon guidance cues in both the subplate and the cortical plate.

The expression of candidate guidance genes in the cortex has been reported previously. EphA family members show area-related pattern in cortical layers in the mouse and monkey (Yun et al., 2003). Especially at birth in mouse brain, EphA gene expression patterns appear in distinctive neocortical compartments. In the study of ephrin/Eph coupled mutant mice, it has been shown that ephrinA5/EphA4 interaction is required for the topographic mapping of TCAs within the somatosensory area (Dufour et al., 2003). Furthermore, from the analysis of mice deficient in Slit1/2, which is expressed in developing telencephalon, suggest a role in axon guidance in cortex (Fig. 2A) (Bagri et al., 2002; Whitford et al., 2002).

Different Response of Pre- AND Post-Subplate Crossing Axons TO Cortical Cues

A comparison of axon trajectories in the two conditions – coordinated shifts in the cortical plate and subplate, versus a mismatch – suggests a required temporal sequence of cues. In the first condition, VB axons shifted their trajectory almost entirely in the subplate, initially ignoring intracortical guidance cues. Thalamocortical axons also ignore intracortical cues after the subplate ablation (Ghosh et al., 1990), and in the *reeler* mouse, where axons cross the inverted cortical plate to an ectopic 'superplate' (Fig. 5C). Only then do axons turn back to innervate their target area (Molnár et al., 1998). Together these observations indicate that thalamic axons encounter the subplate, and then acquire the ability to respond to cortical plate cues. For this scheme to work, growth cones need to change their responsiveness to guidance cues as they pass subplate and enter into cortex. This is reminiscent of findings from Flanagan's lab (Brittis et al., 2002), showing that when spinal axons cross the midline, growth cones synthesize new proteins that allow them to respond to available cues differently.

Other Determinants that Control Projection

When TCAs come close to cortex, they are guided by the cues in subplate and cortex. What are the factors mediating positional information and controlling TCAs till they get to the cortex? First, possible several guidance molecules have been suggested (Braisted et al., 1999; Braisted et al., 2000; Garel et al., 2002). Both in vitro and *in vivo* analysis showed that expression of Slits in hypothalamus force TCAs to make a sharp turn toward telencephalon (Bagri et al., 2002). Recent experiments have also revealed a role of neurogenin2 (ngn2): a basic HLH transcription factor, involved in neuronal determination and differentiation (Seibt et al., 2003). Inactivation of ngn2, which is expressed in developing diencephalon, did not appear to have a major effect on thalamic regional specification but did modify the pathfinding of TCAs projection. This suggest determinants in thalamus is also required for the specific responsiveness of their axon guidance cues encountered in intermediate targets.

Conclusions

Considerable progress has been made in understanding the mechanisms of thalamocortical innervation in the developing brain. TCAs originating from specific thalamic nuclei follow distinct pathways through the subcortex, and are segregated within the major fiber tract, the IC, target a specific cortical area (inter-areal targeting) and form a representation of a sensory or motor field (intra-area targeting). As reviewed here, there is new evidence that there are several check points for TCAs to pass for correct projection. These are 1) Correct cell identity for neurons born in thalamus, 2) Expression of repulsive cues in hypothalamus, and other subcortical cues, 3) Topography of guidance cues in the IC, 4) Area identity in the subplate to navigate TCAs towards the correct cortical entry position, 5) Regional cues in the cortex to navigate TCAs to the final target area. Furthermore, several lines of evidence indicate that neural activity plays a crucial role in conferring presynaptic patterns to postsynaptic cells via neurotransmitter receptor-mediated intracellular signals to form mature barrel fields (Iwasato et al., 2000; Rebsam et al., 2002; Erzurumlu and Kind, 2001, Erzurumlu and Iwasato, 2005, this volume). But the molecular mechanisms for activity dependent refinement are largely unknown in the somatosensory system. Further understanding of how VB axons project to somatosensory area in development will also help to understand neuronal activity-dependent cytoarchitectonic formation of mammalian brain.

Acknowledgement

All figures are taken from the article of Advances in Neurological Science with permission. This work was supported by the RIKEN Brain Science Institute(TS), the National Institutes of Health and The March of Dimes Birth Defects Foundation (to Elizabeth Grove, University of Chicago).

Literature Cited

Auladell, C., Pérez-Sust. P., Supèr. H., and Soriano. E. (2000). The early development of thalamocortical and corticothalamic projections in the mouse, *Anat. Embryol. (Berl.)* 201;169–179.

Bachiller. D., Klingensmith. J., Kemp. C., Belo. J. A., Anderson. R. M., May. S. R., McMahon. J. A., McMahon. A. P., Harland. R. M., Rossant. J., and De Robertis. E. M. (2000). The organizer factors Chordin and Noggin are required for mouse forebrain development. Nature. 403:658–61.

Bagri. A., Marin. O., Plump. A. S., Mak. J., Pleasure. S. J., Rubenstein. J. L., and Tessier-Lavigne. M. (2002). Slit proteins prevent midline crossing and determine the dorsoventral position of major axonal pathways in the mammalian forebrain. *Neuron*. 33:233–48.

Bayer. S. A., and Altman. J. (1990). Development of layer I and the subplate in the rat neocortex. *Exp Neurol*. 107:48–62.

Bishop. K. M., Rubenstein. J. L., and O'Leary. D. D. (2002). Distinct actions of Emx1, Emx2, and Pax6 in regulating the specification of areas in the developing neocortex. *J. Neurosci.* 22:7627–38.

Braisted. J. E., Tuttle. R., and O'leary. D. D. (1999). Thalamocortical axons are influenced by chemorepellent and chemoattractant activities localized to decision points along their path. *Dev Biol*. 208:430–40.

Braisted. J. E., Catalano. S. M., Stimac. R., Kennedy. T. E., Tessier-Lavigne. M., Shatz. C. J., and O'Leary. D. D. (2000). Netrin-1 promotes thalamic axon growth and is required for proper development of the thalamocortical projection. *J Neurosci*. ;20:5792–801.

Briscoe. J., and Ericson. J. (2001) Specification of neuronal fates in the ventral neural tube. *Curr Opin Neurobiol*. 11:43–9.

Brittis. P. A., Lu. Q., and Flanagan. J. G. (2002). Axonal protein synthesis provides a mechanism for localized regulation at an intermediate target. *Cell*. 110:223–35.

Caviness. Jr., V. S. (1982). Neocortical histogenesis in normal and *reeler* mice: a developmental study based upon [3H]thymidine autoradiography. *Dev Brain Res* 4:293–302.

Crossley. P. H., Martinez. S., and Martin. G. R. (1996). Midbrain development induced by FGF8 in the chick embryo. *Nature*. 380:66–8.
Dingwell. K. S., Holt. C. E., and Harris. W. A. (2000). The multiple decisions made by growth cones of RGCs as they navigate from the retina to the tectum in Xenopus embryos. *J Neurobiol* 44:246–59.
Dufour. A., Seibt. J., Passante. L., Depaepe. V., Ciossek. T., Frisen. J., Kullander. K., Flanagan. J. G., Polleux. F., and Vanderhaeghen. P. (2003). Area specificity and topography of thalamocortical projections are controlled by ephrin/Eph genes. *Neuron*. 39:453–65.
Erzurumlu. R. S., and Jhaveri. S. (1990). Thalamic axons confer a blueprint of the sensory periphery onto the developing rat somatosensory cortex. *Brain Res Dev Brain Res*. 56:229–34.
Erzurumlu. R. S., and Kind. P. C. (2001). Neural activity: sculptor of 'barrels' in the neocortex. *Trends Neurosci*. 24:589–95.
Fukuchi-Shimogori. T., and Grove. E. A. (2001). Neocortex patterning by the secreted signaling molecule FGF8. *Science*. 294:1071–4.
Garel. S., Yun. K., Grosschedl. R., and Rubenstein. J. L. (2002). The early topography of thalamocortical projections is shifted in Ebf1 and Dlx1/2 mutant mice. *Development*. 129:5621–34.
Garel. S., and Rubenstein. J. L. (2004). Intermediate targets in formation of topographic projections: inputs from the thalamocortical system. *Trends Neurosci*. 27:533–9.
Ghosh. A., and Shatz. C. J. (1993). A role for subplate neurons in the patterning of connections from thalamus to neocortex. *Development*. 117:1031–47.
Ghosh. A., Antonini. A., McConnell. S. K, and Shatz. C. J. (1990). Requirement for subplate neurons in the formation of thalamocortical connections. *Nature*. 347:179–81.
Ghosh. A., and Shatz. C. J. (1992). Involvement of subplate neurons in the formation of ocular dominance columns. *Science*. 255:1441–3.
Grove. E. A., and Fukuchi-Shimogori. T. (2003). Generating the cerebral cortical area map. *Annu Rev Neurosci*. 26:355–80.
Hevner. R. F, Shi. L., Justice. N., Hsueh. Y., Sheng. M., Smiga. S., Bulfone. A., Goffinet. A. M, Campagnoni. A. T, and Rubenstein. J. L. (2001). Tbr1 regulates differentiation of the preplate and layer 6. *Neuron*. 29:353–66.
Iwasato. T., Datwani. A., Wolf. A. M, Nishiyama. H., Taguchi. Y., Tonegawa. S., Knöpfel. T., Erzurumlu. R. S., and Itohara. S. (2000). Cortex-restricted disruption of NMDAR1 impairs neuronal patterns in the barrel cortex. *Nature*. 406:726–31.
Jeanmonod. D., Rice. F. L., and Van der Loos. H. (1981). Mouse somatosensory cortex: alterations in the barrelfield following receptor injury at different early postnatal ages. *Neuroscience*. 6:1503–35.
Kiecker. C., and Niehrs. C. (2001). A morphogen gradient of Wnt/beta-catenin signalling regulates anteroposterior neural patterning in Xenopus. *Development*. 128:4189–201.
Lee. L. J, Iwasato. T., Itohara. S., and Erzurumlu. R. S. (2005). Exuberant thalamocortical axon arborization in cortex-specific NMDAR1 knockout mice. *J Comp Neurol*. 485:280–92.
Lee. S. M., Tole. S., Grove. E., and McMahon. A. P. (2000). A local Wnt-3a signal is required for development of the mammalian hippocampus. *Development*. 127:457–67.
López-Bendito. G., and Molnár. Z. (2003). Thalamocortical development: how are we going to get there? *Nat. Rev. Neurosci.* 4; 276–289.

Mackarehtschian. K., Lau. C. K, Caras. I., and McConnell. S. K. (1999). Regional differences in the developing cerebral cortex revealed by ephrin-A5 expression. *Cereb Cortex.* 9:601–10.

McQuillen. P. S, DeFreitas. M. F, Zada. G., and Shatz. C. J. (2002). A novel role for p75NTR in subplate growth cone complexity and visual thalamocortical innervation. *J Neurosci.* 22:3580–93.

Miller. B., Chou. L., and Finlay. B. L. (1993). The early development of thalamocortical and corticothalamic projections. *J. Comp. Neurol.* 335;16–41

Miyashita-Lin. E. M., Hevner. R., Wassarman. K. M, Martinez. S., and Rubenstein. J. L. (1999). Early neocortical regionalization in the absence of thalamic innervation. *Science.* 285:906–9.

Molnár. Z., Adams. R., Goffinet. A. M, and Blakemore. C. (1998). The role of the first postmitotic cortical cells in the development of thalamocortical innervation in the reeler mouse. *J Neurosci.* 18:5746–65.

Molnár. Z., Higashi. S., and López-Bendito. G. (2003) Choreography of early thalamocortical development, *Cereb. Cortex* 13;661–669.

Muzio. L., DiBenedetto. B., Stoykova. A., Boncinelli. E., Gruss. P., and Mallamaci. A. (2002). Emx2 and Pax6 control regionalization of the pre-neuronogenic cortical primordium. *Cereb. Cortex* 12:129–39

Nordstrom. U., Jessell. T. M., and Edlund. T. (2002) Progressive induction of caudal neural character by graded Wnt signaling. *Nat Neurosci.* 5:525–32.

O'Leary. D. D, and Wilkinson. D. G. (1999). Eph receptors and ephrins in neural development. *Curr Opin Neurobiol* 9:65–73.

Rakic. P. (2001). Neurobiology. Neurocreationism–making new cortical maps. *Science* **294**, 1011–2.

Rebsam. A., Seif. I., and Gaspar. P. (2002). Refinement of thalamocortical arbors and emergence of barrel domains in the primary somatosensory cortex: a study of normal and monoamine oxidase a knock-out mice. *J Neurosci.* 22:8541–52.

Seibt. J., Schuurmans. C., Gradwhol. G., Dehay. C., Vanderhaeghen. P., Guillemot. F., and Polleux. F. (2003). Neurogenin2 specifies the connectivity of thalamic neurons by controlling axon responsiveness to intermediate target cues. *Neuron.* 39:439–52.

Senft. S. L., and Woolsey. T. A. (1991). Mouse barrel cortex viewed as Dirichlet domains. *Cereb Cortex.* **1**:348–63.

Shamim. H., Mahmood. R., and Mason. I. (1999). In situ hybridization to RNA in whole embryos. *Methods Mol Biol.* 97:623–33.

Shimogori. T., and Grove. E. A. (2005). FGF8 regulates neocortical guidance of area-specific thalamic Innervation. *J Neurosci.* in press.

Tuttle. R., Nakagawa. Y., Johnson. J. E., and O'Leary. D. D. (1999). Defects in thalamocortical axon pathfinding correlate with altered cell domains in Mash-1-deficient mice. *Development.* 126:1903–16.

Van der Loos. H., and Woolsey. T. A. (1973). Somatosensory cortex: structural alterations following early injury to sense organs. *Science.* 179:395–8.

Vanderhaeghen. P., Lu. Q., Prakash. N., Frisen. J., Walsh. C. A, Frostig. R. D., and Flanagan. J. G. (2000). A mapping label required for normal scale of body representation in the cortex. *Nat Neurosci.* 3:358–65.

Whitford. K. L., Marillat. V., Stein. E., Goodman. C. S., Tessier-Lavigne. M., Chedotal. A., and Ghosh. A. (2002). Regulation of cortical dendrite development by Slit-Robo interactions. *Neuron.* 33:47–6.

Williams. S. E., Mason. C. A., and Herrera. E. (2004). The optic chiasm as a midline choice point. *Curr Opin Neurobiol.* 14:51–60.

Woolsey. T. A., and Van der Loos. H. (1970). The structural organization of layer IV in the somatosensory region (SI) of mouse cerebral cortex. The description of a cortical field composed of discrete cytoarchitectonic units. *Brain Res.* 17:205–42.

Wurst. W., and Bally-Cuif. L. (2001). Neural plate patterning: upstream and downstream of the isthmic organizer. *Nat Rev Neurosci.* 2:99–108.

Yun. M. E., Johnson. R. R., Antic. A., and Donoghue. M. J. (2003). EphA family gene expression in the developing mouse neocortex: regional patterns reveal intrinsic programs and extrinsic influence. *J Comp Neurol.* 456:203–16.

Zhou. C., Tsai. S. Y., and Tsai. M. J. (2001). COUP-TFI: an intrinsic factor for early regionalization of the neocortex. *Genes Dev.* 15:2054–59.

4

The Earliest Thalamocortical Interactions

Zoltán Molnár, Guillermina López-Bendito, Daniel Blakey, Alexander Thompson, and Shuji Higashi

Abstract

Developing thalamic projections cross the various segments of the diencephalon and telencephalon to reach appropriate cortical areas without the influence of external sensory input in an autonomous fashion. This initial map requires no major fiber reorganization between the thalamus and cortex: a simple 90 degree rotation can explain the early topographic relations between thalamic volume and the cortical sheet. However, this initially simple layout may be modified close to the cortical target. It has been proposed that the early interactions between thalamic projections and subplate neurons could play a major role in establishing the ultimate innervation pattern. We review evidence for early functional interactions between thalamic projections and cortex in wild-type rat and in the reeler-like mutant Shaking Rat Kawasaki (SRK). In the reeler mouse or SRK the thalamocortical projections ascend to the cortical surface comprising early born cells before descending to their ultimate target cells equivalent to layer 4. This suggests that the initial thalamocortical trajectory has a limited role in final cortical map formation. The normal periphery related pattern is assumed by the thalamic fibers since they are capable of correcting defects in trajectory closer to their targets. A candidate for effecting the realignment of these projections is the neural activity pattern transmitted to the immature cortex via thalamic projections. There is current *in vivo* and *in vitro* evidence for the role of neural activity in area and lamina-specific thalamic targeting. In the *Snap25* mouse all neurons lack action potential-mediated neurotransmitter release. In spite of this, the embryonic development of thalamic projections follows a normal pattern until birth. There is rich evidence for changes in cortical map formation if the sensory input is changed or the size of the cortical sheet is altered and followed beyond postnatal life. The challenge is to relate the complex events of cortical remodeling to area specific thalamic fiber invasion and modality-specific sensory activation patterns.

The thalamus is a major relay structure in the diencephalon, and as such transmits nearly all sensory input destined for the cortex. Each sensory organ (except olfactory) and major subcortical motor center provides input to one or more specific nuclei of the thalamus, and these nuclei have well defined reciprocal interconnections with particular cortical areas (Jones, 1985). Layer specificity is remarkably similar for all cortical territories and is conserved between species. All areas in which the early stages of thalamic innervation have been studied, the majority of thalamocortical axons terminate in layer 4 and to a lesser extent in layers 6, 3 and 1 (Jones, 1985). A high proportion of layer 6 neurons send corticofugal projections back to their appropriate thalamic nucleus. Most input comes from the cortex (Mitrofanis and Guillery, 1993), the pattern of which enables the two types of thalamic nuclei to be distinguished (Guillery and Sherman, 2002). "First order nuclei" receive primary afferent fibers from the periphery and corticothalamic afferents from layer 6 pyramidal cells. "Higher order nuclei" receive input from pyramidal cells in cortical layer 5. Only layer 6 projections send branches to the thalamic reticular nucleus and are believed to have modulatory function (Jones, 2002). Layer 5 projections are proposed to transmit information about the output of one cortical area to another, and thus are thought to be involved in corticocortical communication (Guillery and Sherman, 2002).

Complexities of the Adult Thalamocortical Innervation; Polarity of Maps and Reversals

The representation of the three-dimensional thalamic volume on the continuous planar sheet of cortex has been studied extensively in several species, including human (Caviness and Frost, 1980; Crandall and Caviness, 1984; Jones, 1985; Johansen-Berg, 2005). These studies demonstrated that all areas of neocortex receive a thalamic projection (Caviness and Frost, 1980). In most cases, neighboring neocortical fields receive their projections from adjacent thalamic nuclei, and proximity relationships correspond in the cortex and the thalamus; but occasionally, non-adjacent thalamic nuclei project to adjoining but distinct cortical regions, with no overlap, separated sharply at the borders of their cortical regions (Caviness, 1988). The relative position and neighborhood relationships of the fibers originating from different thalamic regions are, to a considerable extent, preserved in their course to the cortex within the characteristic fan shaped radiation of thalamic fibers (Caviness and Frost, 1980; Caviness, 1988; Agmon and Connors, 1991; Molnár et al., 1998a). This does not appear to be the case at the level of individual fibers, which exhibit mixing and twisting along their path, especially in the thalamic reticular nucleus and in white matter close to the cortex (Bernardo and Woolsey, 1987). Intriguingly, during development these compartments contain largely transient premature neuronal populations, the subplate and the thalamic reticular neurons (Allendoerfer and Shatz, 1994; Mitrofanis and Guillery, 1994), which can host the accumulating thalamocortical or corticofugal projections before they enter cortex or thalamus and establish their connections (Rakic, 1977; Lund and Mustari, 1977; Shatz and Rakic, 1981).

Sites of Fiber Rearrangements, Modifications of Early Connections Coincide with Transient Cell Populations and Waiting Compartments

It has been proposed that the early topographic ordering in the developing thalamocortical projection, when the fibers accumulate in the intermediate zone and subplate, establishes the gross distribution of input to cortical areas (Blakemore and Molnár, 1990; Catalano et al., 1996). This layout might even determine most of the general areal mapping of thalamocortical interconnections. The initial map requires no fiber reorganization between the thalamus and cortex; a simple twist (approximately 90 degrees) of the fiber bundles can explain the relationships between the two representations (Molnár and Blakemore, 1995; Price et al., 2006). Initial simple fiber ordering accomplished during the earliest phases of thalamocortical development might be preserved along the pathway. This attractive notion, that cortical mapping is first determined by a simple ordered projection from thalamus to cortex, has been challenged on the basis of the topography of the visual field representation in the dorsal lateral geniculate nucleus (dLGN) and striate cortex. Conolly and Van Essen (1984), examining the visual field representation in the dLGN and the visual cortex of the macaque monkey, found that the map is mediolaterally but not anteroposteriorly reversed between LGN and cortex. Since this map transformation cannot be explained with a single twist, they concluded that the geniculocortical fibers in the white matter must reverse positions along one major axis but not the orthogonal one.

Nelson and Le Vay (1985), using pairs of tracer injections to reveal fiber order in the adult cat, did indeed report crossing of thalamic axons in the mediolateral but not in the rostrocaudal axis of the optic radiation. This anisotropic decussation occurred only a short distance (2-500 μm) below the visual cortex, at a depth that might well have been in the transient subplate during development. The thalamic fiber side branches in the intermediate zone and subplate described by Naegele et al., (1988) and Ghosh and Shatz (1992) could form the anatomical substrate for this reversal (Krug et al., 1998; Molnár and Hannan, 2000). It was proposed that the decussation pattern seen in the adult represents a modification of the array of thalamocortical fibers that occurs during or at the end of the waiting period (Molnár, 1998).

Adams et al (1997) proposed that the thalamocortical fiber rearrangements which require fiber crossing are very common in the cortex. Sensory maps of the primary and secondary visual (Allman and Kaas, 1971) and somatosensory areas (McCasland and Woolsey, 1988; Catania and Kaas, 1995) are reversed with respect to each other. This implies that only one of these maps can be established with a simple transformation, while the other requires the rearrangement and crossing of the thalamocortical connectivity at some point along the pathway. Interestingly the secondary areas seem to correspond to the initial overall layout of thalamocortical connectivity. At present we do not know the development of these mirror reversals. We suspect that these transformations and map reversals might even operate within the same cortical field based on mechanisms separate from the initial deployment.

Early Fiber Deployment is Autonomous, But Fiber Entry and Branching is Modified by Early Activity

The initial layout of thalamic axons might be very clear and regular in embryonic life, but becomes substantially altered with further development near the termination site. It has been proposed that thalamocortical development has two major phases. The early deployment of thalamocortical connectivity is established in an autonomous fashion before the afferents from the sensory periphery reach the dorsal thalamus. Early outgrowth from the thalamus might be directed by gradients within the dorsal thalamus and ventral pallidum (Molnár and Blakemore, 1995; Seibt et al, 2003; Marín, 2003; Garel and Rubenstein, 2004) or the cortex (Vanderhaegen et al., 2000; Fukuchi-Shimogori and Grove, 2001). Thalamic fibers change fasciculation pattern and growth kinetics as they cross gene expression boundaries along their trajectory towards the cortex. There seem to be at least two especially critical zones for axon outgrowth in the embryonic forebrain. One critical zone is at the diencephalic-telencephalic border, and the other one at the striatocortical junction, the pallial subpallial boundary (PSPB). Thalamic axons prove to be very sensitive indicators of regionalization defects in the developing forebrain. Altered gene expression patterns along the thalamocortical path can arrest or modify their development at specific sites. Having examined the aberrant development of thalamocortical projections in various mutants some basic principles are beginning to emerge (Hevner et al., 2002; Jones et al., 2002; López-Bendito et al., 2003; Garel and Rubenstein, 2004). There seem to be characteristic default pathways where thalamic projections are derailed if the early developmental steps are perturbed (see López-Bendito and Molnár, 2003).

As the thalamic projections arrive at the cortex the early topography is simple. AP-movement on the cortical convexity corresponds to a mediolateral movement in the thalamus, whilst the cortical ventro-dorsal axis corresponds to an antero-posterior axis along a slab of thalamic cells (Molnár and Blakemore, 1995). The thalamic nuclei are still not fully formed at these early stages, yet even in adult the basic pattern can be recognized. However, the initial layout is not the final topography and the maps formed can be altered substantially. Evidence suggests that this rearrangement can occur in subplate and within the cortex closer to the ultimate target cells. However, this rearrangement does not modify the rest of the fibre trajectory, which preserve the initial juvenile arrangement.

In mammals, thalamic fibers arrive at the appropriate cortical regions before their ultimate target neurons are born, and pause before continuing to establish their final innervation pattern within the cortical plate (Rakic, 1977; Shatz and Luskin, 1986). This period has been called the waiting period and is the stage from which the sensory periphery could start to modify its own juvenile cortical representation after initial thalamocortical targeting. In this second phase, during the process of thalamic fiber ingrowth and arborization, activity dependent mechanisms have been implicated. The role of neural activity in the initial arborization of thalamic axons within cortical layer 4 has been demonstrated by Herrmann and Shatz (1995). This has been further examined under *in vitro* conditions by Wilkemeyer and Angelides (1995) and Anderson and Price (2002). Moreover, experiments where activity has been abolished earlier, suggested that it plays a role in the area-specific delivery and refinement of thalamic projections (Catalano and Shatz, 1998). The side branch formation might be regulated by electric fields that are generated by activity along the axons. Interestingly, when TTX (a sodium channel antagonist that blocks action potentials) was delivered into the brain of cat fetuses at the time of arrival of thalamic projections at the subplate, abnormal connections were established by the LGN axons (Catalano and Shatz, 1998). Only a few thalamic fibers entered the visual cortex, and an aberrant topography was formed within the cortical plate. Although this data indicates that even the initial phases of thalamocortical targeting might depend on early activation patterns, the exact nature of the required neural activity is not known. The cornerstone of this hypothesis is that functional interaction is demonstrated by the thalamic projections and the developing cortex at an early stage. We shall review evidence for early thalamocortical interactions, which may elicit neural activity patterns, which in turn refine the area- and lamina-specific thalamic innervations.

Evidence for Functional Transient Thalamocortical Circuitry Prior the Final Layout of the Thalamic Projections

The role of thalamic projections in determining cortical areas and architecture might start at these early stages, therefore, understanding the earliest interactions between thalamic projections and developing cerebral cortex can be important in many respects. Recent *in vitro* studies indicate that thalamic afferents release a diffusible factor that promotes proliferation of neurons and glia in the proliferative zones of the cortex (Dehay et al., 2001). Moreover, neuronal migration in the cortex is facilitated by thalamic fibers in organotypic thalamus-cortex co-cultures

(Edgar and Price, 2001). If a similar mechanism operates *in vivo*, this early influence of thalamocortical axons on corticogenesis might contribute to the cortical cytoarchitectonic differences. Nevertheless, early gene expression patterns are normal until birth if no thalamic projections enter the cortex (Miyashita et al., 1999; Nahagawa et al., 1999; Tuttle et al., 1999) or there is a shifted delivery of thalamic connectivity during embryonic life (Garel et al., 2002).

Evidence for Early Thalamocortical Transmission in the Cerebral Cortex

It was demonstrated that the peripheral sensory organs can generate spontaneous activity patterns at ages when the sensory afferents begin to reach the thalamus (Galli and Maffei, 1988; Meister et al., 1991). These activity patterns could elicit EPSPs on thalamic projection neurons (Mooney et al., 1996) capable of relaying them to cortex (Friauf and Shatz, 1991), and thus these activity patterns may alter the forming terminals within the subplate and cortical plate by controlling side branch formation. The existence of some synapses in the SP at early stages has been demonstrated with electron microscopy (Herrmann et al., 1994). These ideas have triggered further electrophysiological studies on the interactions between thalamic axons and cortex. Single cell recording and current source density (CSD) analysis demonstrated that thalamic axons establish functional synaptic contacts with subplate cells (Friauf and Shatz, 1991; Friauf et al., 1990; Hanganu et al., 2002; Molnár et al., 2003; Higashi et al., 2005).

To gain further insight into the formation of early thalamocortical synapses, we recorded optical images, using voltage sensitive dyes, in the cerebral cortex of prenatal rats by selective thalamic stimulation (Higashi et al., 2002) of thalamocortical slice preparations (Agmon et al., 1993). At E17, thalamic stimulation elicits excitation that rapidly propagates through the internal capsule to the cortex. These responses last less than 10–15ms, and are not affected by the application of glutamate receptor antagonists, suggesting they might reflect presynaptic fiber responses. At E18, long-lasting (more than 300 ms) responses appear in the internal capsule (Higashi et al., 2002) at the site of the cells which possess early thalamic projections (Métin and Godement, 1996; Molnár et al., 1998). These responses in the cortex and internal capsule are both abolished by perfusion of glutamate receptor antagonists, which indicates synapse-mediated activation. At E19, distinct long-lasting responses appeared mainly in the cortical subplate. By E21, shortly before birth, the deep cortical layers are also activated in addition to the subplate. The laminar location of the responses was determined in the same slices by Nissl-staining or birthdating with bromodeoxy-uridine (BrdU) at E13 (Higashi et al., 2002). These results demonstrated that there is a delay of several days between the arrival of thalamocortical axons at the subplate at E16 and the appearance of functional thalamocortical synaptic transmission at E19. Since thalamocortical connections are already functional within subplate and deep cortical plate at embryonic ages, prenatal thalamocortical synaptic connections could influence cortical circuit formation even before birth (Allendoerfer and Shatz, 1994).

The activation patterns revealed with optical responses later extend into and then become confined to layer IV during early postnatal periods (Higashi et al., 2002; 2005; see left panels of Figure 1). These early functional interactions are different from mature postnatal activation, being relatively small, but much longer lasting. There are parallel changes in the receptor compositions and even the intracortical connectivity (Arber, 2004; Higashi et al., 2002; Hoerder et al., 2006). It has been suggested that subplate neurons integrate into the cortical circuitry in various ways (Allendoerfer and Shatz, 2004). The above mentioned anatomical and physiological observations lead to the suggestion that subplate neurons orchestrate the ultimate thalamocortical synapse

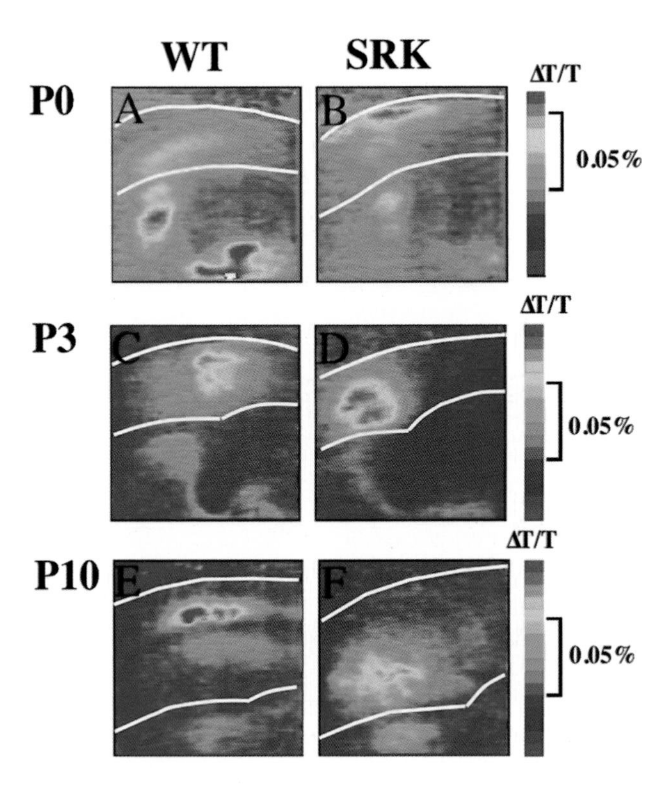

Figure 1 The spatial pattern of optical recording images taken during the beginning of the sustained cortical depolarization in WT (A, C and E) and SRK (B, D and F) at P0, P3 and P10. Each image is an averaged of 5 serial frames taken from the cortex of thalamocortical slices after the offset of initial thalamic fiber volley following direct thalamic stimulation. At P0 the deep cortical layers are activated in WT, in contrast to SRK, where the peak is close to the pial surface. At P3, the peak of the responses ascended in WT and descended in SRK, and then it was refined at P10 to the position corresponding to the position of layer 4 cells. (Modified from Higashi et al., 2005).

formation during development (Kanold et al., 2004). However, it is still not known whether the synaptic responses observed in subplate during late embryonic and early postnatal days are specific to the early-generated neurons or whether the migration of the functional response from the early-generated neurons of subplate to layer IV cells of the cortex is a general process. Would the same developmental steps occur if they were displaced relative to each other?

Reeler *Mutant Mouse and Shaking Rat Kawasaki (SRK) as Models to Study Thalamocortical Development*

The mutant rat, Shaking Rat Kawasaki (SRK), which has reverse cortical layering similar to the phenotype observed in the *reeler* mouse, provides an interesting model system to test whether the functional reorganization is dependent on cell location or on cell type. The phenotype of SRK resembles that of the *reeler* mouse (Aikawa et al, 1988; Ikeda and Terashima; 1997). The role of Reelin in choreographing thalamocortical ingrowth may be complex and is most probably indirect (no direct effect of Reelin on the growth of thalamocortical axons has not been demonstrated). In SRK and *reeler*, this population of cells is abnormally located in the superplate (SP) due to failure of the cortical plate (CP) to split it into marginal zone (MZ) and subplate (SP). Thus, thalamocortical fibers penetrate the CP of SRK and *reeler* prematurely *en route* to the SP (Molnár and Blakemore, 1995). Although morphological studies showed that in *reeler* mutant mice the thalamocortical axons extend toward the SP before the projections descend to the cortical plate (Caviness, 1976; Yuasa et al., 1994; Molnár et al., 1998b), the functional responses elicited through these immature synapses between thalamic projections and these early-generated neurons have only recently been documented in these mutants (Higashi et al., 2002; 2005; Figure 1 right panels).

We have performed birthdating experiments to confirm that the aberrant cortical lamination seen in SRK is similar to that in the *reeler* mutant mouse. We also performed fiber tracing studies using biotinylated dextran amine (BDA) in living slices and carbocyanine dye tracing in fixed brains (Higashi et al., 2005). In thalamocortical slice preparations from SRK the spatial and temporal pattern of excitation was investigated using optical recording with voltage sensitive dyes during the first 10 postnatal days (P0-10). At birth, a strong optical response was elicited within the superplate of the SRK, in the cell layer corresponding to subplate in wild type (WT) rats (Fig. 1B). This response rapidly decreased during postnatal days, as the activation descended into deep cortical layers comprising layer IV cells, (as identified by birthdating with 5-bromo-2′-deoxyuridine at E17). The migration of the optical response occurred during the same postnatal periods in wild type as in SRK, but in different directions (Fig. 1D, F).

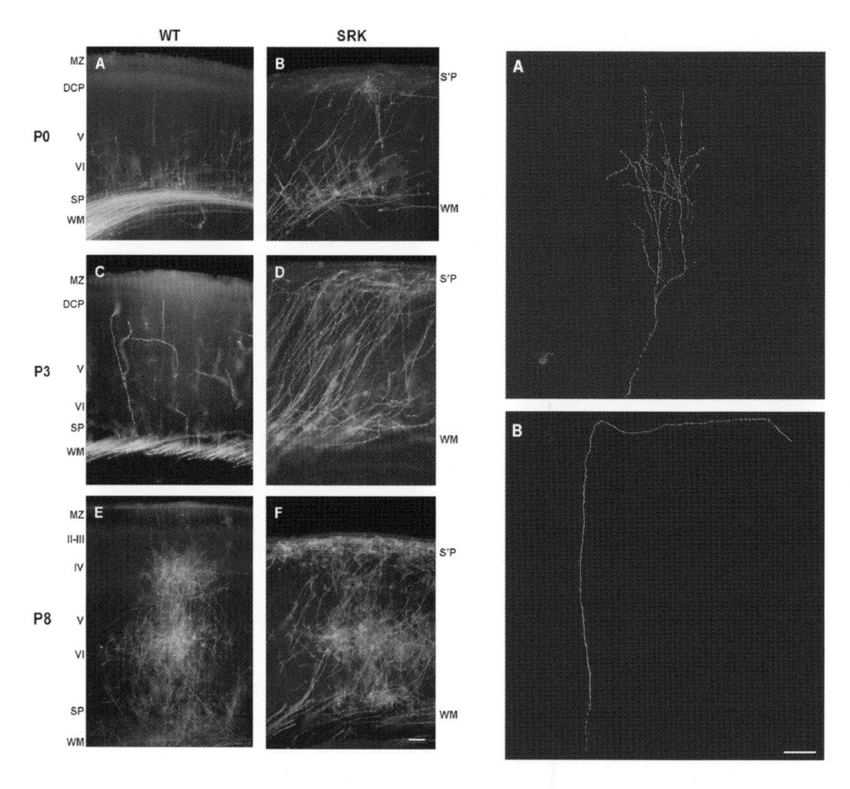

Figure 2 Composite figure of primary somatosensory cerebral cortex of WT (A, C, E) and SRK (B, D, F) at P0, P3 and P8. DiI injection to ventrobasal complex of the thalamus reveals individual axons as they enter the cortical plate. In WT, fibers enter the cortical plate in an ordered fashion, after pausing and aligning in the subplate (A). Fibers penetrate the cortical plate orthogonal to the pial surface. In SRK, axons traverse the cortical plate in oblique fascicles and enter the superplate before turning (B,D). At the p8 panels (E, F) the thalamocortical arbors form at the depth where the peak of the optic responses were observed (see E and F in Fig. 1). To show lamination pattern, sections were counterstained with bisbenzimide. MZ: marginal zone; DCP: dense cortical plate; SP: subplate; WM: white matter; S'P: superplate. Scale bar 100 μm. **(right panels)** Confocal microscopic images of single axons in primary sensory cortex labelled with DiI at P3. In WT (A) fibers typically show branching morphology around layer IV of the cortical plate. This is not observed in SRK (B) where fibers cross the entire breadth of the cortical plate as oblique fascicles and enter the superplate. Here fibers turn and continue parallel to the pial surface before they descend into the cortical plate. Scale bar 50 μm.

Tracing individual axons in SRK revealed that at P0 a large number of thalamocortical axons reach the superplate (Fig. 2), but by P10 only a fraction of them preserve the loop up to the pial surface. The majority of arbors appeared to be pruned back and very often only the arbors to the middle cortical layers were present. It is not clear whether all thalamic axons must extend via this 'loop', or some can establish direct contact with layer IV cells.

Normal Thalamocortical Terminal Clustering in *Reeler* and Shaking Rat Kawasaki Somatosensory Cortex

The periphery related thalamocortical axon patterning was normal in SRK, but the cytoarchitectonic barrels in the SI cortex were not apparent (Higashi et al., 2005). This is very similar to our previous findings in the *reeler* mutant (Bronchti et al., 1999a,b). Nissl sections of the mutant mice did not show clearly defined barrel boundaries, but cytochrome oxidase staining revealed normal periphery related pattern in a region corresponding to S1 (Bronchti et al., 1999b). This suggests that in *reeler* the majority of thalamic fibers assume normal periphery related pattern in the barrel cortex, but that cell patterning in the barrel field might be impaired. We examined the radioactively labeled 2-deoxy-glucose (DG) uptake after clipping all the mystacial whiskers with the exception of the 3 caudalmost of rows B and D. DG-uptake examined on the coronal plane, revealed a columnar activation pattern with a highest DG-uptake in the intermediate layers. In the tangential plane, DG-uptake showed that the cortical activation pattern, and thus the areal distribution of whisker representation in *reeler*, is organized in an identical manner to that in normal mice (Bronchti et al., 1999a,b). Therefore, the abnormal trajectory to the cortex in the *reeler* does not seem to alter the ordered functional whisker representation.

These results suggest that the general developmental pattern of synapse formation between thalamic axons and subplate (superplate) neurons in WT and SRK is very similar, but it follows the altered position of the displaced subplate cells and the individual arbors are considerably re-modeled. In later postnatal ages the latency of layer 4 activation appears identical in normal and SRK (Higashi et al., 2005). The appearance of fibers in P8 specimens that do not loop through the SP but appear to enter the deep CP and arborise, as seen in normal animals, is an interesting finding and may have important implication on the current models for the mechanisms of stopping and branching of thalamocortical axons in the cortical plate. This warrants further investigation to determine how these connections and terminal arbors develop. The extent and rapidity of the thalamocortical arbor remodeling in the *reeler* mutant mouse and in SRK is spectacular. The detailed analysis of individual arbors during late embryonic stages and during the first postnatal week could reveal the sequence or coexistence of synaptic contacts in subplate and cortical plate.

It has been suggested that the CP is not permissive for thalamocortical ingrowth as projections reach its borders (Hubener et al., 1995). This appears to be the case in the dense cortical plate (DCP) of wild-type animals, though the CP appears to become growth-permissive at a later point (Molnár and Blakemore, 1999). Indeed, it has been suggested that growth-promoting factors are upregulated in the cerebral cortex postnatally (Hubener *et al.*, 1995); however, this does not necessarily mean that

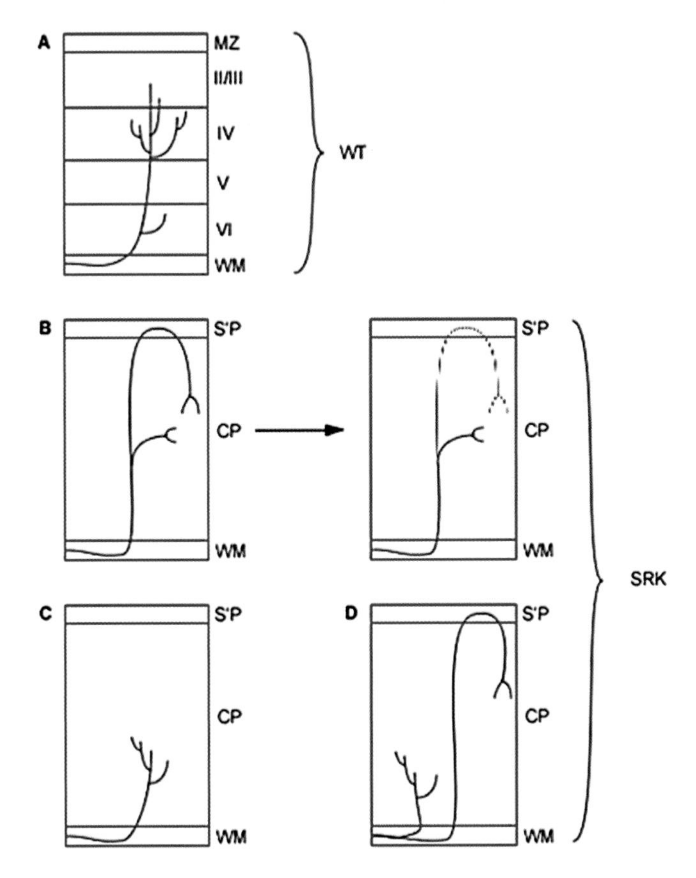

Figure 3 Possible origins of non-looping thalamocortical projections to presumptive cerebral cortical layer IV neurons in SRK. **A:** In WT rat, axon enters the cortex and arborises around layer IV. **B:** Non-looping axon in SRK may result from 'pruning' of originally looped axon after side branch growth. In SRK ingrowth without loop may alternatively occur in some axons (**C**) or as white matter side branches of other pre-existing axons (**D**). MZ: marginal zone; WM: white matter; S'P: superplate; CP: cortical plate.

it is growth permissive. In agreement with the handshake hypothesis (Molnár and Blakemore, 1995), thalamocortical axons follow a scaffold laid down by their pioneer axons from the subplate. In the case of *reeler* mutants, this projection arises in the S'P and traverses the CP in the form of oblique axonal fascicles. If this is also the case for SRK, it could account for the fiber trajectory through the CP whilst maintaining its non-permissive nature until after the majority of fibers enter the S'P. The scaffold originating from the S'P and in close association with thalamocortical axons has been demonstrated in *reeler* (Molnár *et al.* 1998b).

The finding that older mutant animals possess some thalamocortical axons that ascend directly to target cells corresponding to those of layer IV may provide some clue as to the mechanism of fiber ingrowth. There are two possibilities for the appearance of non-looping axons in SRK. Fibers may follow the route observed in most SRK thalamocortical fibers: looping up to the SP and descending to the CP. This loop must then be lost after formation of a side-branch arborising within the CP, followed by 'pruning' of the distal section of the axon (Fig. 3B). Alternatively, the axon could enter the CP once it has become permissive for thalamocortical ingrowth, growing directly to cortical targets. This in turn could arise either as an axonal branch occurring directly from the white matter tract (Fig. 3D), or as an axon terminal (Figure 3C). The proportions of the two forms of arborisations might be related to the fraction of the cells showing normal and inverted positions. Not all cells are inverted in SRK, similarly to reeler mouse there is a significant number assuming inside out polarity (Higashi et al., 2005). With the techniques used, these possibilities are not resolvable. Further studies must be carried out to determine this mechanism. More quantitative single-axon reconstructions at different ages might suggest the mechanism by which correct laminar axonal targeting occurs, but the examination of thalamocortical axon growth using time-lapse imaging for 2–3 days would be ideal (Portera-Cailliau et al., 2005). The mechanism of development of the non-looping projections might represent an important postnatal remodeling in SRK. If the 'pruning' suggested above were to occur, it may have implications in developmental plasticity. Our understanding of this process suggests that perhaps a combination of figure 3 B and D might result in the observed phenotype. A potential mechanism for this would be an activity-dependent pruning whereby the early loop is lost. This could represent a means by which relatively normal topography is maintained in the SRK mutant. This could be tested by blocking neuronal activity in the *reeler* or SRK mutants. Or more directly, the arrival of thalamocortical axons could perhaps be studied by observing cortical slice preparations over time; alternatively the lengths of thalamocortical axons in the cortical plate could be examined at stages in development. Results from such an experiment would require cautious interpretation: more rigorous studies would follow the fate of an identified axon at different developmental stages.

There are numerous mouse models where the thalamocotical axon delivery is altered during the autonomous phase of delivery (L1 KO, Sema6A KO), but then the final cortical topography seems to compensate (see Leighton et al., 2001; Molnár et al., 2003). The layer specific stop and branch signals seem to be distinct from the ones governing the early deployment and entry of the thalamic projections (Yamamoto et al., 2000).

Layer Specific Stop and Branch Characteristics of Thalamocortical Axons

In the mouse somatosensory system thalamocortical axons reach their major target of layer 4 between postnatal days 2 and 4 (Agmon et al, 1993; Rebsam et al, 2002). Once the lamina has been reached the axons cease radial growth, and elaborate their complex arbors. Both molecular and neural activity dependent mechanisms have been shown to direct this behavior, however the degree of redundancy between the regulatory mechanisms is not yet known.

The *in vitro* technique of thalamus-cortex co-culture has been used to unravel some of these mechanisms. By growing late embryonic thalamus explants adjacent to early postnatal cortex explants thalamocortical axon development can be observed with the same lamina specific termination pattern as *in vivo* (Yamamoto et al., 1989, 1992; Molnár and Blakemore, 1991; Bolz et al., 1992; 1993). Yamamoto et al (1997) were able to observe the growth of thalamic axons *in vitro* using confocal video microscopy. Analysis of cultures in which thalamic axons entered the cortex from the ventricular, pial, or lateral side indicated that the "stop" and "branch" signals are essentially independent. Axons growing radially stop, their growth cones collapsing, and branches emerge behind their distal tip, whereas axons entering the cortex laterally in layer 4 continue to grow whilst branches emanate from their length.

Several molecular cues, which may represent the stop/branch signals, have subsequently been identified using this and other tissue culture techniques. The transmembrane glycoprotein N-cadherin is expressed in both layer 4 neurons and the sensory thalamus (Gil et al, 2002), and may mediate axon-cell contacts through homophilic, calcium dependent association (Geiger and Ayalon, 1992). Perturbation of this association by antibody and peptide blockade in co-cultures prevents both the proper growth of axons through the deep cortical layers, and termination in layer 4: axons grow through their target towards the pial margin of the culture (Poskanzer et al, 2003). Ephrin-Eph tyrosine kinase interactions have also been shown to regulate thalamocortical lamination patterns. Mann et al (2002) grew thalamic explants on membranes isolated from either layer 5 or layer 4, they observed great branching and reduced growth on layer 4 derived preparations indicating that membrane bound molecules regulate thalamocortical axon morphology. Furthermore, ephrin A5 expression was shown to be layer specific in the cortex, and cognate EphA receptors expressed in the thalamus, while blockade of the Eph-ephrin signaling prevented the preferential branching phenotype of thalamic axons. Enzymatic disruption of co-cultures has also narrowed down the list of potential candidates for stop and branch signals. Disruption of the neural cell adhesion molecule (NCAM) with endoneuraminidase-N promoted thalamocortical axon branching across all layers of the cortex, implying that this molecule may be a branch inhibitor in non-target layers of the cortex (Yamamoto et al, 2000; and 2006 see in this volume).

Are cell adhesion and signaling mechanisms sufficient for the correct laminar deployment and arborization of thalamocortical axons? Certainly results published by Yamamoto et al (2000) in which thalamic axons grew correctly though co-cultures were grown using chemically fixed cortical explants, imply this is the case; however several studies have indicated that neural activity is also required. Tetrodotoxin (TTX) infusion into embryonic cat brains during the radial growth of thalamocortical axons through the cortex prevented their termination in layer 4, fibers instead projecting to the more superficial cortical layers (Catalano and Shatz, 1998). Repetition of these experiments using the co-culture system allowed a more careful analysis of the axon morphology and investigation of neurotransmitter receptor specific agonist and antagonists (Wilkemeyer and Angelides, 1996; Anderson and Price, 2002) (Figure 4). Addition of TTX prevented layer specific termination (Anderson and Price, 2002 Fig. 4B), but also induced a greater degree of axon branching and presumptive presynaptic bouton development (Wilkemeyer and Angelides, 1996 Fig. 4D). The NMDA specific agonist aminophosphovalerate also induced the grow-through effect (Anderson and Price, 2002). These techniques can be thought of as depressing

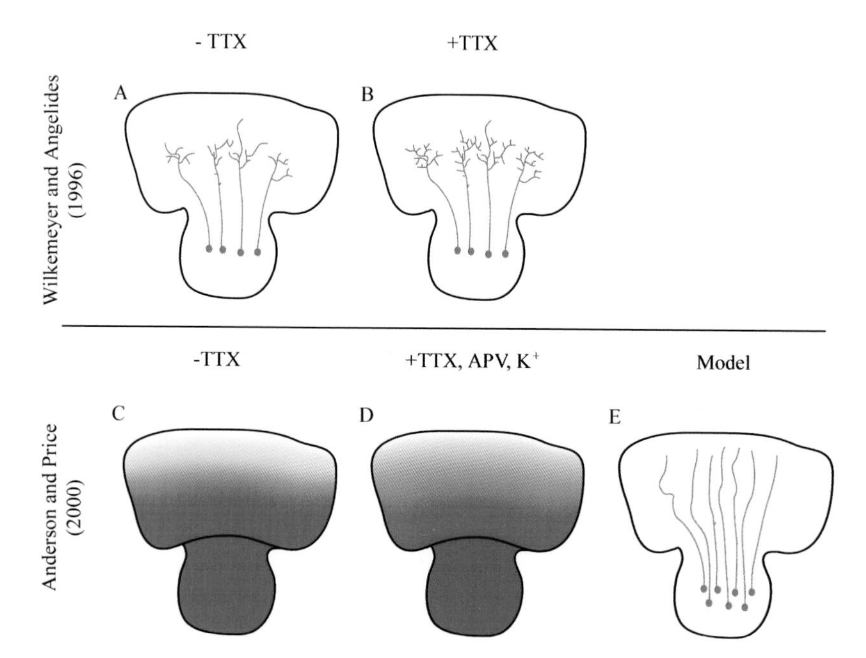

Figure 4 Cartoon illustrates the results of the addition of neural activity modifiers to co-cultures. A and B: co-cultures analysed by single axon tracing and morphometric quantification (Wilkemeyer and Angelides, 1996). A: Controls: axons emanating from the thalamic explant grow into the cortical regions and form characteristic arbours in the central regions of the cortex, shown by vital staining to be layer 4. B: the addition of TTX has no effect on the length of the axon collaterals, or the layer of termination; however fibers have more branch points. C and D: co-cultures analysed by bulk-labelling of fibers and fluorescence intensity differences (Anderson and Price, 2000). C: Controls, fibers labelled with DiI enter the cortical explant, and there is a sharp drop in labelling intensity in the central regions corresponding to layer 4, indicating that many fibers terminate here, NB the cultures were not long enough for extensive arborisation to occur. D: application of TTX, APV or an increase in the concentration of potassium ions results in the abolition of a sharp drop in label intensity in layer 4. Instead the drop is gradual over the width of the cortex. These results were interpreted as axons growing through the slice to reach the pia, E. Abreviations: TTX—tetrodotoxin, APV—aminophosphovalerate, K$^+$—potassium ions

neural activity, however the elevation of potassium ions in the culture medium, increasing the probability of action potentials, also caused the axons to grow through layer 4 (Anderson and Price, 2002). The genetic disruption of NMDA mediated glutamate transmission in the $NR1^{-/-}$ and the cortex specific *NR1* knock-out, does not entirely support the *in vitro* data (Iwasato et al, 2000; Datwani et al, 2002; see Erzurumlu and Iwasato, 2005 in this volume). There was no reported alteration in the thalamocortical axon lamination pattern; however exuberant branching of individual axons has subsequently been reported (Lee et al, 2005).

At present the relative contribution of the periphery driven, activity dependent patterning and the genetically controlled axon-target molecular signaling mechanisms is not clear. Both systems can be shown to affect the lamination and arborisation of thalamocortical axons. Neural activity may alter the expression of molecular factors, which subsequently mediate growth cone collapse causing axon growth termination. Furthermore the correct pattern of regulated neurotransmitter signaling is likely to be required to allow post-synaptic cells to direct the elaboration of axonal arbors.

Debate rages over the plasticity of thalamocortical projections in the arrangement of fiber topography. Experiments of the early 1990s by Agmon and colleagues (Agmon *et al.*, 1993, 1995) suggested that topographically organized projections were apparent from birth, while other experiments provided convincing evidence that there is some form of postnatal remodeling resulting in accurate topographical mapping from thalamus to cortex (Krug *et al.*, 1998; Rebsam *et al.*, 2002; see Shimogori and Grove, 2006 in this volume). Although putative somatosensory and visual cortices show overall general patterns of thalamic innervation they seem to have considerable differences in the innervation density of the projection and their topographic precision. We have compared the number of single and double labeled cells from pairs of single crystals of carbocyanine dyes (DiI and DiAsp) placed at 250 or 500 μm distances from each other into the putative visual and somatosensory cortex on whole brains rather than slices (Molnár et al., 1998). These experiments revealed that at early postnatal ages, a similar low percentage of double-labelled cells were observed in VB and LGN (VB = 0.6%; LGN 0.5%). However, in LGN, the number of labelled cells was 3–5 times smaller than in VB. Moreover the backlabelled cell bodies of thalamic projection neurons in LGN showed more scatter than in VB (Molnár et al., 1996; Molnár 1998). These areal differences in density and topography might be significant in the differences in the extent of possible plastic changes.

The *SNAP-25* KO Mouse as Model Systems to Test the Role of Early Neural Activity in Brain Development

Understanding of the contribution of a: action potential mediated, b: spontaneous and c: paracrine neurotransmitter release during axonal pathway formation and neural migration has very general basic biological importance (Rizo and Südhof, 2002). Our data (Washbourne et al., 2002; Molnár et al., 2002) suggest that *Snap25*$^{-/-}$ mice are a unique resource to study the role of spontaneous synaptic activity in target recognition, synapse maturation and plasticity that are required to develop the effective neural circuitry of the mammalian brain. SNAP-25, together with syntaxin-1 and VAMP-2, form the core SNARE complex, which plays an essential role in exocytotic release of neurotransmitter (Südhof, 1995). *Snap25*$^{-/-}$ mice develop to term, and fetal brain development appears superficially normal, even though action potential-evoked neurotransmitter release is entirely eliminated (Washbourne et al., 2002). SNAP-25 deficient neurons extend axons that terminate in synapses where spontaneous, action potential-independent release still occurs but action potentials do not trigger neurotransmission. Thus, genetic ablation of SNAP-25 expression appears to selectively disable the vesicular processes responsible for evoked synaptic transmission, leaving intact membrane trafficking for axon outgrowth and exocytosis for spontaneous neurotransmitter secretion. *Munc18 1*$^{-/-}$ or *Munc 13-1, 2*, and *3*$^{-/-}$ mice cannot release transmitter from synaptic vesicles at all. The paracrine mechanism whereby transmitter is released out of the cell in a vesicle independent manner is however still present in these animals and it is surprising that they also develop to term with relatively normal brains (Verhage et al., 2000; Varoqueaux et al., 2002).

We have studied both the general developmental pattern of the forebrain and the development of cortical lamination in *Snap25*$^{-/-}$ (Washbourne et al., 2002; Molnár et al., 2002). These experiments showed that, within the resolution of our techniques (Nissl stain, immunohistochemistry for calcium binding proteins etc.) the cerebral cortex develops normally in these knockouts. We also examined the synapses qualitatively in the cerebral cortex at the EM level (Washbourne et al., 2002), but we did not observe any major gross abnormalities within the forebrain of these mutants. Using carbocyanine dye tracing we have examined the development of thalamocortical projections in *Snap25*$^{-/-}$. We were particularly concerned with axon elongation, growth-cone morphology and the kinetics of cortical innervation (Molnár et al., 2002). We paid special attention to the side branch formation during the waiting period and cortical plate innervation and branching. The results showed no significant differences between the SNAP25 deficient animals and their heterozygote and wild-type littermates (Fig. 5).

What Form of Intercellular Communication is Needed for the Remodeling of the Initial Thalamocortical Map?

In contrast with the initial gross deployment of thalamocortical and corticothalamic connections, the remodeling of cortical circuitry during

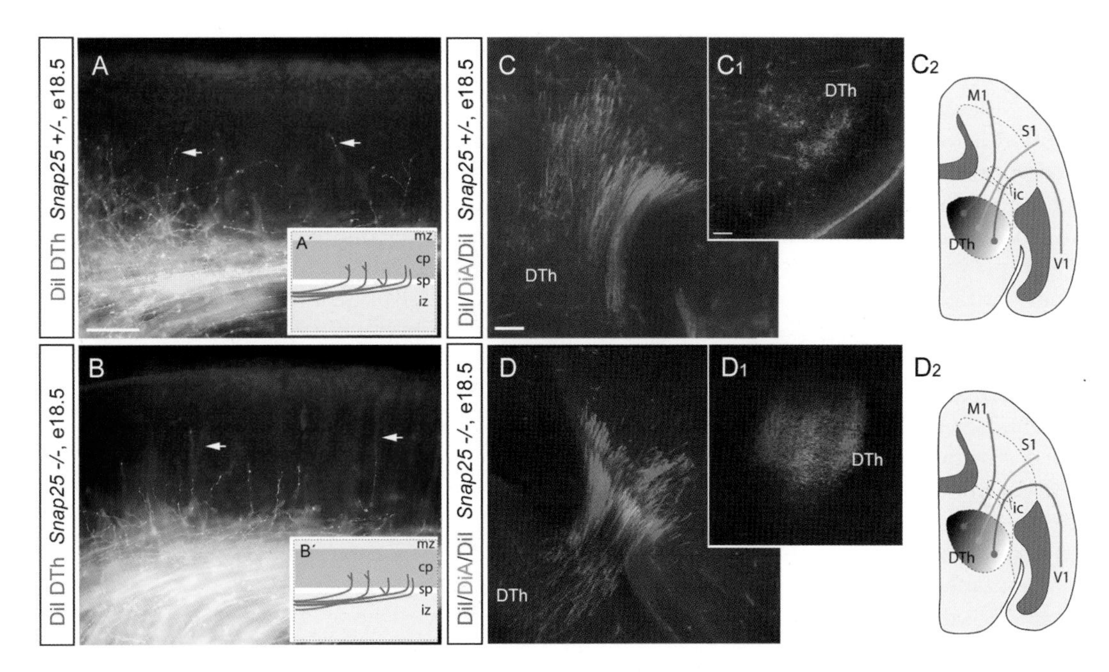

Figure 5 Normal cortical invasion and early topography of thalamic connectivity in the Snap25 knockout mouse. (A-B) Thalamocortical projections, traced with DiI placed in the dorsal thalamus (DTh), show similar ingrowth patterns in heterozygous (A) and *Snap25* deficient mouse (B) at e18.5. Double-exposure photomicrographs of coronal sections showing the DiI-labeled axons (red) and bis-benzimide counterstaining (blue). Note that in both genotypes, axons had started to invade the cortical plate (arrows; see schemas A′ and B′). (C-D_2) Crystals (DiI, DiA, DiI) were implanted along a parasagittal row in putative motor (M1), somatosensory (S1) and visual (V1) areas respectively, in the right hemisphere of heterozygous (C) and *Snap25* deficient mouse (D). Confocal micrographs at the level of the internal capsule (ic; C and D) and dorsal thalamus (C_1 and D_1) showing that in both brains the topographical arrangement of the fibers is maintain not only in the dorsal thalamus but throughout the entire axonal pathway (see schemes C_2 and D_2). mz, marginal zone; sp, subplate; iz, intermediate zone. Scale bar: 100 μm (A-D); 100 μm (C_1-D_1). Data modified from Molnár et al., (2002b).

thalamic fiber invasion is probably a more complex process in which patterns of afferent and local activity, expression of surface molecules and growth factors, and cell death all play crucial roles (Katz and Shatz, 1996). In turn the remodeling may lead to the formation of new synapses and therefore to new distributions of activity. Early thalamic projections are capable of eliciting sustained depolarization patterns in the subplate at the time of their own side-branch formation (Friauf and Shatz, 1991; Higashi et al., 2002; Molnár et al., 2002a). This early interaction is different from the mature postnatal form, being relatively smaller, but much longer. Our data in embryonic *Snap25*$^{-/-}$ mice suggest that axonal growth and early topographic arrangement of these fiber pathways do not rely on activity-dependent mechanisms requiring evoked neurotransmitter release at embryonic stages. We propose that other forms of intercellular communication might still play a part, such as the spontaneous release of neurotransmitter vesicles, and non-vesicular paracrine release of neurotransmitter.

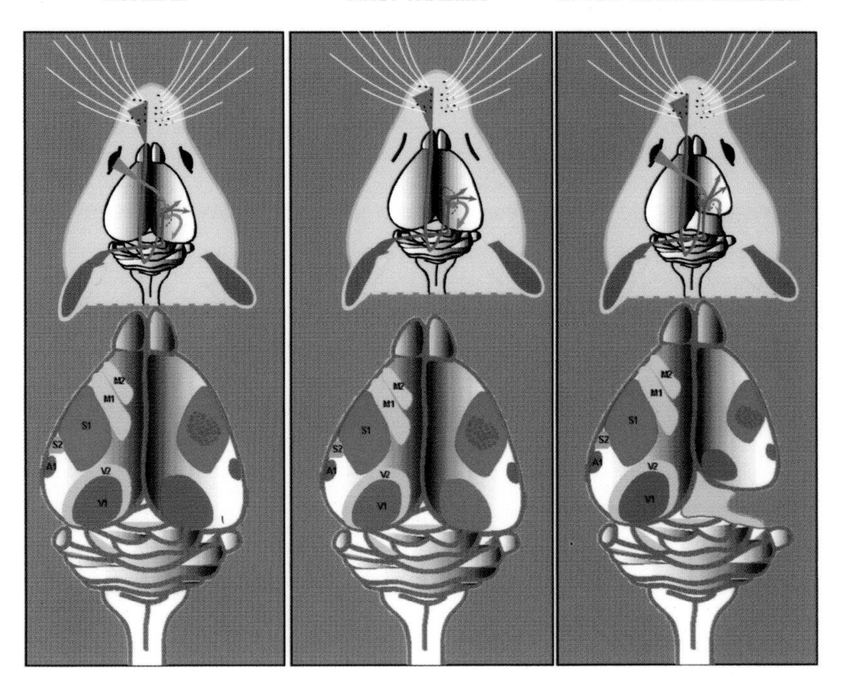

Figure 6 Schematic representation of the sensory pathways from the different sensory modalities (upper row: somatosensory—red, visual—blue, auditory—purple) and the distribution of cortical areas (lower row, S1—primary somatosensory cortex red, V1—primary visual cortex blue, and A1—primary auditory cortex, purpule) in normal, anophthalmic and early cortical lesioned animals. The normal thalamocorical relationship is maintained in anophthalmic mice, but the auditory and somatosensory modalities invade the visual thalamic nuclei (dLGN) and thus the cortex will be responsive to these modalities (Bronchti et al., 2000). The thalamocortical relations do not show major rearrangements. In contrast to this after extremely early cortical lesions in the marsupial, cortical areas showed substantial rearrangements on the remaining cortical sheets, together with considerable rearrangements of the thalamic afferents (Huffmann et al., 1999).

Can Thalamic Connectivity Rearrange Considerably in Experimental Paradigms?

To understand cortical arealization, it is essential to comprehend whether the initial allocation of thalamic projections with different sensory modalities can be shifted during development. There are numerous examples demonstrating that depriving just one sensory modality is not sufficient to change the early thalamocortical allocation although modality specific cortical responses substantially rearrange (Bronchti et al., 2000; Kahn and Krubitzer, 2002; Fig. 6 middle column). Even considerable re-routing surgery had no effect on the early thalamic matching (Sur and Leamy, 2001). In both paradigms the thalamic projections remained in their original place: the cross modal rearrangements were mostly due to changes on a subcortical level (Figure 6). There is however evidence for early alterations of these thalamic projections under extreme conditions. We used a marsupial model to study whether thalamic projections and early thalamocortical matching can realign themselves onto a drastically reduced cortical sheet at a very early stage (Molnár et al., 1998c; Huffmann et al., 1999). This work showed that following an early occipital cortical lesion all cortical areas with their thalamic connections were compressed and a visually responsive area formed on a more rostral region of the neuroepithelium, not originally destined for vision (Huffman et al., 1999; Fig. 6 right column).

Conclusion

Although the adult organization of thalamocortical projections is intriguingly complex, we are beginning to understand how the initial layout of this complicated pathway is constructed with a cascade of simple rules. There are multiple mechanisms involved in establishing thalamocortical connectivity. The early deployment of thalamocortical connectivity is established in an autonomous fashion before the afferents from the sensory periphery reach the dorsal thalamus. However, peripheral signals could begin to modify this juvenile topography after initial cortical targeting, during thalamic fiber ingrowth and arborization. In some cortical areas this rearrangement may be more prominent than in others. With the help of mouse transgenic models we are now able to dissect the mechanisms which are dependent on early neural activity and it will be a major challenge to understand how the pattern of external influence is translated to cellular and molecular signals which organize connectivity within the cortex.

Acknowledgements

We are grateful to Dr Amanda Cheung, Dr Carmen Pinon, Miss Anna Hoerder for their comments on the manuscript. The original work of ZM's laboratory was supported by Grants from Medical Research Council GO300200, European Community (QLRT-1999-30158), The Welcome Trust (063974/B/01/Z), The Royal Society, Swiss National Science Foundation (3100-56032.98), Human Frontier Science Program (RGP0107/2001) and St John's College, Oxford. Optical recording study was supported by grants from the Japanese Ministry of Education, Science and Culture (Grant-in-Aid for Encouragement of Young Scientist No. 08780790 to Shuji Higashi). We are very grateful to Prof Minoru Kimura for his continued support.

References

Adams, N.C., Lozsadi, D.A., and Guillery, R.W. (1997). Complexities in the thalamocortical and corticothalamic pathways. *Eur. J. of Neurosci.*, 9:204–209.

Agmon, A., Yang, L.T., O'Dowd, D. K., Jones, E. G. (1993). Organized growth of thalamocortical axons from the deep tier of terminations into layer IV of developing mouse barrel cortex. *J Neurosci.* 13:5365–82.

Agmon, A., and Connors, B. W. (1991). Thalamocortical responses of mouse somatosensory (barrel) cortex *in vitro*. *Neuroscience,* 41:365–379.

Agmon, A., Yang, L. T., Jones, E. G., and O'Dowd, D. K. (1995). Topological precision in the thalamic projection to neonatal mouse barrel cortex. *J. Neurosci.,* 15:549–561.

Aikawa, H., Nonaka, I., Woo, M., Tsugane, T., Esaki, K. (1988). Shaking rat Kawasaki (SRK): a new neurological mutant rat in the Wistar strain. *Acta Neuropathol (Berl).* 76:366–372.

Allendoerfer, K. L., Shatz, C. J. (1994). The subplate, a transient neocortical structure: Its role in the development of connections between thalamus and cortex. *Ann Rev Neurosci* 17:185–218.

Allman, J. M., and Kaas, J. M. (1971). Representation of the visual field in striate and adjoining cortex of the owl monkey (*Aotus trivirgatus*). *Brain Res.* 35:89–106.

Anderson, G. and Price, D. J. (2002). Layer-specific thalamocortical innervation in organotypic cultures is prevented by substances that alter neural activity. *Eur. J. Neurosci.* 16: 345–349.

Arber, S. (2004). Subplate neurons: bridging the gap to function in the cortex. *Trends Neurosci.* 27:111–113.

Blakemore, C. and Molnár, Z. (1990). Factors involved in the establishment of specific interconnections between thalamus and cerebral cortex. *Cold Spring Harbor Symposia on Quantitative Biology*, 55:491–504.

Bolz, J., Götz, M., Hübener, M., Novak, N. (1993). Reconstructing cortical connections in a dish. *Trend in Neurosci.* 16:310–316.

Bolz, J., Novak, N., Staiger, V. (1992). Formation of specific afferent connections in organotypic slice cultures from rat visual cortex cocultured with lateral geniculate nucleus. *J Neurosci.* 12: 3054–3070

Bronchti, G., Corthesy, M. E., Welker, E. (1999a). Partial denervation of the whiskerpad in adult mice: altered patterns of metabolic activity in barrel cortex. *Eur J Neurosci* 11:2847–55.

Bronchti, G., Katznelson, A., Van Dellen, A., Blakemore, C., Molnár, Z., and Welker, E. (1999b). Deoxyglucose mapping reveals and ordered cortical representation of whiskers in the reeler mutant mouse. *Swiss Soc for Neurosci Meeting, Zürich.* 4:22.

Catalano, S. M., Robertson, R. T., Killackey, H. P. (1996). Individual axon morphology and thalamocortical topography in developing rat somatosensory cortex. *J Comp Neurol.* 367:36–53.

Catalano, S. M., Shatz, C. J. (1998). Activity-dependent cortical target selection by thalamic axons. *Science.* 281:559–62.

Catania, K. C., and Kaas, J. H. (1995). Organization of the somatosensory cortex of the star-nosed mole. *J. Comp. Neurol.,* 351:549–567.

Caviness, V. S. Jr, and Frost, D. O. (1980). Tangential organization of thalamic projections of the neocortex in the mouse. *J. Comp. Neurol.* 194:355–367.

Caviness, V. C., Jr (1988). Architecture and development of the thalamocortical projection in the mouse. *In Cellular Thalamic Mechanisms,* eds. M. Bentivoglio and R. Spreafico, Excetra Medica, Amsterdam-New York, 489–499.

Conolly, M., and Van Essen, D. (1984). The representation of the visual field in parvicellular and magnocellular layers of the lateral geniculate nucleus in the macaque monkey. *J. Comp. Neurol.,* 226:544–564.

Crandall, J. E. and Caviness, V. S., Jr (1984). Thalamocortical connections in newborn mice. *J. Comp. Neurol.* 228:542–556.

Datwani, A., Iwasato, T., Itohara, S., Erzurumlu, R. S. (2002). NMDA receptor-dependent pattern transfer from afferents to postsynaptic cells and dendritic differentiation in the barrel cortex. *Mol. Cell. Neurosci.* 21: 477–492

Dehay, C., Savatier, P., Cortay, V., Kennedy, H. (2001). Cell-cycle kinetics of neocortical precursors are influenced by embryonic thalamic axons. *J Neurosci.* 21:201–214.

Edgar, J. M., Price, D. J. (2001). Radial migration in the cerebral cortex is enhanced by signals from thalamus. *Eur J Neurosci.* 13:1745–1754.

Erzurumlu, R. S., Kind, P. C. (2001). Neural activity: sculptor of 'barrels' in the neocortex. *Trends Neurosci.* 24:589–95.

Friauf, E., Shatz, C. J. (1991). Changing patterns of synaptic input to subplate and cortical plate. J. *Neurophysiology,* 66:2059–2071.

Fukuchi-Shimogori, T., and Grove, E. A. (2001). Neocortex patterning by the secreted signaling molecule FGF8. *Science* 294:1071–1074.

Galli, L., and Maffei, L. (1988). Spontaneous impulse activity of rat retinal ganglion cells in prenatal life. *Science* 242:90–91

Garel, S., and Rubenstein, J. L. (2004). Intermediate targets in formation of topographic projections: inputs from the thalamocortical system. *Trends Neurosci.* 27:533–539.

Garel, S., Yun, K., Grosschedl, R., Rubenstein, J. L. (2002). The early topography of thalamocortical projections is shifted in Ebf1 and Dlx1/2 mutant mice. *Development.* 129:5621–34.

Geiger, B., Ayalon, O. (1992). Cadherins. *Annu Rev Cell Biol* 8: 307–332

Ghosh, A., Shatz, C.J. (1992). Pathfinding and target selection by developing geniculocortical axons. *J. Neurosci.* 12:39–55

Gil, O. D., Needleman, L. A., Huntley, G. W. (2002). Developmental patterns of cadherin expression and localization in relationship to compartmentalized thalamocortical terminations in rat barrel cortex. *J Comp Neurol* 453:372–388.

Guillery, R. W., Sherman, S. M. (2002). Thalamic relay functions and their role in corticocortical communication: generalizations from the visual system. *Neuron.* 33:163–175.

Hanganu, I. L., Kilb, W., Luhmann, H. J. (2002). Functional synaptic projections onto subplate neurons in neonatal rat somatosensory cortex. *J Neurosci.* 22:7165–7176.

Herrmann, K., and Shatz, C. J. (1995). Blockade of action potential activity alters initial arborization of thalamic axons within cortical layer 4. *Proc. Natl. Acad. Sci. USA* 92:11244–11248.

Hevner, R. F., Shi, L., Justice, N., Hsueh, Y., Sheng, M., Smiga, S., Bulfone, A., Goffinet, A. M., Campagnoni, A. T., Rubenstein, J. L. (2001). *Tbr1* regulates differentiation of the preplate and layer 6. *Neuron* 29:353–366.

Hevner, R. F., Miyashita-Lin, E., Rubenstein, J. L. R. (2002). Cortical and Thalamic axon pathfinding defects in Tbr1, Gbx2, and Pax6 mutant mice: Evidence that cortical and thalamis axons interact and guide each other. *J. Comp Neurol* 447:8–17.

Higashi, S., Hioki, K., Kurotani, T., Kasim, N., Molnár, Z. (2005). Functional thalamocortical synapse reorganization from subplate to layer IV during postnatal development in the reeler-like mutant rat (shaking rat Kawasaki). *J Neurosci.* 25:1395–406.

Higashi, S., Molnár, Z., Kurotani, T., Inokawa, H., Toyama, K. (2002). Functional thalamocortical connections develop during embryonic period in the rat: an optical recording study. *Neuroscience* 115:1231–1246.

Hoerder, A., Paulsen, O., Molnár, Z. (2006). Developmental changes in the dendritic morphology of subplate cells with known projections in the mouse cortex. FENS abstracts.

Hübener, M., Götz, M., Klostermann, S., Bolz, J. (1995). Guidance of thalamocortical axons by growth-promoting molecules in developing rat cerebral cortex. *Eur J Neurosci.* 7:1963–1972.

Huffman, K., Molnár, Z., van Dellen, A., Khan, D., Blakemore, C. and Krubitzer, L. (1999). Compression of sensory fields on a reduced cortical sheet. *J. Neurosci.* 19:9939–9952.

Ikeda, Y., and Terashima, T. (1997). Corticospinal tract neurons are radially malpositioned in the sensory-motor cortex of the shaking rat Kawasaki. *J. Comp. Neurol.* 383:370–380.

Iwasato, T., Datwani, A., Wolf, A. M., Nishiyama, H., Taguchi, Y., Tonegawa, S., Knopfel, T., Erzurumlu, R. S., Itohara, S. (2000). Cortex-restricted disruption of NMDAR1 impairs neuronal patterns in the barrel cortex. *Nature* 406:726–731.

Johansen-Berg, H., Behrens, T. E., Sillery, E., Ciccarelli, O., Thompson, A. J., Smith, S. M., Matthews, P. M. (2005). Functional-anatomical validation and individual variation of diffusion tractography-based segmentation of the human thalamus. *Cereb Cortex.* 15:31–39.

Jones, L., López-Bendito, G., Gruss, P., Stoykova, A., Molnár, Z. (2002). Pax6 is required for the normal development of the forebrain axonal connections. *Development* 129:5041–5052.

Jones, E.G. (1985). The thalamus, Plenum Press, New York.

Jones, E.G. (2002). Thalamic organization and function after Cajal. *Progress in Brain Research* 136:333–357.

Kahn, D. M., Krubitzer, L. (2002). Massive cross-modal cortical plasticity and the emergence of a new cortical area in developmentally blind mammals. *Proc Natl Acad Sci U S A.* 99:11429–11434.

Katz, L. C., and Shatz, C. J. (1996). Synaptic activity and the construction of cortical circuits. *Science* 274: 1133–1138.

Krug, K., Smith, A. L., Thompson, I. D. (1998). The development of topography in the hamster geniculo-cortical projection. *J. Neurosci.* 18:5766–5776.

Lee, L. J., Iwasato, T., Itohara, S., Erzurumlu, R. S. (2005). Exuberant thalamocortical axon arborization in cortex-specific NMDAR1 knockout mice. *J Comp Neurol.* 485:280–292.

Leighton, P. A., Mitchell, K. J., Goodrich, L. V., Lu, X., Pinson, K., Scherz, P., Skarnes, W. C., Tessier-Lavigne, M. (2001). Defining brain wiring patterns and mechanisms through gene trapping in mice. *Nature* 410:174–179.

López-Bendito, G. and Molnár, Z. (2003). Thalamocortical Development: How are we going to get there? *Nature Reviews Neuroscience* 4:276–289.

López-Bendito, G., Chan, C.-H., Mallamaci, A., Parnavelas, J., Molnár, Z. (2002). The role of *Emx2* in the development of the reciprocal connectivity between cortex and thalamus. *J. Comp. Neurol* 451:153–169.

Lund, R. D. and Mustari, M. J. (1977). Development of the geniculocortical pathways in rat. *J. Comp. Neurol.* 173:289–305.

Mann, F., Peuckert, C., Dehner, F., Zhou, R., Bolz, J. (2002). Ephrins regulate the formation of terminal axonal arbors during the development of thalamocortical projections. *Development* 129:3945–3955

Marín, O. (2003). Thalamocortical topography reloaded: it's not where you go, but how you get there. *Neuron* 39:388–91.

McCasland, J. S., and Woolsey, T. A. (1988). High resolution 2-deoxyglucose mapping of functional cortical columns in mouse barrel cortex. *J. Comp. Neurol.* 278:555–569.

McConnell, S. K., Ghosh, A., and Shatz, C. J. (1989). Subplate neurons pioneer the first axon pathway from the cerebral cortex. *Science* 245:978–982.

Meister, M., Wong, R. O. L., Baylor, D. A., and Shatz, C.J. (1991). Synchronous bursts of action potentials in ganglion cells of the developing mammalian retina. Science 252:939–943.

Métin, C. and Godement, P. (1996). The ganglionic eminence may be an intermediate target for corticofugal and thalamocortical axons. *J. Neurosci.* 16:3219–3235.

Mitrofanis, J. and Guillery, R. W. (1993). New views of the thalamic reticular nucleus in the adult and the developing brain. *Trends Neurosci* 16:240–245.

Miyashita-Lin, E. M., Hevner, R., Wassarman, K. M., Martinez, S. and Rubenstein, J. L. (1999). Early neocortical regionalization in the absence of thalamic innervation. *Science* 285:906–909.

Molnár,Z., López-Bendito, G., Small, J., Partridge, L. D., Blakemore, C., Wilson, M. C., (2002b). Normal development of embryonic thalamocortical connectivity in the absence of evoked synaptic activity. *J. Neurosci* 22:10313–10323.

Molnár, Z. (1998). *Development of Thalamocortical Connections*, Springer - Heidelberg, 264p.

Molnár, Z., and Blakemore, C. (1999). Development of signals influencing the growth and termination of thalamocortical axons in organotypic culture. *Experimental Neurology.* 156:363–393.

Molnár, Z., Adams, R., Blakemore, C. (1998a). Mechanisms underlying the establishment of topographically ordered early thalamocortical connections in the rat. *J Neurosci* 18:5723–5745.

Molnár, Z., Adams, R., Goffinet, A. M., Blakemore, C. (1998b). The role of the first postmitotic cells in the development of thalamocortical fiber ordering in the *reeler* mouse. *J Neurosci* 18: 5746–5785.

Molnár, Z., Knott, G. W., Blakemore, C., and Saunders, N. R. (1998c). Development of thalamocortical projections in the South American grey short-tailed opossum (Monodelphis Domestica). *J Comp Neurol* 398:491–514.

Molnár, Z., Blakemore, C. (1991). Lack of regional specificity for connections formed between thalamus and cortex in coculture. *Nature* 351: 475–477.

Molnár, Z., Blakemore, C. (1995). How do thalamic axons find their way to the cortex? Trends Neurosci 18:389–397.

Molnár, Z., Hannan, A. (2000) Development of thalamocortical projections in normal and mutant mice. In *Mouse Brain Development*, Eds A. Goffinet and P. Rakic Springer-Verlag, Heidelberg, 293–332.

Molnár, Z., Higashi, S., López-Bendito, G. (2003). Choreography of early thalamocortical development. *Cerebral Cortex* 13:661–669.

Molnár, Z., Bronchti, G., Blakemore, C., and Welker, E. (1996). Initial topological order in thalamocortical projections in the barrelless mutant mouse. *Soc. Neurosci. Abstr.* 22:1013.

Molnár, Z., Kurotani, T., Higashi, S., and Toyama, K. (2002a). Development of functional thalamocortical synapses studied with current source density analysis in whole forebrain slices. *Brain Research Bulletin* 60:355–372.

Mooney, R., Penn, A. A., Gallego, R., Shatz, C. J. (1996). Thalamic relay of spontaneous retinal activity prior to vision. *Neuron.* 17:863–874.

Naegele, J. R., Jhaveri, S., Schneider, G. E. (1988). Sharpening of topographical projections and maturation of geniculocortical axon arbors in the hamster. *J. Comp. Neurol.* 277:593–607.

Nakagawa, Y., Johnson, J. E. and O'Leary, D. D. (1999). Graded and areal expression patterns of regulatory genes and cadherins in embryonic neocortex independent of thalamocortical input. *J. Neurosci.* 19:10877–85.

Nelson, S. B., and LeVay, S. (1985). Topographic organization of the optic radiation of the cat. *J. Comp. Neurol.* 240:322–330.

Portera-Cailliau, C., Weimer, R. M., De Paola, V., Caroui, P., Svoboda, K. (2005). Diverse modes of axon elaboration in the developing neocortex. PLoS Biology, 3(8): e272.

Poskanzer, K., Needleman, L. A., Bozdagi, O., Huntley, G. W. (2003). N-cadherin regulates ingrowth and laminar targeting of thalamocortical axons. *J Neurosci.* 23:2294–2305.

Price D. J., Kennedy H., Dehay C., Zhon L., Mercier M., Jossin Y., Goffinet A. M., Tissir F., Blakey D., Molnár Z. (2006). The development of cortical connections. *Eur. J. Neurosci*: 23: 910–920.

Rakic, P. (1977). Prenatal development of the visual system in the rhesus monkey. *Philos Trans R Soc Lond B Biol Sci* 278:245–260.

Rebsam, A., Seif, I., Gaspar, P. (2002). Refinement of thalamocortical arbors and emergence of barrel domains in the primary somatosensory cortex: a study of normal and monoamine oxidase a knock-out mice. J Neurosci. 22:8541–8552.

Rizo, J. and Südhof, T. C. (2002). Snares and munc18 in synaptic vesicle fusion. Nat Rev Neurosci. 3:641–653.

Seibt, J., Schuurmans, C., Gradwhol, G., Dehay, C., Vanderhaeghen, P., Guillemot, F., Polleux, F. (2003). Neurogenin2 specifies the connectivity of thalamic neurons by controlling axon responsiveness to intermediate target cues. *Neuron* 39:439–452.

Shatz, C. J. and Rakic, P. (1981). The genesis of efferent connections from the visual cortex of the fetal rhesus monkey. *J. Comp. Neurol.* 196:287–307.

Shatz, C. J. and Luskin, M. B. (1986). Relationship between the geniculocortical afferents and their cortical target cells during development of the cat's primary visual cortex. *J. Neurosci.* 6:3655–3668.

Shimogori, T. and Grove, E. A. (2006) Subcortical and neocortical guidance of Area-specific thalamic innervation, In: Development and plasticity in sensory thalamus and cortex, Eds: Erzurumlu, R., Guido, W., Molnár, Z., Springer, Heidelberg. p. 68.

Südhof, T. C. (1995). The synaptic vesicle cycle: a cascade of protein-protein interactions. *Nature* 375:645–653.

Sur, M., Leamey, C. A. (2001). Development and plasticity of cortical areas and networks. *Nat Rev Neurosci.* 2:251–262.

Tuttle, R., Nakagawa, Y., Johnson, J. E. and O'Leary, D. D. (1999). Defects in thalamocortical axon pathfinding correlate with altered cell domains in Mash-1-deficient mice. *Development* 126:1903–1916.

Vanderhaegen, P., Lu, Q., Parakash, N., Frisen, I., Walsh, C. A., Frostig, R. D., Flanagan, J. G. (2000) A mapping label required for normal scale of body representation in the cortex. *Nat Naurosci* 3:358–365.

Varoqueaux, F., Sigler, A., Rhee, J.-S., Brose, N., Enk, C., Reim, K., Rosenmund, C. (2002). Total arrest of spontaneous and evoked synaptic transmission but normal synaptogenesis in the absence of Munc13-mediated vesicle priming. *Proc Natl Acad Sci* 99:9037–9042.

Verhage, M., Maia, A. S., Plomp, J. J., Brussaard, A. B., Heeroma, J. H., Vermeer, H., Toonen, R. F., Hammer, R. E., van den Berg, T. K., Missler, M., Geuze, H. J., Südhof, T. C. (2000). Synaptic assembly of the brain in the absence of neurotransmitter secretion. *Science* 287:864–869.

Washbourne, P., Thompson, P. M., Carta, M., Costa, E. T., Mathews, J. R., López-Bendito, G., Molnár, Z., Becher, M. W., Valenzuela, C. F., Partridge, L. D., Wilson, M. C. (2002). Genetic ablation of the t-SNARE SNAP-25 distinguishes mechanisms of neuroexocytosis. *Nature Neuroscience* 5:19–26.

Wilkermeyer, M. F., Angelides, K. J. (1996). Addition of tetrodotoxin alters the morphology of thalamocortical axons in organotypic cocultures. *J. Neurosci. Res.* 43: 707–718.

Yamamoto, N., Higashi, S., Toyama, K. (1997). Stop and branch behaviors of geniculocortical axons: a time-lapse study in organotypic cocultures. *J Neurosci.* 17:3653–3663.

Yamamoto, N., Yamada, K., Kurotani, T., Toyama, K. (1992). Laminar specificity of extrinsic cortical connections studied in coculture preparations. *Neuron* 9:217–228.

Yamamoto, N., Inui, K., Matsuyama, Y., Harada, A., Hanamura, K., Murakami, F., Ruthazer, E.S., Rutishauser, U., Seki, T. (2000). Inhibitory mechanism by polysialic acid for lamina-specific branch formation of thalamocortical axons. *J. Neurosci.* 20:9145–9151.

Yamamoto, N., Kurotani, T., Toyama, K., (1989). Neural connectionsbetween the lateral geniculate nucleus and visual cortex *in vitro*. *Science* 245:192–194.

Yamamoto, N., Takemoto, M., Hattori, Y., Hanamura, K. (2006). Molecular basis for the formation of lamina-specific thalamocortical projection. In: Development and plasticity in sensory thalamus and cortex, Eds: Erzurumlu, R., Guido, W., Molnár, Z., Springer, Heidelberg. pp. 78–90

5

Molecular Basis for the Formation of Lamina-Specific Thalamocortical Projection

Nobuhiko Yamamoto, Makoto Takemoto, Yuki Hattori, and Kenji Hanamura

The thalamocortical (TC) projection is one of fundamental neural circuits in the neocortex. The most characteristic feature of TC projection is layer specificity: Sensory thalamic neurons primarily project to layer 4 of corresponding neocortical areas (Jones, 1981; Gilbert, 1983). An intriguing problem is how lamina-specific TC projections are formed during development, since laminar specificity contains not only a principle of cortical circuit formation but also a common feature in axonal targeting mechanisms in the CNS.

Neurobiologists have revealed that growing axons are guided by local and long-range molecular cues (Tessier-Lavigne and Goodman, 1996): Chemoattractive and chemorepulsive molecules released from target cells are able to act as long-range cues, whereas membrane-associated molecules such as cell surface or extracellular matrix (ECM) molecules act as contact-mediated cues. In fact, a number of molecules with these properties have been identified during the past decade. However, what molecular mechanisms govern pathway guidance and target recognition processes in a given neural system remained unknown. Moreover, regulation by multiple molecules, which must work *in vivo*, is vague.

We have explored how TC axons recognize their target layer, focusing on axonal termination and branching. In this chapter, we describe cellular and molecular mechanisms underlying TC axonal targeting mechanisms by demonstrating TC axon behavior *in vitro* and gene expression pattern in the developing cortex.

1. TC Axon Termination and Branching in the Target Layer

During development TC axons originating from sensory thalamic nuclei travel in the intermediate zone and reach appropriate cortical areas. TC axons then grow into the cortical plate (CP) and form branches primarily

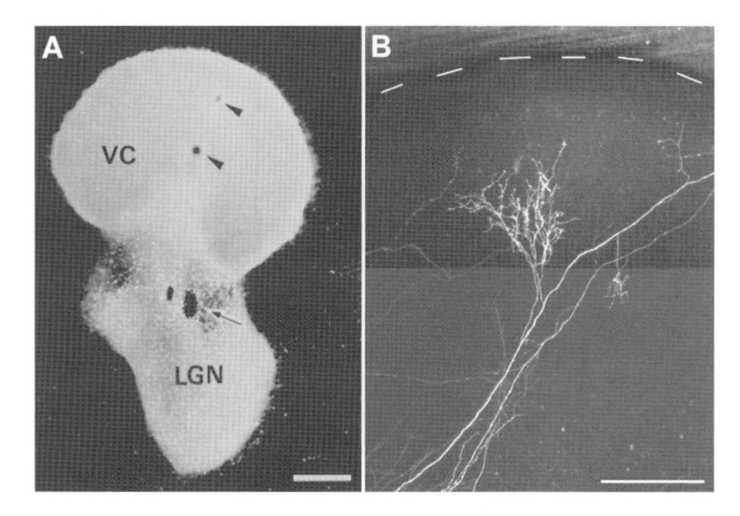

Figure 1 Coculture preparations of the thalamus and cortex. A, organotypic coculture of thalamic explant (lateral geniculate nucleus, LGN) with cortical explant (visual cortex, VC) after one week *in vitro* (From Yamamoto et al., 1989). Bar represents 0.5 mm. B, fluorescent dye-labeled TC axon arbor in cortical explant after two weeks *in vitro* (From Yamamoto et al., 1997). Bar represents 0.25 mm.

in layer 4 without extending to the more superficial layers (Lund and Mustari, 1977; Ghosh and Shatz, 1992; Agmon et al., 1993; Kageyama and Robertson, 1993; Catalano et al., 1996; Molnár et al., 1998).

The question is how TC axons can recognize their target layer. To gain some insights into this issue, we developed organotypic coculture preparations in which specific neuronal connections were reconstructed with the cytoarchitecture preserved (Yamamoto et al., 1989, 1992). In fact, in an organotypic coculture of the rat cortex with the thalamus, the TC projection was formed with essentially the same laminar specificity as that found *in vivo* (Fig. 1) (Yamamoto et al., 1989, 1992; Molnár and Blakemore, 1991; Bolz et al., 1992). This accessible preparation has been useful to study the characteristic axon behaviors that reflect the cellular mechanisms responsible for laminar specificity of the TC projection.

Based on observations of the projection pattern in the culture system, Molnár and Blakemore (1991) proposed that a stop signal for TC axons is present in the cortex. This idea is supported by the fact that thalamic axons entering from the pial surface of cortical slices terminate in the same target layer as do axons entering in the normal direction from the ventricular surface, in spite of the different distances and orientations (Bolz et al., 1992; Yamamoto et al., 1997). We examined TC axon behavior more directly in a time-lapse study (Yamamoto et al., 1997). The majority of the axons traveled at a constant growth rate in layers 6 and 5, but suddenly stopped in and around layer 4. An interesting feature is that the stop behavior was observed in only the axons that traveled perpendicularly to cortical layers but not in those running along layer 4 (Fig. 2).

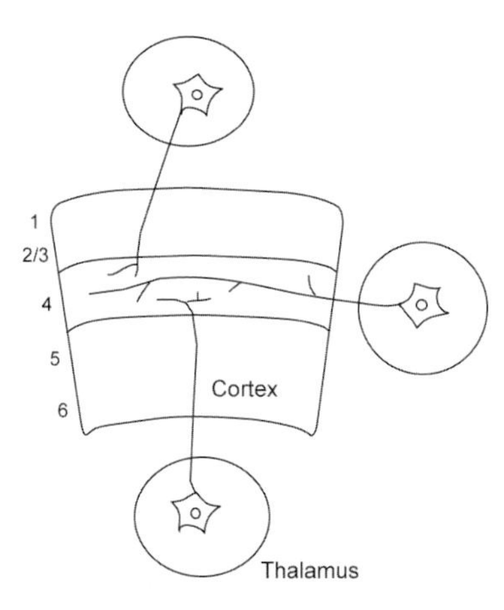

Figure 2 Stop and branch behaviors observed in cocultures of the thalamus and cortex. Thalamic explant is juxtaposed to the ventricular surface (normal ingrowth), pial surface or the lateral edge of the cortical explant. The stop behavior in layer 4 is observed in TC axons which enter the ventricular and pial surface, but not in those entering from the lateral edge. On the other hand, TC axons exhibit the branch behavior regardless of the orientation of ingrowth (From Yamamoto, 2002).

This implies that the stop signal is detected by TC axons as a relative difference in the environmental cues rather than an absolute cue (see below). Noctor et al. (2001) have demonstrated that TC axons extend to the superficial layers after disruption of layer 4 cells in developing ferret cortex, indicating that the stop signal is not simply due to an artifact *in vitro*. The existence of the stop signal has also been demonstrated in neuronal circuits in the cerebellum and the spinal cord, by using *in vitro* preparations (Baird et al., 1992; Sharma et al., 1994). Thus, axonal stop behavior may reflect a general aspect of axonal targeting processes.

Branch formation is another key behavior for axonal targeting. In the coculture of the thalamus with cortex, TC axonal branching was confined to layer 4, indicating the existence of the cellular interactions that induce branch formation (Yamamoto et al., 1989, 1992; Bolz et al., 1992). The time-lapse study further demonstrated that TC axonal branching mostly took place in layer 4 without any transient branch appearance in other layers (Yamamoto et al., 1997). One may suppose that axonal branching simply follows the stop behavior. Indeed, branching was often observed just after TC axons stopped growing. In accordance with this result, in cortical neurons *in vitro* and frog retinal ganglion cells *in vivo*, there is an obvious association between axonal stopping and branching (Szebenyi et al., 1998; Campbell et al., 2001). However, stopping and branching of TC axons did not always occur in the time-lapse study. Moreover, the branching behavior was found regardless of the orientation of ingrowth, which is contrast to stopping behavior (Fig. 2). These findings imply that branching is regulated by the mechanisms that are distinct from those of axonal termination.

Thus, the two independent mechanisms of axonal termination and branching appear to be involved in the formation of lamina-specific TC connections.

2. Mechanisms for TC Axon Termination

In general, axonal termination can be regulated by some molecular mechanisms. One possible mechanism is that axons may stop growing by detecting a molecule of the "stop signal". At the neuromuscular junction, s-laminin, which is concentrated in the endplate is involved in termination of axonal growth (Porter et al., 1995). In culture, growth cones of motor axons stop for up to several hours when they contact s-laminin, without retracting or turning, indicating that s-laminin acts as a stop signal.

A spatial difference of growth-regulating molecules can also produce axonal stop behavior. It is likely that this mechanism acts in TC axonal targeting in the cortex, that is, the laminar difference in growth-promoting or growth-inhibitory molecules could produce termination of TC axon growth. This regulatory mechanism was studied by growing thalamic axons on chemically fixed cortical slices. Why fixed cortical slices were used is because membrane-associated molecules rather than diffusible molecules were thought to underlie TC axon growth. Indeed, TC axons were found to stop growing suddenly in a restricted region in the time-lapse study, indicating that local molecular cues regulate axonal termination. Moreover, this *in vitro* system in combination with enzymatic perturbation permits a direct assessment of activity and biochemical properties of membrane-associated molecules which may affect TC axon behaviors without the confounding influence of diffusible factors that would be released from living cortical cells (Yamamoto et al., 2000b).

TC axons entering the lateral edge of the fixed cortical slice exhibited an obvious difference in axonal extension between cortical layers: TC axons grow more extensively in layers 5 and 6 than layers 2/3 and 4 (Fig. 3). One possible explanation for this phenomenon is that growth-permissive or promoting activity is weaker in the upper layers than in the deep layers. Alternatively, growth-inhibitory components may be expressed in the upper layers. To test these possibilities, fixed cortical slices were subjected to several enzymatic treatments prior to culturing with thalamic explants (Yamamoto et al., 2000b). The result showed that the suppression of axonal growth in the upper layers is reduced considerably by phosphatidylinositol-specific phospholipase C pretreatment of cortical slices (Fig. 3). Therefore, the laminar difference in axonal growth can be attributed to growth-inhibitory activity, in large part due to glycosylphosphatidylinositol-linked molecules in the upper layers, which supersedes some growth-promoting activity (Götz et al., 1992) in all cortical layers. The existence of growth-inhibitory components in the upper layers has also been suggested from the point of view of cell

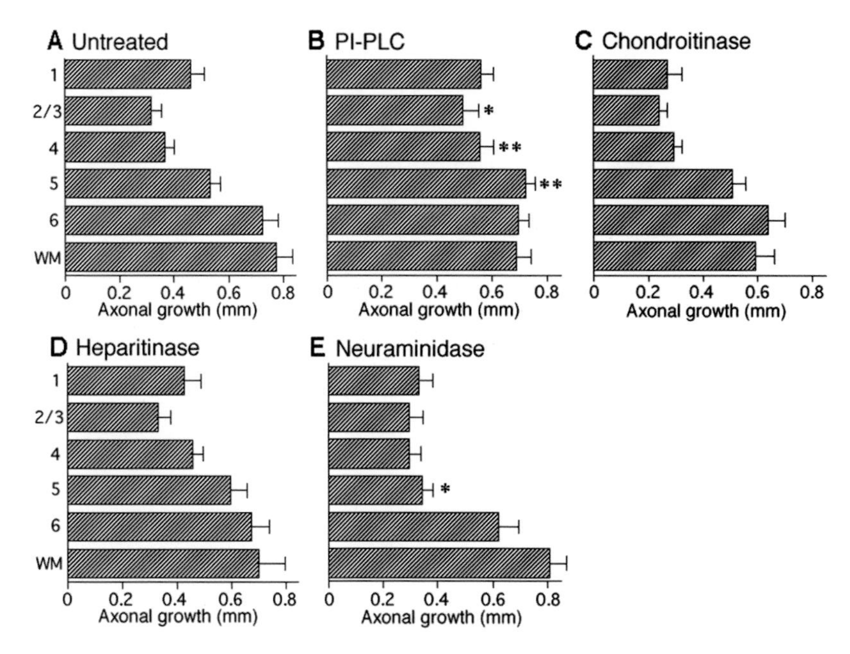

Figure 3 Lamina-specific growth of TC axons on fixed cortical slices (From Yamamoto et al., 2000b). Axonal growth was analyzed in untreated slices (A) or slices treated with phosphatidylinositol-specific phospholipase C (B), chondroitinase (C), heparitinase (D) or neuraminidase (E). Note that the suppression of axonal growth in the upper layers is reduced significantly (asterisks) by phosphatidylinositol-specific phospholipase C treatment.

adhesion. Emerling and Lander (1994) have shown that dissociated thalamic cells adhere better to the deep cortical layers of brain slices than to the upper layers. If cell attachment is mechanistically related to axonal elongation (Forster et al., 1998), this finding further supports the notion that inhibitory activity in the upper layers is involved in axonal targeting.

It should be noted that the growth-inhibitory activity in the upper layers is not strong enough to suppress TC axon invasion into the CP, as TC axons are able to enter the CP from the pial surface (Yamamoto et al., 2000b). Therefore, the relative difference between the upper and deep layers rather than absolute concentration of growth-regulating molecules could contribute to axonal stop behavior.

3. Mechanisms of TC Axonal Branching in the Target Layer

Molecular mechanisms for axonal branching are poorly understood, but several ECM or cell surface molecules have been reported to be able to affect axonal branching. N-cadherin (Inoue and Sanes, 1997), ephrin-A5, one of the EphA ligands (Castellani et al., 1998), reelin, which is

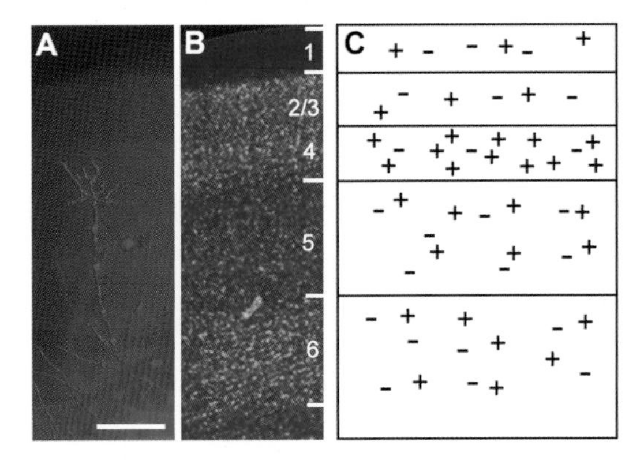

Figure 4 Lamina-specific TC axon branching. A, DiI-labeled TC axon forms specifically in layer 4 of fixed cortex (From Yamamoto et al., 2000a). Bar represents 0.2 mm. B, Cortical lamination is shown by bisbenzamide staining. C, Distribution of branch-regulating molecules in the developing cortex. There are a branch-inducing factor (+) and a branch-inhibitory factor (−), PSA in the developing cortex.

responsible for the reeler phenotype (Del Rio et al., 1997), and Slit-2, which was originally identified as a repellent molecule for a subset of neurons (Wang et al., 1999; Ozdinler and Erzurumlu, 2002) promote axonal branching in CNS and PNS neurons. Although one difficulty is isolation of branch-inducing factors from growth-promoting activity, all of the factors listed act as positive regulators.

We have demonstrated that an inhibitory mechanism also plays an important role in TC axonal branching in the target layer. This was revealed by examining axonal branching on fixed cortical slices as described above (Yamamoto et al., 2000a): Living thalamic explants were juxtaposed to fixed cortical slices, and the laminar location of axonal branching was analyzed. Interestingly, most axonal branches formed in layer 4 of even fixed cortical slices, indicating that lamina-specific branching is regulated by membrane-associated molecules (Figs. 4A and B). Moreover, pretreatment of fixed cortical slices with endoneuraminidase N, the enzyme that specifically digests polysialic acid (PSA) on neural cell adhesion molecule, resulted in axonal branches emerging across broader laminar locations. This role for PSA would be easiest to explain if PSA expression was lowest in layer 4. However, PSA expression was nearly uniform across cortical layers (Seki and Arai, 1991; Yamamoto et al., 2000a). A plausible explanation is that a positive cue whose activity is suppressed by PSA is distributed in all layers but with a peak in layer 4. If the threshold of activity required for branching is only attained in layer 4, then in effect PSA serves as a filter that increases the signal-to-noise ratio during innervation (Fig. 4C). Thus, PSA in the cortex is considered to prevent TC axons from forming branches in inappropriate layers rather than eliminate aberrant branches that appear in layers other than the target layer (Seki and Rutishauser, 1998).

However, the fact that laminar specificity is reduced but still retained after the enzymatic treatment implies the presence of a branch-inducing factor, which is localized in layer 4.

4. Identification of Lamina-Specific Growth- and Branch-Regulating Molecules

Axonal stopping and branching are crucial for TC axonal targeting. Furthermore, it is likely that both positive and negative molecules regulate these axon behaviors. Thus, multiple molecules could be involved in this process. There is now strong evidence for growth-inhibitory factor(s) in layers 2/3–4, branch-inducing factor(s) in layer 4, and growth-promoting and branch-inhibitory factors in all cortical layers.

The great challenge that now stands before us is to identify these molecules. It has been reported that several cell surface and ECM molecules, such as Eph receptors and ligands (Donoghue and Rakic, 1999; Mackarehtschian et al., 1999; Vanderhaeghen et al., 2000; Yabuta et al., 2000), and chondroitin and heparan sulfate proteoglycans (Litwack et al., 1994; Oohira et al., 1994; Maeda and Noda, 1996; Watanabe et al., 1996), are expressed with varying degrees of laminar specificity. As for the upper layers, Sema-7A, a semaphorin family member, is expressed in layers 2/3–4 of the neonatal rodent cortex (Xu et al., 1998). *Cadherin-6* and *rCNL3*, a G-protein-coupled receptor are also expressed in the upper layers of the developing cortex (Suzuki et al., 1997; Chenn et al., 2001). These distributions match the laminar expression of the growth-inhibitory activity for thalamic axons. It would be worthwhile to examine whether these molecules affect TC axonal behaviors.

However, it is not certain whether TC axon termination and branching can be explained by only the known molecules. We attempted to search for the molecules expressed specifically in the upper layers, in particular, in layer 4, by constructing a subtraction cDNA library in which cDNAs derived from layer 4 strips of P7 rat somatosensory cortex were enriched by subtracting cDNAs from layer 5 strips (Zhong et al., 2004). Differential screening and *in situ* hybridization demonstrated that several clones were expressed strongly in layer 4 or layer 2/3–4 of P7 rat cortex. One of the obtained clones was expressed rather specifically in layer 4 (Zhong et al., 2004) (Fig. 5). Sequence analysis demonstrated its features of a transmembrane protein, including a signal peptide sequence, two immunoglobulin and thrombospondin domains. Its cytoplasmic region consists of ZU5 and death domains, which are common to unc5-like netrin receptors. We designated this novel member of the unc5 family as unc5h4.

The expression profile during development showed a migrating behavior of *unc5h4* expression. Layer 4 cells are born at E16-17 in the ventricular zone and migrate to the subventricular and intermediate zones at E18. At P0, they move to the most superficial part of the CP, and gradually settle in layer 4 by P6. *Unc5h4* expression closely

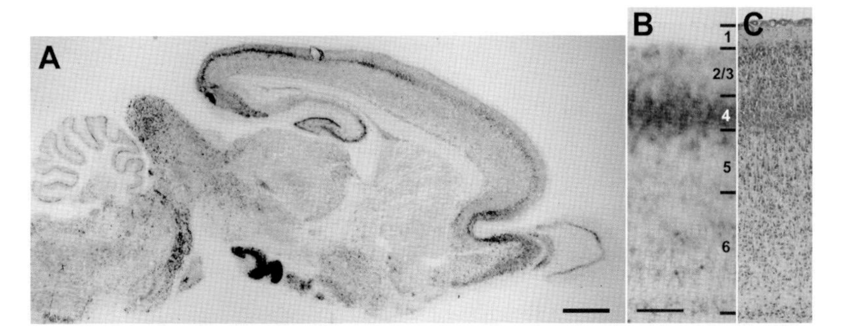

Figure 5 Existence of the gene that is specifically expressed in layer 4 of P7 rat cortex (From Zhong et al., 2004). A, Expression pattern of *unc5h4*, one of the genes obtained from a subtraction cDNA library. Bar represents 1 mm. B, layer 4-specific expression of *unc5h4* in the somoatosensory cortex. Bar represents 0.2 mm. C, Nissl-stained section.

resembles this migration pattern, suggesting that *unc5h4* is expressed during development by the cells destined to form layer 4 of the cortex. Moreover, *unc5h4* expression in embryonic stages was rather uniform across all cortical areas, while it is weak in the motor cortex in postnatal stages. This finding is consistent with the view that layer 4 cells express *unc5h4*, as granular cells, the major population in layer 4, are scarce in the motor cortex. Our recent study further showed that thalamic cells tended to form axonal fasciculation on unc5h4-coated dishes (our own unpublished observations). Thus, its molecular characteristics raise the possibility that it is involved in the interactions between layer 4 cells, or between layer 4 cells and TC fibers.

5. Regulatory Mechanisms by Secreted Factors

As described above, TC axonal targeting is primarily regulated by membrane-associated molecules. However, secreted molecules could also affect these axon behaviors. Indeed, some secreted factors from living cortical cells have been shown to promote TC axon growth (Lotto and Price, 1995; Rennie et al., 1994). Neurotrophins are plausible candidate molecules, since a number of studies have demonstrated that they are involved in the regulation of axonal growth and branching in the CNS as well as PNS (Davies et al., 1986; Morfini et al., 1994; Segal et al., 1995; Cohen-Cory and Fraser, 1995).

First, we studied the precise time course of neurotrophin expression in the developing cortex (Hanamura et al., 2004). A quantitative enzyme immunoassay demonstrated that neurotrophin-3 (NT-3) expression was strong between E18 and P7 with a peak value at P3, when TC axons are invading the CP and begin to form branching. On the contrary, brain-derived neurotrophic factor (BDNF) expression was negligible until P7 but afterwards increased dramatically. Similar developmental

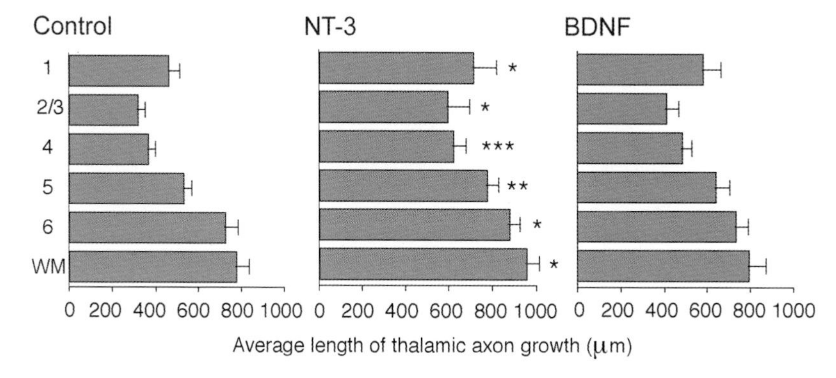

Figure 6 TC axon growth on fixed cortical slices in the presence of NT-3 and BDNF. E15-16 thalamic explants were juxtaposed to the lateral side of fixed P7 cortical slices in the presence of either NT-3 or BDNF (From Hanamura et al., 2004). Quantitative analysis shows that NT-3 enhances axonal growth.

time courses of *BDNF* and *NT-3* mRNA expressions are found in the cat visual cortex (Lein et al., 2000). Thus, it is likely that NT-3 rather than BDNF contributes to TC axon targeting processes.

Whether NT-3 affects TC axon behavior was examined in the culture system where living thalamic explants were juxtaposed to the chemically fixed cortical slices (Yamamoto et al., 2000b; see above). This experimental condition allows us to examine direct action of neurotrophins, since endogenous neurotrophins are no longer secreted from the target cortical cells. Furthermore, the possibility that dendritic changes of cortical cells by the neurotrophin may bring some influences on TC axons can also be excluded. The result showed that in NT-3 containing medium TC axon growth dramatically increased in all cortical layers of fixed cortex (Hanamura et al., 2004) (Fig. 6). In agreement with this, geniculocortical axons have been demonstrated to fail to invade the CP in NT-3 knockout mice (Ma et al., 2002). However, the growth-promoting activity of NT-3 was not found on collagen-coated dishes. This finding implies that NT-3 enhances TC axon growth by cooperating with membrane-associated molecules on cortical cells but not with collagen.

Although the component that cooperates with NT-3 has not been identified, it should be present in the membrane fraction (Castellani and Bolz, 1999). Moreover, a quantitative analysis demonstrated that enhancement of axonal growth in the upper layers including layer 4 was higher than that in the deep layers, indicating that the membrane-bound component should appear in higher concentrations in the upper layers. This cooperative activity of NT-3 may contribute to TC axon growth in layer 4 by overcoming the inhibitory activity that is distributed in the target layer (Fig. 7). Such a neutralization effect of NT-3 for growth inhibitory molecules has been demonstrated in regenerating axon growth in the spinal cord (Schnell et al., 1994). In addition, *NT-3* messages are localized in layer 4 of the cat visual cortex (Lein et al., 2000), although the protein expression pattern is not obvious. Such localization of NT-3

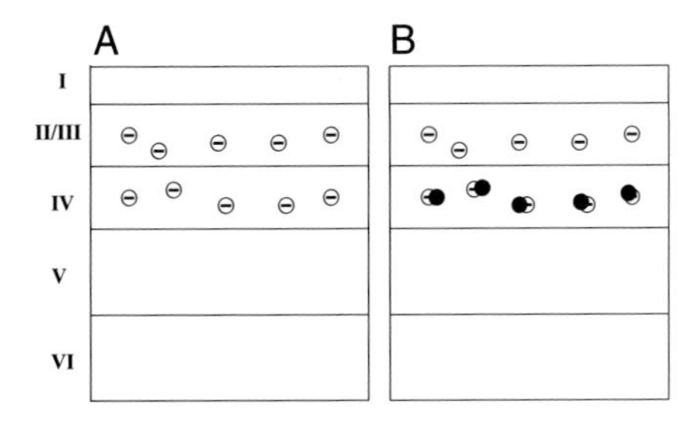

Figure 7 NT-3 action in the developing neocortex. In the developing cortex, a growth-inhibitory molecule (−) is present in the upper layers (A). NT-3 (closed circle) can promote TC axon growth in the target layer, by neutralizing the inhibitory activity (B).

expression can produce more restricted action on TC axon growth in the target layer (Fig. 7).

6. Conclusion

Our findings in *in vitro* preparations have demonstrated that TC axonal targeting involves axonal termination and branching processes. Evidence further indicates that these processes are primarily regulated by the cell surface molecules and/or ECM molecules that are expressed in a lamina-specific manner in the developing cortex, although secreted factors such as neurotrophins also play a role in these axon behaviors. Our extensive molecular screening has further shown that some membrane-associated factors including unc5h4 are expressed specifically in layer 4, the target layer. A future study would be to reveal the role of the candidate molecules and their regulatory mechanisms.

References

Agmon, A., Yang, L. T., O'Dowd, D. K., Jones, E. G. (1993). Organized growth of thalamocortical axons from the deep tier of terminations into layer IV of developing mouse barrel cortex. *J. Neurosci.* 13:5365–5382.

Baird, D. H., Hatten, M. E., Mason, C. A. (1992). Cerebellar target neurons provide a stop signal for afferent neurite extension in vitro. *J. Neurosci.* 12:619–634.

Bolz, J., Novak, N., Staiger, V. (1992). Formation of specific afferent connections in organotypic slice cultures from rat visual cortex cocultured with lateral geniculate nucleus. *J. Neurosci.* 12:3054–3070.

Campbell, D. S., Regan, A. G., Lopez, J. S., Tannahill, D., Harris, W. A., Holt, C. E. (2001). Semaphorin 3A elicits stage-dependent collapse, turning, and branching in Xenopus retinal growth cones. *J. Neurosci.* 21:8538–8547.

Castellani, V., Bolz, J. (1999). Opposing roles for neurotrophin-3 in targeting and collateral formation of distinct sets of developing cortical neurons. *Development* 126:3335–45.

Castellani, V., Yue; Y., Gao, P. P., Zhou, R., Bolz, J. (1998). Dual action of a ligand for Eph receptor tyrosine kinases on specific populations of axons during the development of cortical circuits. *J. Neurosci.* 18:4663–4672.
Catalano, S. M., Robertson, R. T., Killackey, H. P. (1996). Individual axon morphology and thalamocortical topography in developing rat somatosensory cortex. *J. Comp. Neurol.* 366:36–53.
Chenn, A., Levin, M. E., McConnell, S. K. (2001). Temporally and spatially regulated expression of a candidate G-protein-coupled receptor during cerebral cortical development. *J. Neurobiol.* 46:167–177. Erratum in: *J. Neurobiol.* 47:244.
Cohen-Cory, S., Fraser, S. E. (1995). Effects of brain-derived neurotrophic factor on optic axon branching and remodelling in vivo. *Nature* 378:192–196.
Davies, A. M., Thoenen, H., Barde, Y. A. (1986). The response of chick sensory neurons to brain-derived neurotrophic factor. *J. Neurosci.* 6:1897–1904.
Del Rio, J. A., Heimrich, B., Borrell, V., Forster, E., Drakew, A., Alcantara, S., Nakajima, K., Miyata, T., Ogawa, M., Mikoshiba, K., Derer, P., Frotscher, M. and Soriano, E. (1997) A role for Cajal-Retzius cells and reelin in the development of hippocampal Connections. *Nature* 385:70–74.
Donoghue, M. J., Rakic, P. (1999). Molecular evidence for the early specification of presumptive functional domains in the embryonic primate cerebral cortex. *J. Neurosci.* 19:5967–79
Emerling, D. E., Lander, A. D. (1994). Laminar specific attachment and neurite outgrowth of thalamic neurons on cultured slices of developing cerebral cortex. *Development* 120:2811–2822.
Forster, E., Kaltschmidt, C., Deng, J., Cremer, H., Deller, T., Frotscher, M. (1998). Lamina-specific cell adhesion on living slices of hippocampus. *Development* 125:3399–3410.
Ghosh, A., Shatz, C. J. (1992). Pathfinding and target selection by developing geniculocortical axons. *J. Neurosci.* 12:39–55.
Gilbert, C. D. (1983). Microcircuitry of the visual cortex. *Ann. Rev. Neurosci.* 6:217–247.
Götz, M., Novak, N., Bastmeyer, M., Bolz, J. (1992). Membrane-bound molecules in rat cerebral cortex regulate thalamic innervation. *Development* 116:507–519.
Hanamura, K., Harada, A., Katoh-Semba, R., Murakami, F., Yamamoto, N. (2004). BDNF and NT-3 promote thalamocortical axon growth with distinct substrate and temporal dependency. *Eur. J. Neurosci.* 19:1485–1493.
Inoue, A., Sanes, J. R. (1997). Lamina-specific connectivity in the brain: regulation by N-cadherin, neurotrophins, and glycoconjugates.*Science* 276:1428–1431.
Jones, E. G. (1981). Anatomy of cerebral cortex: Columnar input-output organization. In: Schmitt, F. O.,Worden, F. G., Adelman, G., Dennis, S. G. (eds.), The Organization of the Cerebral Cortex, pp. 199–235. Cambridge, Massachusetts, London, England: The MIT press.
Kageyama, G. H., Robertson, R. T. (1993). Development of geniculocortical projections to visual cortex in rat: evidence for early ingrowth and synaptogenesis. *J. Comp. Neurol.* 335:123–148.
Lein, E. S., Hohn, A., Shatz, C. J. (2000). Dynamic regulation of BDNF and NT-3 expression during visual system development. *J. Comp. Neurol.* 420:1–18.
Litwack, E. D., Stipp, C. S., Kumbasar, A., Lander, A. D. (1994). Neuronal expression of glypican, a cell-surface glycosylphosphatidyl-inositol-anchored heparan sulfate proteoglycan, in the adult rat nervous system. *J. Neurosci.* 14:3713–3724.

Lotto, R. B., Price, D. J. (1995). The stimulation of thalamic neurite outgrowth by cortex-derived growth factors in vitro: the influence of cortical age and activity. *Eur. J. Neurosci.* 7:318–328.

Lund, R. D., Mustari, M. J. (1977). Development of geniculocortical pathway in rats. *J. Comp. Neurol.* 173:289–306.

Ma, L., Harada, T., Harada, C., Romero, M., Hebert, J. M., McConnell, S. K., Parada, L. F. (2002). Neurotrophin-3 is required for appropriate establishment of thalamocortical connections. *Neuron* 36:623–34.

Mackarehtschian, K., Lau, C. K., Caras, I., McConnell, S. K. (1999). Regional differences in the developing cerebral cortex revealed by ephrin-A5 expression. *Cereb. Cortex* 9:601–10.

Maeda, N., Noda, M. (1996). 6B4 proteoglycan/phosphacan is a repulsive substratum but promotes morphological differentiation of cortical neurons. *Development* 122:647–658.

Molnár, Z., Blakemore, C. (1991). Lack of regional specificity for connections formed between thalamus and cortex in coculture. *Nature* 351:475–477.

Molnár, Z., Adams, R., Blakemore, C. (1998). Mechanisms underlying the early establishment of thalamocortical connections in the rat. *J. Neurosci.* 18:5723–5745.

Morfini, G., DiTella, M. C., Feiguin, F., Carri, N., Caceres, A. (1994). Neurotrophin-3 enhances neurite outgrowth in cultured hippocampal pyramidal neurons. *J. Neurosci. Res.* 39:219–232.

Noctor, S. C., Palmer, S. L., McLaughlin, D. F., Juliano, S. L. (2001). Disruption of layers 3 and 4 during development results in altered thalamocortical projections in ferret somatosensory cortex. *J. Neurosci.* 21:3184–3195.

Oohira, A., Matsui, F., Watanabe, E., Kushima, Y., Maeda, N. (1994). Developmentally regulated expression of a brain specific species of chondroitin sulfate proteoglycan, neurocan, identified with a monoclonal antibody IG2 in the rat cerebrum. *Neuroscience* 60:145–157.

Ozdinler, P. H., Erzurumlu, R. S. (2002). Slit2, a branching-arborization factor for sensory axons in the Mammalian CNS. *J. Neurosci.* 22:4540–4549.

Porter, B. E., Weis, J., Sanes, J. R. (1995). A motoneuron-selective stop signal in the synaptic protein s-laminin. *Neuron* 14:549–559.

Rennie, S., Lotto, R. B., Price, D. J. (1994). Growth-promoting interactions between the murine neocortex and thalamus in organotypic co-cultures. *Neuroscience* 61:547–564.

Schnell, L., Schneider, R., Kolbeck, R., Barde, Y. A., Schwab, M. E. (1994). Neurotrophin-3 enhances sprouting of corticospinal tract during development and after adult spinal cord lesion. *Nature* 367:170–173.

Segal, R. A., Pomeroy, S. L., Stiles, C. D. (1995). Axonal growth and fasciculation linked to differential expression of BDNF and NT3 receptors in developing cerebellar granule cells. *J. Neurosci.* 15:4970–4981.

Seki, T., Arai, Y. (1991). Expression of highly polysialylated NCAM in the neocortex and piriform cortex of the developing and the adult rat. *Anat. Embryol.* (Berl) 184:395–401.

Seki, T., Rutishauser, U. (1998). Removal of polysialic acid-neural cell adhesion molecule induces aberrant mossy fiber innervation and ectopic synaptogenesis in the hippocampus. *J. Neurosci.* 18:3757–3766.

Sharma, K., Korade, Z., Frank, E. (1994). Development of specific muscle and cutaneous sensory projections in cultured segments of spinal cord. *Development* 120:1315–23.

Suzuki, S. C., Inoue, T., Kimura, Y., Tanaka, T., Takeichi, M. (1997). Neuronal circuits are subdivided by differential expression of type-II classic cadherins in postnatal mouse brains. *Mol. Cell. Neurosci.* 9:433–447.

Szebenyi, G., Callaway, J. L., Dent, E. W., Kalil, K. (1998). Interstitial branches develop from active regions of the axon demarcated by the primary growth cone during pausing behaviors. *J. Neurosci.* 18:7930–7940.

Tessier-Lavigne, M. and Goodman, C. S. (1996). The molecular biology of axon guidance. *Science* 274:1123–1133.

Vanderhaeghen, P., Lu, Q., Prakash, N., Frisén, J., Walsh, C. A., Frostig, R. D., Flanagan, J. G. (2000). A mapping label required for normal scale of body representation in the cortex. *Nature Neurosci.* 3:358–365.

Wang, K. H., Brose, K., Arnott, D., Kidd, T., Goodman, C. S., Henzel, W., Tessier-Lavigne, M. (1999). Biochemical purification of a mammalian slit protein as a positive regulator of sensory axon elongation and branching. *Cell* 96:771–784.

Watanabe, E., Matsui, F., Keino, H., Ono, K., Kushima, Y., Noda, M., Oohira, A. (1996). A membrane-bound heparan sulfate proteoglycan that is transiently expressed on growing axons in the rat brain. *J. Neurosci. Res.* 44:84–96.

Xu, X., Ng, S., Wu, Z. L., Nguyen, D., Homburger, S., Seidel-Dugan, C., Ebens, A., Luo, Y. (1998). Human semaphorin K1 is glycosylphosphatidylinositol-linked and defines a new subfamily of viral-related semaphorins. *J. Biol. Chem.* 273:22428–22434.

Yabuta, N. H., Butler, A. K., Callaway, E. M. (2000). Laminar Specificity of Local Circuits in Barrel Cortex of Ephrin-A5 Knockout Mice. *J. Neurosci.* 20:RC88.

Yamamoto, N. (2002). Cellular and Molecular basis for the formation of lamina-specific thalamocortical projections. *Neurosci. Res.* 42:167–173.

Yamamoto, N., Kurotani, T., Toyama, K. (1989). Neural connections between the lateral geniculate nucleus and visual cortex in vitro. *Science* 245:192–194.

Yamamoto, N., Yamada, K., Kurotani, T., Toyama, K. (1992). Laminar specificity of extrinsic cortical connections studied in coculture preparations. *Neuron* 9:217–228.

Yamamoto, N., Higashi, S., Toyama, K. (1997). Stop and Branch behaviors of geniculocortical axons: A time-lapse study in organotypic cocultures. *J. Neurosci.* 17:3653–3663.

Yamamoto, N., Inui, K., Matsuyama, Y., Harada, A., Hanamura, K., Murakami, F., Ruthazer, E. S., Rutishauser, U. and Seki, T. (2000a). Inhibitory mechanism by polysialic acid for lamina-specific branch formation of thalamocortical axons. *J. Neurosci.* 20:9145–9151.

Yamamoto, N., Matsuyama, Y., Harada, A., Inui, K., Murakami, F., Hanamura, K. (2000b). Characterization of factors regulating lamina-specific growth of thalamocortical axon. *J. Neurobiol.* 42:56–68.

Zhong, Y., Takemoto, M., Fukuda,T., Hattori,Y., Murakami,F., Nakajima, D., Nakayama,M., Yamamoto, N. (2004). Identification of the genes that are expressed in the upper layers of the neocortex. *Cereb. Cortex* 14:1144–1152.

6

Role of Citron K in the Development of Cerebral Cortex

Patrizia Muzzi, Paola Camera, Ferdinando Di Cunto, and Alessandro Vercelli

Abstract

Citron-K (CIT-K) is a target molecule for activated Rho which is expressed at high levels in the proliferating areas of the CNS from E10.5 to E16. CIT-K −/− mice display severe defects in neurogenesis, due to altered cytokinesis and aptoptosis: these cellular alterations result in severe microencephaly and in death of the animal due to fatal seizures between the 2nd and the 3rd week of age. We have analysed the development of somatosensory cortex in the CIT-K −/− mice, showing i) a decrease in the barrelfield area and in the size of single whisker-related barrels, ii) a decrease in the cortical thickness, especially in supragranular layers, iii) a decrease in the density of myelinated fibers. We also report cellular changes in cortical neurons: i) both pyramidal neurons and interneurons show altered dendritic development and frequent polyploidy, and ii) the distribution of interneurons is affected by CIT-K deletion. CIT-K −/− mice are a useful tool to study the role of this molecule in the cellular development of cerebral cortex in vivo, and the development and plasticity of cortical areas and connections in a microencephalic animal. Moreover, they represent an interesting model of neonatal epilepsy, in which to study the role of changes in cellular morphology and interneuron distribution in the genesis of epileptic seizures.

From a cellular point of view, the development of cerebral cortex consists of a series of progressive events including cell proliferation, migration, emission and growth of dendrites and axons and synaptogenesis, associated to regressive events such as normally occurring, developmentally-regulated apoptosis, dendritic pruning, elimination of axon collaterals and of inactive synapses.

Small GTPases of the Rho family have been implicated in the regulation of several of these phenomena in vitro and, more recently, in vivo. These proteins act as critical regulators of cytoskeletal structures, and have been involved in a wide variety of cellular events, including polarization, establishment of cell-to-cell contacts, motility, migration, membrane trafficking, cell growth control, cytokinesis and transcriptional activation by growth factors and environmental stress (Van Aelst and Souza-Schorey, 1997; Hall, 1998; Bishop and Hall, 2000).

Rho GTPases exert their complex functions through a network of effector proteins, which physically interact with the GTP-bound conformation and change their biologic activity upon binding (Narumiya et al., 1997; Hall, 1998). These molecules could work as integration points in Rho-dependent signal transduction pathways and play more restricted roles in the regulation of cytoskeletal dynamics (Van Aelst and Souza-Schorey, 1997). Therefore, inactivation of specific effectors is even more informative than targeting of any given GTPase. Abnormalities in Rho GTPase signalling are a prominent cause of mental retardation (Ramarkers, 2002).

Citron-N (CIT-N) and Citron-K (CIT-K) are two target molecules for activated Rho, produced by the same transcription unit (Di Cunto et al., 1998; Madaule et al., 1998). CIT-N was first identified for its ability to interact with GTP-bound Rho and Rac (Madaule et al., 1995). It is specifically expressed in the postnatal and adult nervous system and is localized to postsynaptic densities, where it forms a stable complex with the membrane-associated guanylate kinase PSD95 (Furuyashiki et al., 1999; Zhang et al., 1999). The functions of CIT-N are unknown, although it has been hypothesized that it may link the Rho signalling cascades to NMDA receptor complexes (Furuyashiki et al., 1999; Zhang et al., 1999). Recently, Di Cunto and coll. (2003) have shown that CIT-N is associated with the Golgi apparatus and that inhibition of its expression results in the dispersion of the Golgi apparatus in hippocampal neuron cultures. On the contrary, CIT-K is expressed at high levels in the proliferating areas of the CNS from E10.5 to E16, and in external granular layer of the cerebellum postnatally.

Recently (Di Cunto et al., 2000), a CIT-K −/− mouse has been generated, bearing severe defects in neurogenesis, due to altered cytokinesis and aptoptosis: these cellular alterations result in severe microencephaly and in death of the animal due to fatal seizures between the 2nd and the 3rd week of age. At the same time, a spontaneous mutation has been shown to produce similar morphological and functional effects in *flathead* rats (Sarkisian et al., 1999; Roberts et al., 2000): this mutation causes a frameshift in the second coding exon, resulting in complete absence of the CIT-K protein (Sarkisian et al., 2002). BrdU labeling studies have shown a normal DNA synthesis in the proliferative layers of the cerebral cortex and of the cerebellum (Di Cunto et al., 2000). On the contrary, TUNEL staining and activated caspase 3 immunohistochemistry showed a marked increase of apoptotic neurons in the intermediate zone of the cerebral cortex and at the interface between the external granular layer and the Purkinje cell layer of the CIT-K −/− mouse at the time of cell proliferation and migration (Di Cunto et al., 2000). Flow cytometry showed a marked increase in tetraploidy in the cerebellum: this observation was accompanied by the finding of frequent binucleated cells in

Table 1 Volumes (mm^3) of different brain areas in the wild type and CIT-K −/− mouse, as measured in three-dimensional reconstructions at the computer using the Neurolucida and Neurorotate programs.

	+/+	−/−	Variation (%)
Hemisphere	80.155 ± 1.54	38.915 ± 0.09	−51
Cerebral cortex	22.1 ± 0.41	14.48 ± 0.11	−34
Corpus callosum	1.655 ± 0.12	0.52 ± 0.03	−69
Striatum	4.775 ± 0.21	2.695 ± 0.19	−44
Lateral ventricles	0.045 ± 0.04	0.13 ± 0	+189
Third ventricle	0.14	0.14	no
Hippocampus	8.14 ± 0.01	2.625 ± 0.08	−68

cerebral cortex, and lead to the hypothesis that increased apoptosis is due to defective cytokinesis (Di Cunto et al., 2000).

1. Cerebral Cortex in the CIT-K −/− Mouse

CIT-K −/− mice are markedly microencephalic, showing striking reduction in the size of cerebral cortex and cerebellum. Three dimensional reconstructions of the brain (from coronal sections, between a plane through the caudal olfactory bulbs and a plane through the caudal end of the superior colliculus) led to the observation that the decrease in size at P13 is consistent (34 to 70%) in all structures considered (see Table 1). The reduction in size of brain structures can be easily observed in Fig. 1A-I vs Fig. 1a-i. In parallel, there is a remarkable increase in size of the lateral ventricles (189%), probably due to cell loss in the brain. In particular, the volume of cerebral cortex was reduced by 34%, and so are the cortical surface (at P13 the surface is 69 mm^2 in the control and 47 mm^2 in the CIT-K −/− mice) and the cortical thickness (from 940 ± 86 μm in controls to 752 ± 120 μm in the CIT-K −/− mice) (Fig. 1J-K vs 1j-k).

1.1. Pattern Formation in the CIT-K −/− Mouse

Notwithstanding microencephaly, pattern formation occurs around P4 as in controls: in the posteromedial subfield of the somatosensory cortex the vibrissae-related barrels are regularly expressed, and all whiskers are represented. In parallel with cortical shrinkage, barrelfield area is reduced, both as overall area and as single barrel area, whereas the barrel septa are larger than in controls (Fig. 2). Barrel formation can still be influenced by manipulating the periphery: neonatal ablation of a whisker row results in the enlargement of the barrels of the neighbour whiskers and in the fusion of the barrels representing the whiskers which have been ablated. Many investigators have shown that during the first few days of postnatal life, cortical barrel patterns are exquisitely sensitive to the state of the periphery (Killackey and Belford, 1980; Andrés and Van

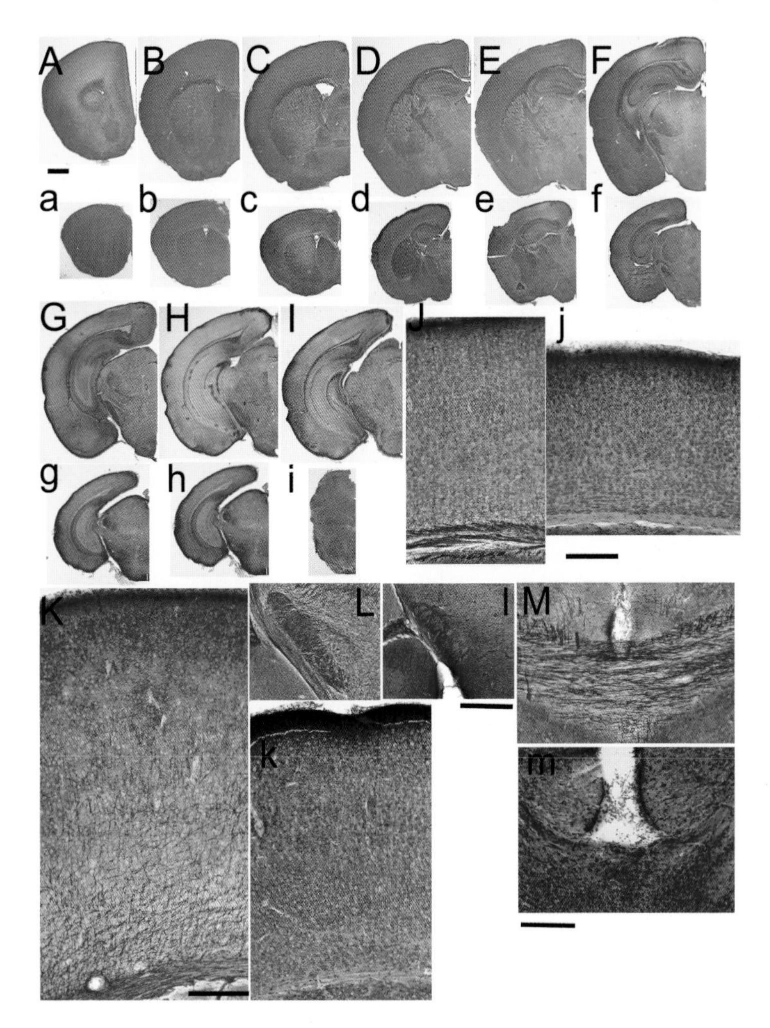

Figure 1 Rostrocaudal serial coronal sections through the encephalon of wild type (A-I) and of CIT-K −/− mouse (a-i) at P13 (scale bar = 1 mm). Note the reduction in thickness of visual (J vs j) and somatosensory (K vs k) cortex, especially evident for the supragranular layers, and the decreased density of myelinated fibers entering the cortex (scale bar = 200 μm). Details show the reduction in size of the cortical peduncle (L-l, scale bar = 500 μm) and of the corpus callosum (M-m, scale bar = 200 μm) in the CIT-K −/− mouse.

der Loos, 1982; Rhoades et al., 1990; Vercelli et al., 1999). Transection of the infraorbital nerve (ION) during this period results in the formation of five continuous stripes (whisker row representations) in layer IV of the posteromedial barrel subfield, with no differentiation of individual whisker-specific clusters within the stripes. Disruption of a single row of vibrissae at birth leads to the development of a single, thin band of cells and axons corresponding to the damaged whisker row (Killackey and Belford, 1980; Welker and Van der Loos, 1986). Our results with CO histochemistry in the CIT-K −/− mice confirm these reports (data not shown).

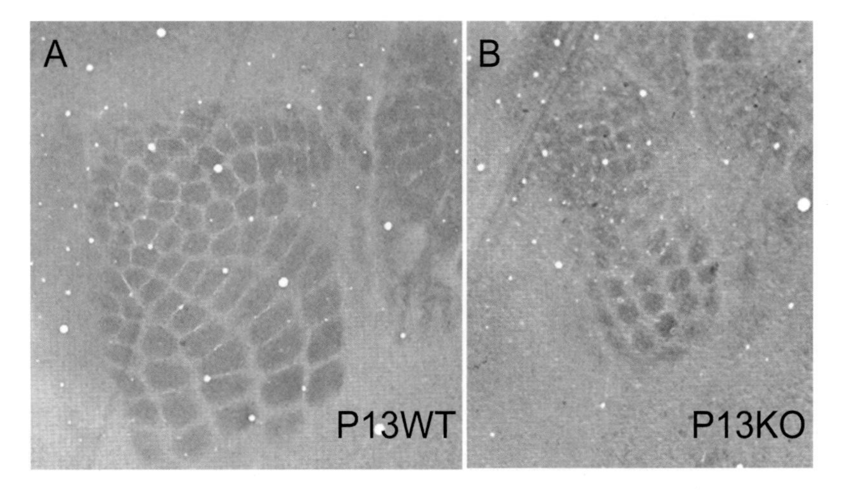

Figure 2 Tangential sections through the barrelfield of somatosensory cortex in P13 wild type (A) and CIT-K −/− (B) mice, revealed by CO histochemistry. The barrelfield is reduced in area, and single barrels are reduced in size, whereas barrel septa are expanded in the CIT-K −/− mice.

1.2. Cortical Layers

In the CIT-K −/− mice, cortical layering is conserved, but layers are thinner than in controls, especially in supragranular layers. The density of neuronal profiles is decreased, and dendritic bundles of pyramidal neurons are disorganised. The density of myelinated fibers is strikingly reduced: they are rare within the layers and the major fiber systems originating from cerebral cortex (corpus callosum, internal capsule and cerebral peduncles) are absent or strongly reduced in size compared to controls. Myelinated fiber depletion can be observed both in blackgold stained material (Figures 1L-l, cerebral peduncle; 1M-m, corpus callosum) and in MBP-immunoreacted sections (data not shown), and is probably due not only to the decreased number of projecting neurons, but also to olygodendrocyte depletion. Changes in fiber systems reminds of some alterations in the organisation of cortical connections in the somatosensory cortex observed after E15 X-ray irradiation (Funahashi et al., 1997) and in MAM-treated microencephalic rats (Ueda et al., 1999)

In all cortical layers, but more frequent in infragranular layers, both pyramidal neurons and interneurons may be binucleated (Fig. 3I and 4E): this aspect is striking in infragranular NADPH-diaphorase positive interneurons (Fig. 4E).

1.3. Development of Cortical Neurons

Cerebral cortex consists of two major types of neurons: pyramidal neurons, representing 60–70 % of total neurons, and GABAergic inhibitory interneurons. Both populations are affected in their number due to the

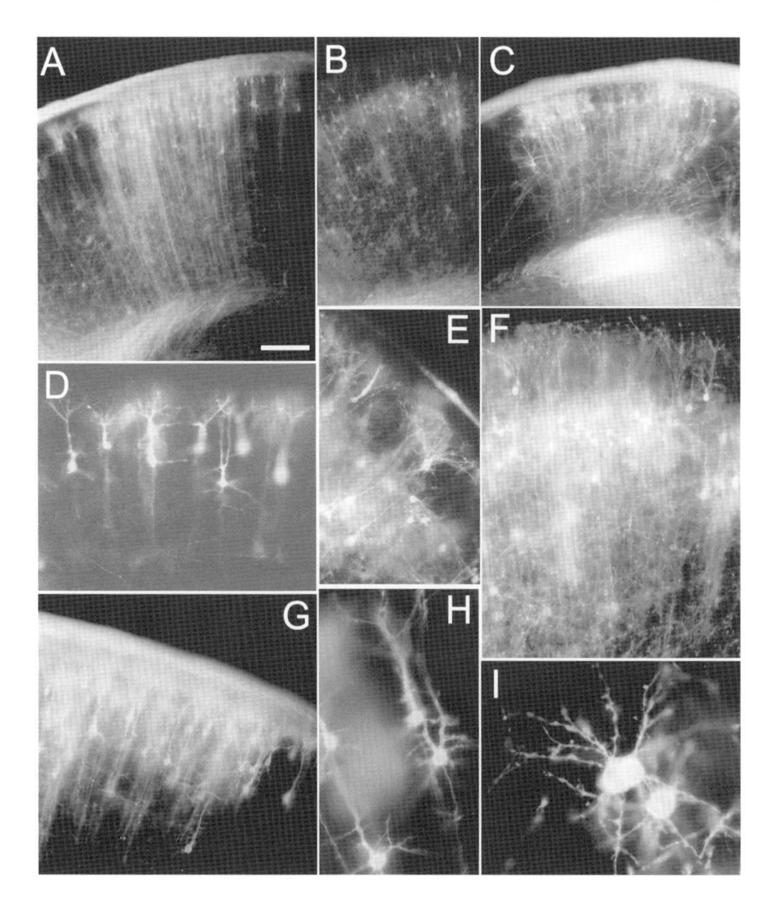

Figure 3 DiI (1,1_-dioctadecyl-3,3,3_,3_-tetramethylindocarbocyanine perchlorate)-labeled cellular elements: Callosally-projecting cortical neurons were retrogradely labelled by inserting a crystal of DiI into the corpus callosum in paraformaldehyde-fixed P4 brains of control (A, D, G and H) and CIT-K −/− (B, C, E, F and I) mice. In A-C, the distribution of callosally-projecting neurons in cerebral cortex can be compared in wild type and CIT-K −/− mice: they are similar, but especially in the supragranular layers in formation callosally-projecting neurons seem to be less frequent, and sometimes their dendrites are not properly oriented. In infragranular layers, binucleated cell bodies are frequently observed (I). Scale bar = 200 μm in A-C, 100 in D-G and 50 in H-I.

increase in neuronal death during development: Di Cunto and coworkers (2000) showed that in the the CIT-K −/− mice neural progenitors proliferate regularly in the subventricular zone, but die soon after having been generated; similarly, LoTurco and coworkers (Sarkisian et al., 1999, 2001, 2002; Lo Turco et al., 2003) showed an increased number of dying neurons in the ganglionic eminence of the developing *flathead* rat. The outcome is the reduced cortical thickness due to depletion of both neuron types, more abundant in supragranular layers than in infragranular ones: i) this could be explained by a diverse sensitivity of later generated neurons to defects in cytokinesis, as suggested by the observation of an higher density of binucleated nuclei in the infragranular layers. ii) Alternatively, and more likely, the major involvement of supragranular layers could be ascribed to the progressive depletion of progenitors due to cell death of newborn cells progressively reducing the number of neurons that are generated in a time unit.

We have shown that, in both pyramidal neurons and in interneurons of the CIT-K −/− mice, dendritic development is altered. A great number of *in vitro* and *in vivo* studies have recently implicated Rho GTPases and their effectors in neurite extension and remodeling. In particular, activation of RhoA induced neurite retraction in cultured mammalian neuronal cells, while the activity of Rac1 and Cdc42 was required in the same experimental models for neurite extension (Kozma et al., 1977; Gallo and Letourneau, 1998). Accordingly, it has been shown that Drosophila Rho is required in vivo to limit the extension of neuronal dendritic arborizations (Lee et al., 2000). Among the effector molecules of Rho, p160-ROCK seems to play a particularly important role in mediating neurite collapse (Hirose et al., 1998; Maekawa et al., 1999). Time-lapse microscopy of single optic tectal neurons in Xenopus tadpoles has shown that enhanced Rac and Cdc42 activity selectively increase branch addictions and retractions, and dominant-negative RhoA increases branch extension (Li et al., 2000). Expression of dominant –negative Rac1 results in progressive elimination of dendritic spines, whereas hyperactivation of RhoA causes a drastic simplification of dendritic branch pattern dependent on the activity of the downstream kinase ROCK (Nakayama et al., 2000). Rho GTPases modulate dendritic spine formation, plasticity and function; moreover, at least three genes (for Oligophrenin 1, PAK3 and αPIX) involved in X-linked mental retardation participate directly in cellular signalling through Rho GTPases (reviewed in Ramakers, 2002). In addition, other genes involved in mental retardation are linked to Rho signalling indirectly, such as in Aarskog-Scott syndrome, in Williams syndrome and in fraX syndrome (Ramakers, 2002). However, despite the evidence obtained with cell culture systems and with invertebrate animal models, the exact roles played by Rho-GTPases and by their effectors during development of the mammalian CNS are still largely obscure.

In order to investigate the effects of deleting CIT-K gene on dendritic development, we have performed a quantitative analysis of neuronal processes in cortical pyramidal cells of CIT-K −/− mice and of their littermate +/+ controls. We studied the development of cortico-collicular and of callosally-projecting neurons at P8 and P14 (after which age the CIT-K −/− usually die), after retrograde labeling with DiI crystals inserted in the superior colliculus and in the corpus callosum (Di Cunto et al., 2003). The length and complexity of neuronal processes were quantitatively analyzed by fluorescence microscopy using the Neurolucida imaging system (Glaser and Glaser, 1990). The distribution of cortico-collicular neurons did not change compared to controls, but the mean distance between their cell bodies and the pial surface was significantly decreased (454.66 ± 118.69 μm in ko vs 541.25 ± 89.8 μm in wt at

P8 – $p = 0.02$, and 466.74 ± 38.29 μm vs 642.97 ± 124.62 μm respectively, at P14 – $p < 0.001$). Callosally-projecting neurons were found both in supragranular and in infragranular layers, but, due to the strong decrease in thickness and cellularity of supragranular layers (Di Cunto et al., 2000), it was difficult to discriminate supra- and infragranular neurons. Nevertheless, supragranular neurons projecting to the corpus callosum were still present. Quantitative analysis (Table 2) showed that, in P8 CIT-K −/− mice, the apical dendrites of the cortico-collicular pyramidal neurons were significantly shorter than in controls. Interestingly, no significant differences could be observed in the same cells at P14; if anything, apical dendrites tended to be longer in the knockouts. The same trend was observed in the callosally-projecting pyramidal neurons, even if in this case the differences were not significant at both the stages which were considered in the study.

In contrast, basal dendrites followed a different developmental pattern in corticocollicular and callosally-projecting neurons. In corticocollicular neurons, the total dendritic length was significantly shorter than in controls in both P8 and P14 knockout mice. The mean length of the terminal segments was similar at P14, thus indicating that the decrease in total length was due mainly to a slightly lower degree of arborization (reflected by the higher number of nodes in the wild type). Basal dendrites of callosally-projecting neurons were of the same length in +/+ and −/− mice at P8, but tended to be longer in −/− mice at P14. This increase in length became statistically significant when we considered terminal segments, since the degree of arborization was almost equal in the two groups (Table 2).

At both ages considered, the size of the somata of corticocollicular neurons was larger in control mice, whereas that of callosally-projecting ones was larger in −/− mice (Table 2).

1.4. Development of Cortical Interneurons

GABA interneurons currently account for 1/6-1/4 of all cortical neurons (15–25% in Jones, 1993; 14.6–22.8% in Gonchar and Burkhalter, 1997) probably close to the upper limit due to the low GABA expression of some interneurons: they are anatomically (Cobas et al., 1987), physiologically and molecularly heterogeneous (Fairen et al., 1984; Jones and Hendry, 1986; Naegele and Barnstable, 1989; Kawaguchi, 1995). They colocalize several different peptides such as somatostatin (SOM), cholecystokinin (CCK), neuropeptide Y (NPY), vasointestinal peptide (VIP) and the neurochemical markers choline acetyltransferase (ChAT) and nitric oxide synthase (NOS) or NADPH-diaphorase (Vincent and Kimura, 1992; Valtschanoff et al., 1993). In addition, different populations of cortical GABAergic neurons may express different calcium binding proteins, such as parvalbumin (PV), calbindin-D28 (CB) and calretinin (CR). GABAergic interneurons probably serve to sharpen the selectivity of sensory responses (in the visual system, Somers et al., 1995) and to prevent excessive firing (Benardo and Wong, 1995) by responding

Table 2 Quantitative analysis of pyramidal neurons of the visual cortex projecting to the corpus callosum or to the superior colliculus in wild type (WT) and CIT-K −/− (KO) mice. Each parameter is expressed as mean ± standard deviation of three animals, five neurons each. Statistical analysis was performed using the unpaired Student T test, two tails (From Di Cunto *et al.*, 2003).

Age	Cell type	Genotype	Cell body	Basal			Apical	
			Area (μm^2)	Total length (μm)	# of nodes	Length of terminal segments (μm)	Total length (μm)	# of nodes
P8	Callosally-projecting	WT	191.12 ± 43.89	378.43 ± 284.35	2.62 ± 1.74		1304.24 ± 797.76	8.73 ± 6.06
		KO	232.14 ± 77.36	380.84 ± 216.23	2.89 ± 1.96		1074.7 ± 617.62	7 ± 3.89
		P	0.05	0.966	0.494		0.217	0.216
	Cortico-collicular	WT	268.44 ± 49.34	340.57 ± 227.58	2.36 ± 1.37		2573.04 ± 1076.13	15.55 ± 3.95
		KO	220.05 ± 48.92	264.64 ± 164.57	2 ± 1.52		1569.2 ± 697.7	14 ± 7.31
		P	0.006	0.016	0.603		0.005	0.447
P14	Callosally-projecting	WT	232.17 ± 30.52	418.17 ± 243.68	2.75 ± 1.65	77.04 ± 64.28	1360.82 ± 840.47	8.06 ± 4.6
		KO	279.57 ± 48.88	513.94 ± 382.63	2.68 ± 1.66	106.88 ± 80.77	1545.71 ± 833.69	8.12 ± 4.11
		P	0.001	0.073	0.775	< 0.001	0.259	0.967
	Cortico-collicular	WT	322.9 ± 71.18	574.83 ± 323.84	3.19 ± 2.02	101.14 ± 75.89	3199.81 ± 1247.84	13.47 ± 3.49
		KO	269.68 ± 91.97	452.34 ± 228.25	2.5 ± 1.45	97.11 ± 67.77	3510.01 ± 1758.2	15.08 ± 5.55
		P	0.1	0.022	0.039	0.55	0.597	0.354

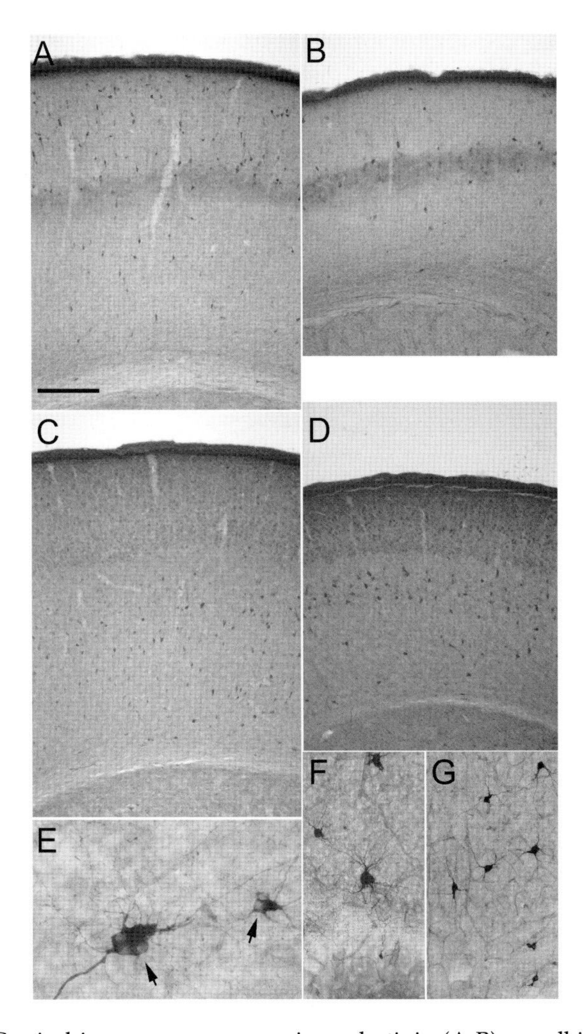

Figure 4 Cortical interneurons expressing calretinin (A-B) or calbindin (C-D) in coronal sections through the somatosensory cortex of control (A and C) and CIT-K −/− (B and D) P13 mice. In E-G, NADPH-d positive interneurons in tangential (E) and coronal (F-G) sections of somatosensory cortex at P7 (E) and P13 (F-G): note the enormous increase in size of the cell body in F (CIT-K −/−) vs G (control), and the polyploidy of the nuclei (arrows) and the increased number of dendrites in E (CIT-K −/−).

with opposing inhibition to the excitation in spiny neurons (Ferster and Jagadeesh, 1992). On the other hand, GABAergic interneurons are in turn inhibited by other GABAergic interneurons (Gonchar and Burkhalter, 1999A and 1999B), thus disinhibiting spiny neurons and synchronizing pyramidal cell activity (Jefferys et al., 1996).

We have recently studied the development of cortical interneurons in the somatosensory cortex of CIT-K −/− mice (Muzzi et al., 2005, submitted). To map their distribution and obtain their density, we have labelled interneurons immunohistochemically with antibodies (Fig. 4) against gamma aminobutyric acid decarboxylase 67 (GAD67), CB, PV and CR; in addition, nitric oxide synthesizing interneurons were labelled by ß nicotinamide dinucleotide phosphate-diaphorase (NADPH-d) histochemistry.

The density of NADPH-d-positive interneurons was significantly decreased by 59% in the supragranular layers, whereas in the infragranular layers their density was comparable to controls. Similar results were obtained with GAD67 and CR immmunohistochemistry, whereas PV-IR interneurons were significantly decreased both in supra- and in infragranular layers, and CB-IR were decreased in infragranular layers. These findings suggest that the syndrome affecting CIT-K mice involves interneuron types differently. Disruptions of the developmental plans for the generation of interneuronal diversity, both of genetic and epigenetic origin (Santhakumar and Soltesz, 2004), lead to pathological alterations in cerebral cortex. For example, i) in schizophrenia PV-positive (Benes and Berretta, 2001) and CB-positive (Reynolds et al., 2001) interneurons are selectively depleted, or ii) in mutations of homeobox genes, related to Dlx1 and Dlx2, associated with infantile spasms, mental retardation and autism the GABAergic system is altered (Sherr, 2003). iii) A decrease in PV immunoreactivity can be observed in the binocular area of the rat visual cortex following monocular deprivation (Cellerino et al., 1992). Deletion of several genes may affect interneuron migration and distribution into the developing cerebral cortex. In the somatosensory cortex of *Otx1* (the murine homolog of the Drosophila orthodenticle gene) −/− mice (which is thinner than in controls, and shows a decreased density of layer V large pyramidal neurons) PV-expressing interneurons are distributed in patches, thus indicating that chandelier and basket cells (the most important inhibitory input to pyramidal neurons) are unevenly distributed (Cipelletti et al., 2002; Pantò et al., 2004). This situation reminds of patients with intractable epilepsy associated with architectural dysplasia (DeFelipe et al., 1994; Spreafico et al., 2000). In the motor cortex of *mdx* (dystrophin deficient) mice, cortical interneurons expressing calcium-binding proteins are increased in numbers, possibly due to an increased intracellular calcium secondary to lack of dystrophin that stimulates the expression of calcium binding proteins or changes in neuronal activity (Carretta et al., 2003).

The morphology of cortical interneurons of the different types was altered from normal, mostly in infragranular layers: their somata were significantly larger then normal, and bore more dendrites, such that we called these interneurons "cockroach cells". Some of these findings remind of morphological aspects in human cortical dysplasias and disorders of development and might explain the recurrence of epileptic seizures that bring these mice to death by the end of the 2nd week of life. Alterations in neuronal morphology, size and in number and morphology of dendrites affect especially in interneurons: in fact, alterations in pyramidal neurons are much more subtle (Di Cunto et al., 2003). An interesting hypothesis has been raised by Sarkisian et al. (2001), who suggest that interneuron hypertrophy might be due to an increased availability of neurotrophins, especially of BDNF. Similar changes in interneuron morphology have been reported in patients bearing cortical malformations and epilepsy (Ferrer et al., 1992; Garbelli et al., 1999; Thom et al., 2000). As for pyramidal neurons, these effects could be the outcome of a complex interplay between genetic and epigenetic constraints. The cerebral cortex in the CIT-K −/− mice is strikingly altered, with an overall decrease in thickness due to deprivation of supragranular layers. This might change the availability of trophic factors, which are reported to be very important in dendritic growth (McAllister et al., 1995). Also, afferents are capable to influence PV expression in visual cortex (Cellerino et al., 1992).

2. Conclusion

CIT-K −/− mice are a useful tool to study the role of this molecule in the cellular development of cerebral cortex in vivo, and the development and plasticity of cortical areas and connections in a microencephalic animal. Moreover, they represent an interesting model of neonatal epilepsy, in which to study the role of changes in cellular morphology and interneuron distribution in the genesis of epileptic seizures.

Acknowledgements

Supported by grants from Fondation LeJeune, Paris to FDC and AV, and from Fondazione Mariani, Milano, and Fondazione Cassa di Risparmio, Torino, to AV (Grant R-04-40). PM is a recipient of a fellowship from Fondazione Cassa di Risparmio, Torino. The authors are grateful to D. Martinuzzi for technical assistance.

References

1. Andrés, F. L., and Van der Loos, H. (1982). The dependence of sensory brain maps on the configuration, early in development, of the periphery. *Neuroscience* 7[suppl.]:S6–7.
2. Benardo, L.S., and Wong, R.K.S. (1995). Inhibition in the cortical network. In: The cortical neuron (Gutnick, M.J., Mody, I., eds) pp 141–155. New York: Oxford University Press.
3. Benes, F.M., and Berretta, S. (2001). GABAergic interneurons: implications for understanding schizophrenia and bipolar disorder. *Neuropsychopharmacology* 25:1–27.
4. Bishop, A.L., and Hall, A. (2000). Rho GTPases and their effector proteins. *Biochem. J.* 348:241–255.
5. Carretta, D., Santarelli, M., Vanni, D., Ciabatti, S., Sbriccoli, A., Pinto, F., and Minciacchi, D. (2003). Cortical and brainstem neurons containing calcium-binding proteins in a murine model of Duchenne's muscular dystrophy: selective changes in the sensorimotor cortex. *J. Comp. Neurol.* 456:48–59.

6. Cellerino, A., Siciliano, R., Domenici, L., and Maffei, L. (1992). Parvalbumin immunoreactivity: a reliable marker for the effects of monocular deprivation in the rat visual cortex. *Neuroscience* 51:749–753.
7. Cipelletti, B., Avanzino, G., Vitellaro-Zuccarello, L., Franceschetti, S., Sancini, G., Gavazza, T., Acampora, D., Simeone, A., Spreafico, R., and Frassoni, C. (2002). Morphological organization of somatosensory cortex in Otx1(−/−) mice. *Neuroscience* 115:657–667.
8. Cobas, A., Welker, E., Fairen, A., Kraftsik, R., and Van der Loos, H. (1987). GABAergic neurons in the barrel cortex of the mouse: an analysis using neuronal archetypes. *J. Neurocytol.* 16:843–870.
9. DeFelipe, J., Huntley, G.W., del Rio, M.R., Sola, R.G., and Morrison, J.H. (1994). Microzonal decreases in the immunostaining for non-NMDA ionotropic excitatory amino acid receptor subunits GluR 2/3 and GluR 5/6/7 in the human epileptogenic neocortex. *Brain Res.* 657:150–158.
10. Di Cunto, F., Calautti, E., Hsiao, J., Ong, L., Topley, G., Turco, E., and Dotto, G.P. (1998). Citron rho-interacting kinase, a novel tissue-specific ser/thr kinase encompassing the Rho-Rac-binding protein Citron. *J. Biol. Chem.* 273:29706–29711.
11. Di Cunto, F., Imarisio, S., Hirsch, E., Broccoli, V., Bulfone, A., Migheli, A., Atzori, C., Turco, E., Triolo, R., Dotto, G.P., Silengo, L., and Altruda, F. (2000). Defective neurogenesis in citron kinase knockout mice by altered cytokinesis and massive apoptosis. *Neuron* 28:115–127.
12. Di Cunto, F., Ferrara, L., Curtetti, R., Imarisio, S., Guazzane, S., Broccoli, V., Bulfone, A., Altruda, F., Vercelli, A., and Silengo, L. (2003). Role of citron kinase in dendritic morphogenesis of cortical neurons. *Brain Res Bull.* 60:319–327.
13. Fairen, A., DeFelipe, J., and Regidor, J. (1984). Nonpyramidal neurons: General account. In: Cerebral cortex, Vol 1, Cellular components of the cerebral cortex (Peters A, Jones EG, eds), pp 201–245, New York, Plenum.
14. Ferrer, I., Pineda, M., Tallada, M., Oliver, B., Russi, A., Oller, L., Noboa, R., Zujar, M.J., and Alcantara, S. (1992). Abnormal local-circuit neurons in epilepsia partialis continua associated with focal cortical dysplasia. *Acta Neuropathol. (Berl)* 83:647–652.
15. Ferster, D., and Jagadeesh, B. (1992). EPSP-IPSP interactions in cat visual cortex studied with in vivo whole-cell patch recording. *J. Neurosci.* 12:1262–1274.
16. Funahashi, A., Darmanto W., and Inouye, M. (1997). Cortical fiber distribution in the somatosensory cortex of rats following prenatal exposure to X-irradiation. *Environ. Med.* 41:37–39.
17. Furuyashiki, T., Fujisawa, K., Fujita, A., Madaule, P., Uchino, S., Mishina, M., Bito, H., and Narumiya S. (1999). Citron, a Rho-target, interacts with PSD-95/SAP-90 at glutamatergic synapses in the thalamus. *J. Neurosci.* 19:109–118.
18. Gallo, G., and Letourneau, P.C. (1998). Axon guidance: GTPases help axons reach their targets. *Curr. Biol.* 8:R80–82.
19. Garbelli, R., Munari, C., De Biasi, S., Vitellaro-Zuccarello, L., Galli, C., Bramerio, M., Mai, R., Battaglia, G., and Spreafico, R. (1999). Taylor's cortical dysplasia: a confocal and ultrastructural immunohistochemical study. *Brain Pathol.* 9:445–461.
20. Glaser, J.R., and Glaser, E.M. (1990). Neuron imaging with Neurolucida–a PC-based system for image combining microscopy. *Comput. Med. Imaging Graph.* 14:307–317.

21. Gonchar, Y., and Burkhalter, A. (1997). Three distinct families of GABAergic neurons in rat visual cortex. *Cereb. Cortex* 7:347–358.
22. Gonchar, Y., and Burkhalter, A. (1999A). Connectivity of GABAergic calretinin-immunoreactive neurons in rat primary visual cortex. *Cereb. Cortex* 9:683–696.
23. Gonchar, Y., and Burkhalter, A. (1999B). Differential subcellular localization of forward and feedback interareal inputs to parvalbumin expressing GABAergic neurons in rat visual cortex. *J. Comp. Neurol.* 406:346–360.
24. Hall, A. (1998). Rho GTPases and the actin cytoskeleton. *Science* 279:509–514.
25. Hirose, M., Ishizaki, T., Watanabe, N., Uehata, M., Kranenburg, O., Moolenaar, W.H., Matsumura, F., Maekawa, M., Bito, H., and Narumiya, S. (1998). Molecular dissection of the Rho-associated protein kinase (p160ROCK)-regulated neurite remodeling in neuroblastoma N1E-115 cells. *J. Cell Biol.* 141:1625–1636.
26. Jefferys, J.G., Traub, R.D., and Whittington, M.A. (1996). Neuronal networks for induced '40 Hz' rhythms. *Trends Neurosci.* 19:202–208.
27. Jones, E.G., and Hendry, S.C.H. (1986). Co-localization of GABA and neuropeptides in neocortical neurons. *Trends Neurosci.*, 9:71–76.
28. Jones, E.G. (1993). GABAergic neurons and their role in cortical plasticity in primates. Cereb. Cortex 3:361–372.
29. Kawaguchi, Y. (1995). Physiological subgroups of nonpyramidal cells with specific morphological characteristics in layer II/III of rat frontal cortex. *J. Neurosci.* 15:2638–2655.
30. Killackey, H. P., and Belford, G. R. (1980). Central correlates of peripheral pattern alterations in the trigeminal system of the rat. *Brain Res.*, 183:205–210.
31. Kozma, R., Sarner, S., Ahmed, S., and Lim, L. (1997). Rho family GTPases and neuronal growth cone remodelling: relationship between increased complexity induced by Cdc42Hs, Rac1, and acetylcholine and collapse induced by RhoA and lysophosphatidic acid. *Mol. Cell. Biol.* 17:1201–1211.
32. Lee, T., Winter, C., Marticke, S.S., Lee, A., and Luo, L. (2000). Essential roles of Drosophila RhoA in the regulation of neuroblast proliferation and dendritic but not axonal morphogenesis. *Neuron* 25:307–316.
33. Li, Z., Van Aelst, L., and Cline, H.T. (2000). Rho GTPases regulate distinct aspects of dendritic arbor growth in Xenopus central neurons in vivo. *Nat. Neurosci.* 3:217–225.
34. LoTurco, J.J., Sarkisian, M.R., Cosker, L., and Bai, J. (2003). Citron kinase is a regulator of mitosis and neurogenic cytokinesis in the neocortical ventricular zone. *Cereb. Cortex* 13:588–591.
35. Madaule, P., Furuyashiki, T., Reid, T., Ishizaki, T., Watanabe, G., Morii, N., and Narumiya, S. (1995). A novel partner for the GTP-bound forms of rho and rac. *FEBS Lett.* 377:243–248.
36. Madaule, P., Eda, M., Watanabe, N., Fujisawa, K., Matsuoka, T., Bito, H., Ishizaki, T., and Narumiya, S. (1998). Role of citron kinase as a target of the small GTPase Rho in cytokinesis. *Nature* 394:491–494.
37. Maekawa, M., Ishizaki, T., Boku, S., Watanabe, N., Fujita, A., Iwamatsu, A., Obinata, T., Ohashi, K., Mizuno, K., and Narumiya, S. (1999). Signaling from Rho to the actin cytoskeleton through protein kinases ROCK and LIM-kinase. *Science* 285:895–898.
38. McAllister, A.K., Lo, D.C., and Katz, L.C. (1995). Neurotrophins regulate dendritic growth in developing visual cortex. *Neuron* 15:791–803.

39. Muzzi, P., Di Cunto, F., and Vercelli, A. (2005). Deletion of Citron K gene selectively affects the number and distribution of cortical interneurons. Submitted.
40. Naegele, J.R., and Barnstable, C.J. (1989). Molecular determinants of GABAergic local-circuit neurons in the visual cortex. *Trends Neurosci.* 12:28–34.
41. Nakayama, A.Y., Harms, M.B., and Luo, L. (2000). Small GTPases Rac and Rho in the maintenance of dendritic spines and branches in hippocampal pyramidal neurons. *J. Neurosci.* 20:5329–5338.
42. Narumiya, S., Ishizaki, T., and Watanabe, N. (1997). Rho effectors and reorganization of actin cytoskeleton. *FEBS Lett.* 410:68–72.
43. Pantò, M.R., Zappala, A., Tuorto, F., and Cicirata, F. (2004) Role of the Otx1 gene in cell differentiation of mammalian cortex. *Eur. J. Neurosci.*, 19:2893–2902.
44. Ramakers, G.J. (2002). Rho proteins, mental retardation and the cellular basis of cognition. *Trends Neurosci.* 25, 191–199.
45. Reynolds, G.P., Zhang, Z.J., and Beasley, C.L. (2001). Neurochemical correlates of cortical GABAergic deficits in schizophrenia: selective losses of calcium binding protein immunoreactivity. *Brain Res. Bull.* 55:579–584.
46. Rhoades, R. W., Bennet-Clarke, C.A., Chiaia, N.L., White, F.A., Macdonald, G.J., Haring, J.H., and Jacquin, M.F. (1990). Development and lesion-induced reorganization of the cortical representation of the rats body surface as revealed by immunocytochemistry for serotonin. *J. Comp. Neurol.* 293:190–207.
47. Roberts, M.R., Bittman, K, Li, W.W., French, R., Mitchell, B., LoTurco, J.J., and D'Mello, S.R. (2000). The flathead mutation causes CNS-specific developmental abnormalities and apoptosis. *J. Neurosci.* 20:2295–2306.
48. Santhakumar, V., and Soltesz, I. (2004). Plasticity of interneuronal species diversity and parameter variance in neurological diseases. *Trends Neurosci.* 27:504–510.
49. Sarkisian, M.R., Rattan, S., D'Mello, S.R., and LoTurco, J.J. (1999). Characterization of seizures in the flathead rat: a new genetic model of epilepsy in early postnatal development. *Epilepsia* 40:394–400.
50. Sarkisian, M.R., Frenkel, M., Li, W., Oborski, J.A., and LoTurco, J.J. (2001). Altered interneuron development in the cerebral cortex of the flathead mutant. *Cereb. Cortex* 11:734–743.
51. Sarkisian, M.R, Li, W., Di Cunto, F., D'Mello, S.R., and LoTurco, J.J. (2002). Citron-kinase, a protein essential to cytokinesis in neuronal progenitors, is deleted in the flathead mutant rat. *J. Neurosci.* 22:RC217.
52. Sherr, E.H. (2003). The ARX story (epilepsy, mental retardation, autism, and cerebral malformations): one gene leads to many phenotypes. *Curr. Opin. Pediatr.* 15:567–571.
53. Somers, D.C., Nelson, S.B., and Sur, M. (1995). An emergent model of orientation selectivity in cat visual cortical simple cells. *J. Neurosci.* 15:5448–5465.
54. Spreafico, R., Tassi, L., Colombo, N., Galli, M.B., Garbelli, R., Ferrario, A., Lo Russo, G., and Munari, C. (2000). Inhibitory circuits in human dysplastic tissue. *Epilepsia* 41 (Suppl.):168–173.
55. Thom, M,, Holton, J.L., D'Arrigo, C., Griffin, B., Beckett, A., Sisodiya, S., Alexiou, D., and Sander, J.W. (2000). Microdysgenesis with abnormal cortical myelinated fibres in temporal lobe epilepsy: a histopathological study with calbindin D-28-K immunohistochemistry. *Neuropathol. Appl. Neurobiol.* 26:251–257.

56. Ueda, S., Nishimura, A., Kusuki, T., Takeuchi, Y., and Yoshimoto, K. (1999). Delayed 5-HT release in the developing cortex of microencephalic rats. *Neuroreport* 10:1215–1209.
57. Valtschanoff, J.G., Weinberg, R.J., Kharazia, V.N., Schmidt, H.H.H.W., Nakane, M., and Rustioni, A. (1993). Neurons in rat cerebral cortex that synthesize nitric oxide: NADPH-diaphorase histochemistry, NOS immunocytochemistry, and colocalization with GABA.*Neurosci. Lett.* 157:157–161.
58. Van Aelst, L., and Souza-Schorey, C. (1997). Rho GTPases and signaling networks. *Genes Dev.* 11:2295–2322.
59. Vercelli, A., Repici, M., Biasiol, S., and Jhaveri S. (1999). Maturation of NADPH-d activity in the rat's barrel-field cortex and its relationship to cytochrome oxidase activity. *Exp. Neurol.* 156:294–315.
60. Vincent, S. R., and Kimura, H. (1992). Histochemical mapping of nitric oxide synthase in the rat brain.*Neuroscience* 46:755–784.
61. Welker, E., and Van der Loos, H. (1986). Is areal extent in sensory cerebral cortex determined by peripheral innervation density? *Exp. Brain Res.* 63:650–654.
62. Zhang, W., Vazquez, L., Apperson, M., and Kennedy, M.B. (1999). Citron binds to PSD-95 at glutamatergic synapses on inhibitory neurons in the hippocampus. *J. Neurosci.* 19:96–108.

7

The Absence of Layer 4 Dramatically Alters Cortical Development in Ferret Somatosensory Cortex

Debra F. McLaughlin, Sylvie Poluch, Beata Jablonska, and Sharon L. Juliano

Anatomical Findings in the Model of Cortical Dysplasia

The failure of neurons to migrate into the neocortex properly causes many problems. Dysplastic cortex can result from genetic causes, which produces dramatic failures of neurons to migrate properly, such as lissencephaly or double cortex. These topics have been reviewed recently by Crino (2004) and Bielas et al. (2004). More subtle abnormal migration patterns can also occur, resulting in human disorders such as dyslexia or epilepsy. Epigenetic factors that interfere with normal developmental mechanisms also contribute to malformations of cerebral cortex (Castro et al., 2002; Ross, 2002). These can include exposure to toxic substances or radiation, or ingesting substances such as alcohol or cocaine during pregnancy. Altered GABAergic systems are a common finding in models of human cortical dysplasia (Benardete and Kreigstein, 2002). GABAergic interneurons may be exceptionally vulnerable to trauma and errors in neurodevelopment partly because they are generated remotely from the developing cortical plate; the long distance they migrate to their target site may leave them particularly susceptible (Santhakumar and Soltesz, 2004). In schizophrenia, which is likely due to a combination of genetic and environmental factors, the number of parvalbumin-positive interneurons decrease, which could reflect an actual loss of GABAergic neurons or decreased expression of this GABAergic cell marker (Benes and Berretta, 2001).

We developed a model of cortical dysplasia that interrupts the birth of cells populating layer 4 in ferret somatosensory cortex. Most neurons that eventually reside in layer 4 are born on E33 in the ferret (Noctor et al., 1997); the gestational period for ferrets is 41–42 days. To do this, we use a toxin that prevents mitosis for a restricted period of time.

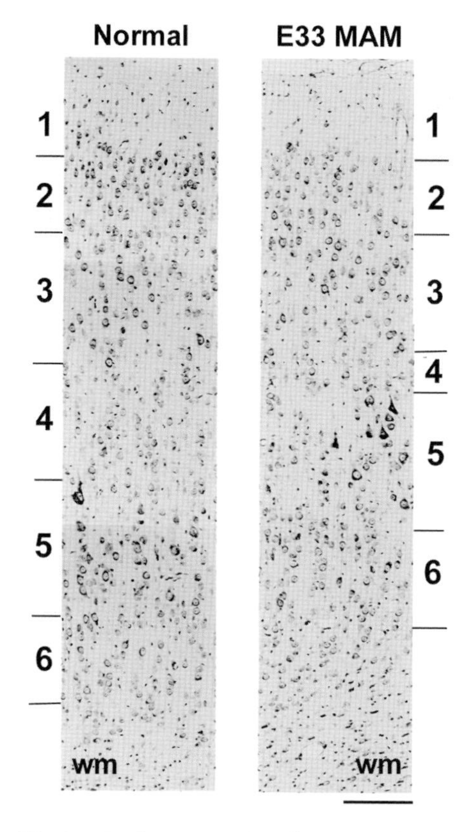

Figure 1 This is a Nissl stain demonstrating the cytoarchitecture in normal and E33 MAM treated somatosensory cortex. The normal cortex clearly demonstrates six layers. The architecture and lamination are relatively normal in the MAM treated section, but layer 4 is very thin. The MAM treated cortex is slightly thinner than the normal cortex, due to a deficit of layer 4 neurons. Scale = 100 μm.

When methylazoxy methanol (MAM) is injected into pregnant ferrets on embryonic day 33 (E33), layer 4 is dramatically thinner than normal, while leaving the remaining layers relatively free of deficits (Noctor et al., 2001) (Figure 1). In the model we developed, the cell size of the remaining layers and the cell density outside of layer 4 are similar in normal and MAM treated cortex (Noctor et al., 2001). Although many features after E33 MAM treatment appear normal, other cortical properties diverge from typical somatosensory cortex. An obvious feature to assess after the diminishment of layer 4 is the trajectory of projections from the thalamus. In somatosensory cortex, the bulk of thalamic afferents usually terminate in layer 4, which is largely absent in the E33 MAM-treated animals. We found that rather than the normal, centrally located, distribution of thalamic afferents, the projection to somatosensory cortex was widespread, with axons from the thalamus distributing nearly equally in all cortical layers (Figure 2).

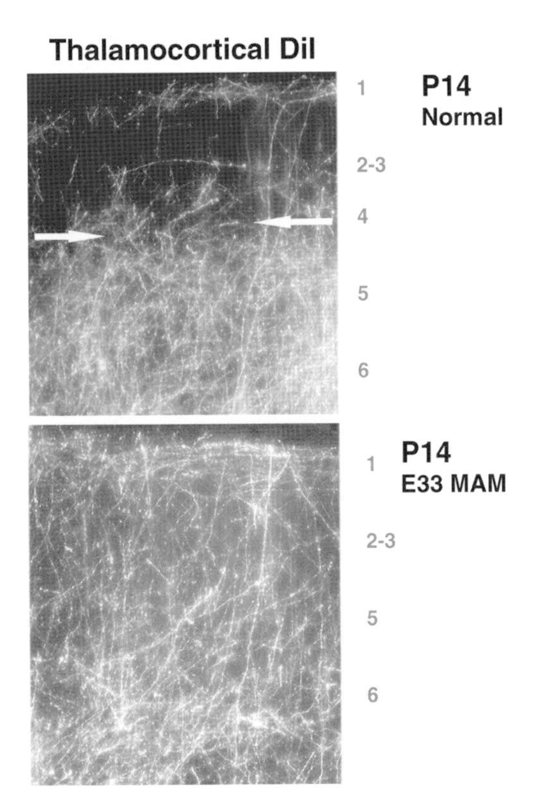

Figure 2 Shown are examples of thalamic axons labeled with DiI terminating in the somatosensory cortex of normal and MAM treated cortex at P14. In the normal cortex (top) the axons do not extend above layer 4 (arrows). In the MAM treated cortex (bottom) the axons extend through and terminate in all cortical layers.

Cortical Information Processing in the Relative Absence of Cortical Layer 4

The widespread distribution of thalamic afferents implied that the cortical responses to stimulation might be altered in the MAM-treated animals. The treated ferrets appear normal for daily activities, although we did not specifically test them on behavioral tasks. They interact with other ferrets, appear well groomed, and live to a normal life span. To evaluate neuronal activity in response to stimulation, we recorded extracellular responses from normal and MAM treated somatosensory cortex. First, we determined that the map of the body in E33 MAM-treated cortex is normal. The position of the limbs and other body parts are similar to those in normal cortex as are the receptive field size and distribution. This observation suggests that although the thalamic projections are not focused regarding their laminar distribution, they are guided appropriately to form a topographical map in the cortex, so that the body loci represented in the thalamus project to their corresponding proper sites in the somatosensory cortex (Noctor et al., 2001).

In further analysis, we determined that in contrast to the normal somatotopic arrangement, the pattern of activity evoked through the cortical layers after stimulation was disturbed in MAM treated somatosensory cortex. Examination of current source density (CSD) profiles in response to a single tap to a digit, the normal pattern of evoked activity was similar to those generally reported for sensory cortex (Aizenman et al., 1996; Di et al., 1990; Kenan-Vaknin and Teyler, 1994; Mitzdorf, 1988; Schroeder et al., 1995) (Figure 3). This includes initial current sinks in layer 4, which progress to the upper and lower layers (McLaughlin and Juliano, 2005). The same general pattern can be observed in the laminar distribution of multiunit responses (MUR). In the somatosensory cortex of MAM treated animals, however, there was no distinct pattern of activity, and the CSD profiles appeared to have little organization with very few distinct sinks (Figure 3). The laminar distribution of MURs in MAM treated cortex is likely to underlie this pattern; they display simultaneous activation across all cortical layers (McLaughlin and Juliano, 2005). This pattern of synchronized activation may reflect the finding that thalamic afferent fibers distribute equally in all layers, resulting in concomitant layer activation.

The concurrent activation pattern across layers could account for the observation that the body map in E33 MAM-treated animals was relatively normal, since the somatosensory cortex appears able to respond strongly to a single tap to the skin. That is, a single tap to the skin elicits a strong response that conveys where the body was stimulated. When we tested the ability of normal and MAM treated somatosensory cortex to respond to a more complex or richer stimulus, however, distinct differences emerged. For these richer stimuli we used intermittent taps delivered at 20 Hz. In normal animals, the cortical response entrained to the stimulus within 100–150 milliseconds, so that all cortical layers exhibited responses with periodicity in the range of the stimulation rate. The initial activity occurred in layer 4 and then transferred to the upper and lower layers (Figure 3). In the MAM treated animals, however, an initial response to the stimulus occurred simultaneously across all cortical layers, but then failed to sustain or entrain, most likely due to an inability to transfer information properly from layer 4 to other layers and alterations in GABAergic mechanisms as discussed below (McLaughlin and Juliano, 2005).

Distribution of Excitatory and Inhibitory Receptors

The observation that the flow of activity through the cortical layers was disturbed after layer 4 disruption suggested that the balance of excitation and inhibition in treated somatosensory cortex may be altered. To test this idea we studied the distribution of selected excitatory and

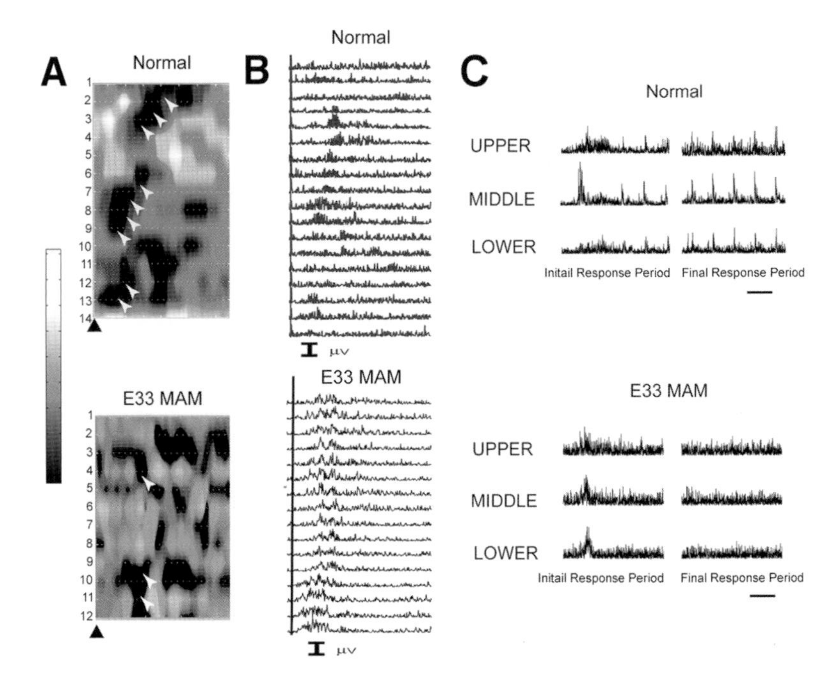

Figure 3 **A.** Illustrated are current source density (CSD) and multiunit response (MUR) profiles obtained after single taps to a digit on a ferret forepaw of a normal (top) and E33 MAM treated (bottom) animal. One-dimensional CSDs were computed for sequential recording levels through the full thickness of the cortical layers. The smoothed contour surface maps represent current source density measures over space (the y axis) and time (x axis). The gray scale represents values that are normalized across treatment groups. The arrowheads follow the activation pattern of the sinks through different cortical layers. In normal cortex, the initial response occurs in layer 4 and then progresses to the upper and lower layers. The time axis is 80 msec. **B.** Multiunit activity evoked by a single tap to a digit in normal (top) and MAM treated cortex (bottom). Each trace represents activity recorded through the cortical layers, beginning at the surface. In the normal somatosensory cortex, the response initiates in layer 4 and progresses to other layers. In the MAM treated cortex, a response to the stimulus occurs almost simultaneously in all layers. **C.** Cortical responses recorded at 3 different cortical levels and shown on a different time scale from those shown in B. The stimuli were intermittent taps delivered at 20 Hz. In normal cortex, the response develops over a period of 100–150 msec, is initiated in central cortical regions and then transferred to the upper and lower layers, and becomes entrained to the stimulus. The entrainment is best in the central layers. In MAM treated cortex, a strong initial response occurs simultaneously across layers and does not successfully entrain at the stimulation rate. Scale = 50 msec.

inhibitory receptors in normal and MAM treated ferret cortex. The pattern of binding for NMDA, AMPA, and kainate receptors showed little difference between the normal and MAM treated somatosensory cortex. Although there were subtle distinctions, the overall laminar patterns were similar. When we assessed the distribution of $GABA_A$ receptors, however, the binding shifted and expanded into the upper and lower layers compared with the normal pattern, heavily concentrated in layer 4 (Jablonska et al., 2004). To assess this pattern shift in more detail we used immunohistochemistry to evaluate the distribution of $GABA_A\alpha$ receptors. In normal animals, these receptors are dense in layer 4, and sparsely represented in the other layers, whereas in MAM treated animals the $GABA_A\alpha$ receptors are dense centrally and expand into the upper and lower layers. This expanded distribution also parallels the widespread distribution of thalamic afferent fibers. Several studies suggest that there is an association between termination of the thalamic afferents and $GABA_A$ receptors although questions remain regarding the exact relationship between thalamic terminations and these receptors (Studler et al., 2002; Meier et al., 2003; Paysan and Fritschy, 1998). To determine the relation between $GABA_A$ receptors and thalamic projections in our system, we labeled both the afferent fibers and $GABA_A\alpha$ receptors in the same sections. This revealed that the thalamic afferents and $GABA_A\alpha$ receptors are often colocalized, suggesting that they may follow similar cues in determining their ultimate locations. They also may respond to parallel cues in creating an altered distribution, such as that seen in E33 MAM treated somatosensory cortex.

Our findings indicate that layer 4 plays a key role in orchestrating overall development of sensory cortex. In our ferret model of cortical dysplasia, the relative absence of cells in layer 4, a major target for thalamic afferent fibers, is most likely related to a broad termination pattern of the ingrowing thalamocortical axons that terminate on the cells remaining in the E33 MAM treated cortex. Since the remaining pyramidal cells are not the most effective in transferring information to other layers, and there are only a few cells in layer 4, the appropriate sequence of projection through the cortex does not occur. This leads to the failure of columns of cells in the somatosensory cortex to process complex stimuli. The inability to process information because layer 4 fails to develop may be similar to deficits in humans where simple incoming information can be processed adequately, but breakdowns occur in attempts to manage more complex stimuli.

Distribution of GABAergic Neurons

Because we found evidence of interruption to the GABAergic system and disorganization of the balance of excitation and inhibition, we studied the distribution and migration of GABAergic interneurons into the cerebral cortex of normal and MAM-treated ferrets. We first assessed the distribution of GABAergic interneurons by evaluating the positions of neurons expressing calcium binding proteins (parvalbumin, calretinin, calbindin) or synthesizing enzymes of GABA, GAD65-67. Calcium binding proteins colocalize with GABA and are often used as markers for GABAergic cells in the cerebral cortex (Jones and Hendry, 1989). We also determined the distribution of MAP2- and GluR2/R3-positive cells, which are presumed to be excitatory. The MAP2-labeled cells were not different in the MAM treated *vs* normal cortex in regard to cell morphology or size of the soma. The overall distribution of

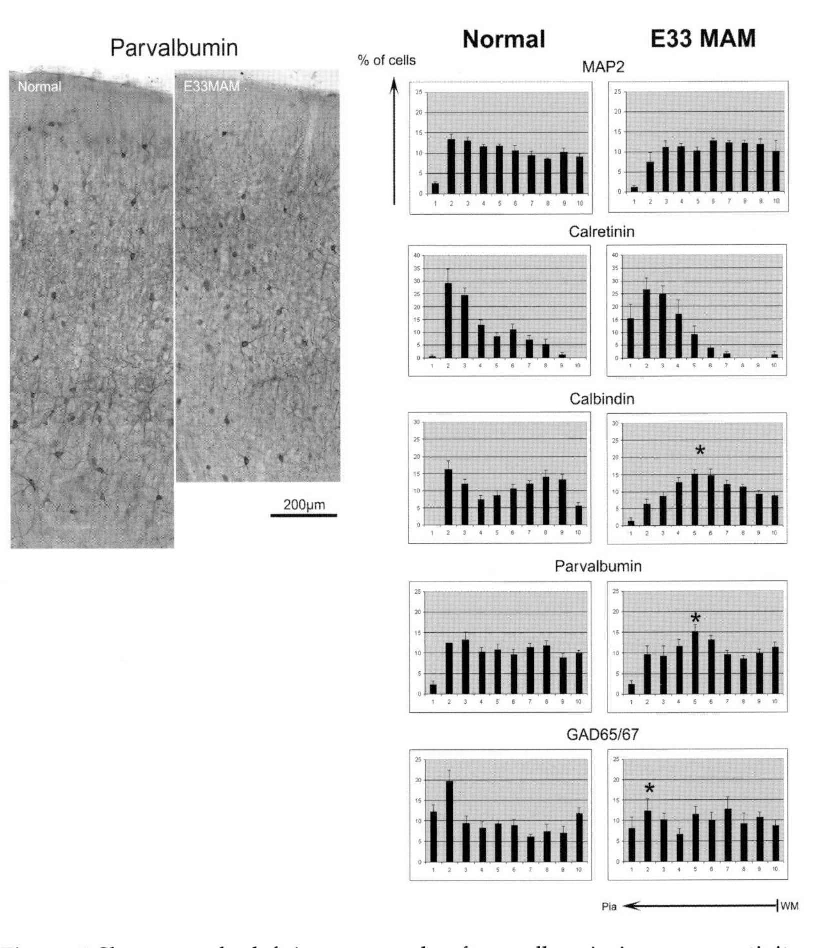

Figure 4 Shown on the left is an example of parvalbumin immunoreactivity in normal and MAM treated somatosensory cortex on P28. The size and shape of the labeled neurons are similar in both cortices. The overall distributions for cells labeled with different markers are shown on the right. The distance between the pia and the bottom of layer 6 was divided into 10 equal bins for each cortical region and the percent of the total cells in each bin indicated on the y-axis, averaged for 4 animals. There are no statistical differences for the distributions of MAP2 and calretinin in normal and E33 MAM treated cortex. The stars indicate regions of statistical significance between normal and MAM treated distributions as measured by ANOVA and post hoc analyses.

MAP2- (Figure 4) and GluR2/R3-positive cells was similar in the normal and MAM treated somatosensory cortex. For a subset of inhibitory markers, however, the labeled cells were distributed differently in MAM treated compared with normal cortex. These include the distributions of parvalbumin, calbindin, and GAD65/67. In the MAM treated cortex, distributions of cells labeled with these markers concentrate around deep or central cortical regions, rather than the more distinct laminar patterns seen in normal cortex. For example, calbindin immunoreactivity is almost bimodal in the normal distribution, with two peaks of increased numbers of neurons, one located in the upper layers, and one in layers 5–6. In the MAM treated cortex, however, more calbindin expressing cells are located centrally, without clear peaks in the distribution. In the parvalbumin distribution, although the peaks in upper and lower layers are subtler, similar patterns of cell dispersal occur, with a large concentration of cells in the middle region of cortex in the MAM treated brains. For the GAD65/67 distribution in normal cortex, the distribution peaks in the upper layers, which is not seen in the MAM-treated cortex. The calretinin patterns of immunoreactivity are comparable for the normal and MAM treated brains. Since the calretinin pattern was relatively normal, this suggests that, although a population of neurons expressing GABA has difficulty reaching its cortical target, this difficulty may be restricted to certain subtypes, since not all neurons expressing GABA have trouble reaching their cortical target. This also corresponded to our earlier receptor binding data, which revealed alterations in the GABAergic receptor system but not in glutamatergic ones (Jablonska et al., 2004). Taken together, these findings suggest that specific subtypes of GABAergic neurons are not able to successfully migrate to their target location in the somatosensory cortex after treatment with MAM on E33, while the system of excitatory neurons remains relatively unaffected.

Origin of GABAergic Neurons

In recent years, it has become evident that the majority of cortical inhibitory neurons in rats and mice originate in the lateral and medial ganglionic eminence and migrate tangentially into the cerebral cortex (for review see Corbin et al., 2001; Marin and Rubenstein, 2003). The same feature had not been specifically demonstrated in ferrets, but we investigated this process in organotypic cultures of normal and MAM-treated neonatal ferrets. To do this, we prepared organotypic cultures of normal ferret cortex at ages ranging from E27 to postnatal day 2 (P2). The cultured slices were injected in the ganglionic eminence with DiI or other dyes taken up by cells migrating away from this site. The slices remained in culture for 2–5 days. We observed that large numbers of neurons leave the ganglionic eminence in normal and treated neonatal ferrets. Because ferrets are altricial animals that mature slowly during development, large numbers of neurons are migrating from the ganglionic eminence at birth and continue to migrate up to P2 (probably even later, but these dates were not tested). At younger ages, the neurons leaving the ganglionic eminence follow overall trajectories similar to those described by others in rodents. There is a deep route of migration that runs just above the cortical ventricular zone before turning to enter the neocortex and a more superficial route in which neurons travel in the subplate or in layer 1 before joining the cortical layers (Marin and Rubenstein, 2003). In rats and mice, the majority of these neurons express GABA and are interneurons in mature cortex. We tested this in our ferret model by immunoreacting the slices containing migrating neurons with antibodies directed against GABA; 68% of the labeled neurons leaving the ganglionic eminence are GABAergic, showing similar properties to these neurons in rodents.

The same set of experiments in MAM treated cortex (examined at E38-P2) indicated that similar numbers of neurons leave the ganglionic eminence and migrate into the cerebral cortex (Figure 5). In MAM-treated animals, however, the neurons migrating from the ganglionic eminence appeared more disorganized on their route to the cerebral cortex. To verify this observation, we determined the angle of orientation of each leading process of a migrating cell for normal and MAM-treated organotypic cultures. The leading process was measured in relation to the pia. An angle was calculated and placed into a bin that contained the 90° orientation (out of 360°), i.e., either oriented radially (toward the pia or the ventricle) or oriented tangentially (medially or laterally). Orientations either radially toward the pia or tangentially in the dorsal direction were considered the "proper" directions. After 2 days in culture, cells migrating from the ganglionic eminence of either normal or MAM treated brains were oriented properly, while after 5 days in culture, the migrating neurons were less likely oriented in the proper direction than those originating from the normal ganglionic eminence (Figure 5). In addition, when the labeled neurons migrating from the ganglionic eminence were immunoreacted for GABA, fewer of them were double labeled (37%), suggesting that the environment in E33 MAM-treated cortex was not conducive to maintaining a GABAergic phenotype.

This led us to wonder whether the source of disorientation was in the migrating neurons themselves, or due to cues originating from the route of migration or target site. To evaluate these questions we prepared mixed organotypic co-cultures obtained at P0, which included explants of normal ganglionic eminence paired with MAM-treated cortical explants and *vice versa*. DiI was injected into the ganglionic eminence and the paired explants remained in culture for 5 days. If the cortical explant was normal, the neurons migrating from the ganglionic eminence were oriented in the designated "proper" directions, even if the ganglionic eminence explant originated from a MAM-treated animal. This suggests that a property of E33 MAM treated cortex impaired proper migration of neurons leaving the ganglionic eminence, while features of normal cortex encouraged proper tangential migration (Figure 5).

Conclusion

MAM treatment on E33 leads to dramatic diminution of layer 4. The relative absence of layer 4 leads to further changes that include improper termination of thalamic afferent fibers, widespread distribution of $GABA_A\alpha$ receptors, and the failure of information transfer in cortical responses to sensory stimulation (Noctor et al., 2001; Palmer et al., 2001; Jablonska et al., 2004; McLaughlin and Juliano, 2005). Further

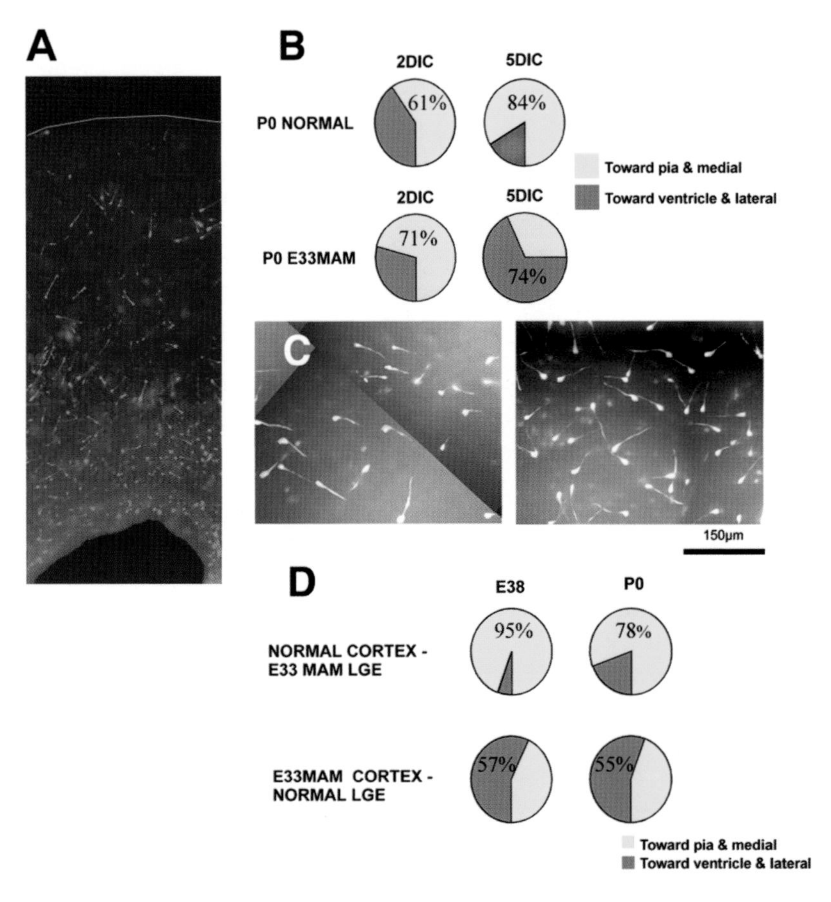

Figure 5 A. An example of cells leaving the ganglionic eminence after an injection of DiI into a normal organotypic culture. The injection was made on P0 and the slice remained in culture for 2 days. The red line indicates the border of the pia. **B.** These pie charts indicate the percent of neurons that left the ganglionic eminence and their direction of migration as indicated by the angle of the leading process after 2 days in culture (DIC) or after 5 days in culture. The "proper" orientation is indicated in yellow, either toward the pia or medially. The other orientations are indicated in gray. **C.** Examples of migrating neurons originating from the ganglionic eminence and labeled with DiI in E33 MAM treated animals. Neurons oriented correctly and migrating for 2 days in culture are on the left, neurons that are more disoriented and migrating for 5 days in culture are indicated on the right. **D.** Pie charts indicating the orientation of migrating neurons from cocultures of normal cortex and E33 MAM treated ganglionic eminence (LGE) or E33 MAM treated cortex and normal LGE. Indicated in the graphs on the left the orientations of the migrating neurons injected on E38 and maintained in culture for 5 days. On the right are graphs of the directions of neurons migrating after injection on P0 after 5 days in culture. In each case, more neurons are oriented in the "proper" direction (yellow) when the cortex is normal. Normal ganglionic eminence did not result in a normal pattern of migration.

analysis shows that cells expressing GABA do not migrate properly and neurons expressing distinct types of calcium binding proteins accumulate in the lower and central layers, rather than reaching their proper sites in the upper layers. Presumed excitatory cells do not show abnormal distributions in MAM treated cortex. Further support for the failure of GABAergic cells to migrate properly is seen in the disorientation of cells leaving the ganglionic eminence and heading toward the cortex in E33 MAM treated animals. We suggest that the relative absence of layer 4 leads to a cascade of effects that result in the inability of cells originating in the ganglionic eminence (presumptive GABAergic cells) to migrate effectively. This leads to a mature cortex in which many GABAergic cells fail to reach their proper targets. As a result, the cortex is not able to appropriately respond to somatic stimulation, probably due to improperly placed GABAergic cells and $GABA_A$ receptors combined with widespread thalamic afferents, which results in an inability to transfer information through the cortical layers. Findings from our model of cortical dysplasia coincide with many observations for human cortical dysplasia, including the consistent finding of altered GABAergic systems (Baraban et al., 2000; Castro et al., 2002; Jablonska et al., 2004; Luhmann et al., 1998).

References

Aizenman, C.D., Kirkwood, A., and Bear, M.F. (1996). A current source density analysis of evoked responses in slices of adult rat visual cortex: Implications for the regulation of long-term potentiation. *Cereb. Cortex* 6:751–758.

Baraban, S.C., Wenzel, H.J., Hochman, D.W., and Schwartzkroin, P.A. (2000). Characterization of heterotopic cell clusters in the hippocampus of rats exposed to methylazoxymethanol in utero. *Epilepsy Res.* 39:87–102.

Benardete, E.A., and Kriegstein, A.R. (2002). Increased excitability and decreased sensitivity to GABA in an animal model of dysplastic cortex. *Epilepsia* 43:970–982.

Benes, S.M. and Berretta, S. (2001). GABAergic interneurons: implications for understanding schizophrenia and bipolar disorder. *Neuropsychopharmacology* 25:1–27.

Bielas, S., Higginbotham, H., Koizumi, H., Tanaka, T., and Gleeson, J.G. (2004). Cortical neuronal migration mutants suggest separate but intersecting pathways. *Annu. Rev. Cell Dev. Biol.* 20:593–618.

Castro, P.A., Pleasure, S.J., and Baraban, S.C. (2002). Hippocampal heterotopia with molecular and electrophysiological properties of neocortical neurons. *Neuroscience* 114:961–972.

Corbin, J.G., Nery, S., and Fishell, G. (2001). Telencephalic cells take a tangent: non-radial migration in the mammalian forebrain. *Nat. Neurosci.* 4 Suppl:1177–1182.

Crino, P. (2004). Malformations of cortical development: molecular pathogenesis and experimental strategies. *Adv. Exp. Med. Biol.* 548:175–191.

Di, S., Baumgartner, C., and Barth, D.S. (1990). Laminar analysis of extracellular field potentials in rat vibrissa/barrel cortex. *J. Neurophysiol.* 63:832–840.

Jablonska, B., Smith, A.L., Palmer, S.L., Noctor, S.C., and Juliano, S.L. (2004). GABAA receptors reorganize when layer 4 in ferret somatosensory cortex is disrupted by methylazoxymethanol (MAM). *Cereb. Cortex* 14:432–440.

Jones, E.G., and Hendry, S.H. (1989). Differential calcium binding protein immunoreactivity distinguishes classes of relay neurons in monkey thalamic nuclei. *Eur. J. Neurosci.* 3:222–246.

Kenan-Vaknin, G., and Teyler, T.J. (1994). Laminar pattern of synaptic activity in rat primary visual cortex: Comparison of in vivo and in vitro studies employing the current source density analysis. *Brain Res.* 635: 37–48.

Luhmann, H.J., Karpuk, N., Qu, M., and Zilles, K. (1998). Characterization of neuronal migration disorders in neocortical structures. II. Intracellular in vitro recordings. *J. Neurophysiol.* 80:92–102.

Marin, O., and Rubenstein, J.L. (2003). Cell migration in the forebrain. *Annu. Rev. Neurosci.* 26:441–483.

McLaughlin, D.F., and Juliano, S.L. (2005). Disruption of layer 4 development alters laminar processing in ferret somatosensory cortex. *Cereb. Cortex*, 15:1791–1803.

Meier, J., Akyeli, J., Kirischuk, S., and Grantyn, R. (2003). GABA(A) receptor activity and PKC control inhibitory synaptogenesis in CNS tissue slices. *Mol. Cell Neurosci.* 23:600–613.

Mitzdorf, U. (1988). Evoked Potentials and their physiological causes: an access to delocalized cortical activity. *Springer Series in Brain Dymanics 1.*

Noctor, S.C., Scholnicoff, N.J., and Juliano, S.L. (1997). Histogenesis of ferret somatosensory cortex. *J. Comp. Neurol.* 387:179–193.

Noctor, S.C., Palmer, S.L., McLaughlin, D.F., and Juliano, S.L. (2001). Disruption of layers 3 and 4 during development results in altered thalamocortical projections in ferret somatosensory cortex. *J. Neurosci.* 21:3184–3195.

Palmer, S. L., Noctor, S.C., Jablonska, B., and Juliano, S.L. (2001). Laminar specific alterations of thalamocortical projections in organotypic cultures following layer 4 disruption in ferret somatosensory cortex. *Eur. J. Neurosci.* 13:1559–1571.

Paysan, J., and Fritschy, J.M. (1998). GABAA-receptor subtypes in developing brain. Actors or spectators? *Perspect. Dev. Neurobiol.* 5:179–192.

Ross, M.E. (2002). Brain malformations, epilepsy, and infantile spasms. *Int. Rev. Neurobiol.* 49:333–352.

Santhakumar, V., and Soltesz, I. (2004). Plasticity of interneuronal species diversity and parameter variance in neurological diseases. *Trends Neurosci.* 27:504–510.

Schroeder, C.E., Seto, S., Arezzo, J.C., and Garraghty, P.E. (1995). Electrophysiological evidence for overlapping dominant and latent inputs to somatosensory cortex in squirrel monkeys. *J. Neurophysiol.* 74:722–732.

Studler, B., Fritschy, J.M., and Brunig, I. (2002). GABAergic and glutamatergic terminals differentially influence the organization of GABAergic synapses in rat cerebellar granule cells in vitro. *Neuroscience* 114:123–133.

8

Influence of Thalamocortical Activity on Sensory Cortical Development and Plasticity

Sarah L. Pallas, Mei Xu, and Khaleel A. Razak

Abstract

The cerebral cortical hemispheres are organized into multiple structurally and functionally distinct areas. The positioning of these areas is nearly invariant across individuals within a species and even between closely related species. We are interested in determining how these cortical areas are specified during development. Another main area of interest is how one cortical area might be induced to take on the identity of, and thus substitute for, another cortical area. We have been taking several different approaches to this long-standing issue. This chapter will report on some of our most recent findings. A more complete summary of our previous work can be found in several other review articles (Pallas, 2001, 2002, 2005, and in press).

Thalamocortical afferent (TCA) targeting is relatively specific during development, in that TCAs do not exhibit the level of exuberancy and subsequent pruning seen in corticofugal projections. The mechanism underlying TCA targeting, however, is unknown. We are exploring the relative roles of gene expression patterns and neuronal activity in directing thalamocortical targeting specificity and thus in specifying cortical areas. We summarize data from three studies of cortical development and plasticity in ferrets, an altricial species with protracted postnatal development. In one study, we investigated the temporal relationship between the targeted ingrowth of TCAs and opposing expression gradients of *Pax6*, *Emx2*, *Cad6*, and *Cad8*. Using real-time PCR coupled with tracing of TCA projections, we found that *Pax6* and *Emx2* expression gradients are declining during TCA ingrowth. They may orchestrate gradients of neurogenesis and/or provide regional patterning signals. On the other hand, differential expression of *Cad6* and *Cad8* is maximal during TCA targeting and synapse formation, and could play a causal role. In the second set of studies described here, we examined the effect of sensory deprivation on cortical specification. If normal levels or sources of neural activity are necessary for targeting, then manipulations of activity during TCA ingrowth should disrupt targeting. Consistent with this prediction, we find that bilateral cochlear ablation in P14 ferrets results in mistargeting of LGN axons to primary auditory cortex. The third section discusses results from cross-modal plasticity studies. In contrast to sensory deprivation, redirection of retinal axons into auditory thalamus does not cause TCA mistargeting, but does respecify several aspects of auditory cortical structure and function. Finally, we examine the relationship between gene expression and the changes seen in cross-modal plasticity. Together these experiments support the idea that early gene expression patterns provide positional information contributes to gradients of neurogenesis and regional identity, but activity plays an essential role in later patterning and plasticity at the level of TCA targeting to functionally defined cortical areas.

Cortical Parcellation is a Stepwise Process

Cortical parcellation, or areal specification, occurs through several different stages, and at each step, the fate of the cells or tissue is further restricted (Fig. 1). Like development of other brain regions, cortical development doubtless depends upon a multitude of critically timed factors. The undifferentiated cortical epithelium may contain information about positional identity and polarity, which then informs the regional specification process, during which gradients and boundaries of gene expression differentiate large regions of cortex as distinct from one another. After this regionalization process, and approximately simultaneous with thalamocortical ingrowth, the cortical areal boundaries are established. To date, no genes have been found to express in a way that

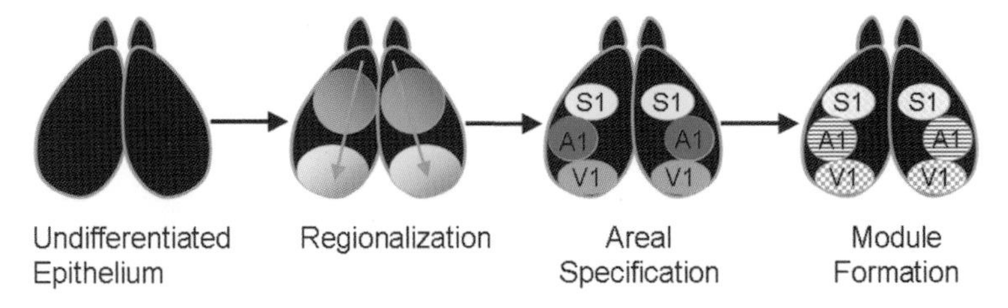

Figure 1 Specification of functionally defined cortical areas occurs through several stages. First, the undifferentiated cortical epithelium develops regional specializations, manifested as local variations in gene expression. At approximately the time when thalamocortical axons enter the cortical plate, which occurs at birth in rats and mice but at P14 in ferrets, areas become distinguishable. Subsequently the unique features of each cortical area appear including formation of modular processing units such as barrels, bands, and blobs for in S1, A1, and V1, respectively.

defines the sharp boundaries between areas, yet the thalamocortical axons (TCAs) project to these areas in a very directed fashion (Crandall and Caviness, 1984), as if there were road markers along the way. After the areas are specified, processing modules are formed that perform area-specific computations on the information received. The long-term goal is to define the mechanisms that are responsible for this stepwise cortical parcellation process at these multiple levels.

Intrinsic and Extrinsic Factors Direct Cortical Parcellation

Some specification events depend on factors intrinsic to cortex, such as patterned expression of molecular cues, and others depend on extrinsic information, either from other brain regions or from the outside world (Fig. 2). The relative contribution of these intrinsic and extrinsic factors to the parcellation of cerebral cortex into different functional areas has been under debate for some time (Pallas, 2001, for review). At one extreme, we could imagine that the location and functional identity of cortical areas are preprogrammed from the beginning, perhaps by intrinsic patterns of gene expression (Bishop et al., 2000; Mallamaci et al., 2000; Huffman et al., 2004). At the other extreme, cortex could be a *tabula rasa*, entirely specified by the extrinsic information it receives. The answer probably lies somewhere in between. Intrinsic factors are the only information available early in development, but once the thalamocortical pathway is formed, cortical development can be influenced

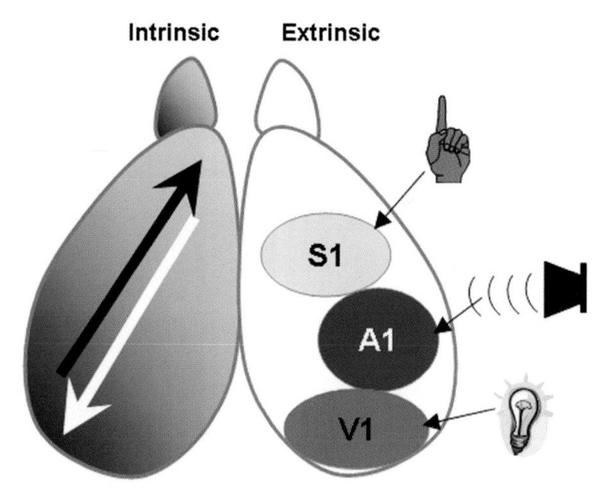

Figure 2 Both intrinsic gradients of gene expression (left) and modality-specific extrinsic neural activity (right) influence the specification of cortical areas. S1-primary somatosensory cortex; A1-primary auditory cortex; V1-primary visual cortex.

by sensory experience as well as by spontaneous afferent activity (Weliky and Katz, 1999; Chiu and Weliky, 2001) or molecular cues present in the afferent fibers (e.g. Dufour et al., 2003).

Thalamocortical projections are obviously an important source of extrinsic information, and could provide a scaffold for parcellation; they are accurately targeted throughout their trajectory and do not exhibit the exuberant overgrowth to extraneous targets seen in corticofugal projections (Innocenti et al., 1977; Olavarria and Van Sluyters, 1985; O'Leary and Stanfield, 1986). It seems simple enough to propose that, for example, visual cortex becomes visual cortex simply because it receives input from the eyes via the visual thalamus. This begs the question of how the TCAs know what path to follow. Do they find their way through intrinsic programming, or does a pre-specified cortical area direct their pathfinding from some distance? Are there other, intermediate cues that help them stay on track? Clearly ventral telencephalon and the cortical subplate contain important guidance information (Ghosh et al., 1990; Garel and Rubenstein, 2004; Molnár, this volume). Given that there are no known molecular markers that correspond with areal boundaries, how are the sharp demarcations between the TCA projections to adjacent cortical areas formed? Projections from visual thalamus (LGN) never stray into auditory or somatosensory cortex at any time during development. Another conundrum is that any mechanism proposed to underlie areal specification during development must be compatible with the fact of cortical evolution– not only has cerebral cortex expanded in size, but there has also been a tremendous increase in the number of cortical areas during mammalian evolution (Felleman and Van Essen, 1991; Kaas, 1993; Rosa and Krubitzer, 1999), requiring concomitant shifts in thalamocortical targeting. The answers to these intriguing questions have been elusive, but it is our hope that our unique experimental approaches can inform future investigations.

Role of Cortical Patterning Genes in TCA Targeting

The first series of experiments that we will discuss in this review address the question of whether there is a direct causal link between known cortical gene expression gradients and TCA targeting. For this work we chose to examine two transcription factor genes that had been suggested as cortical patterning genes by other investigators. The homeobox genes *Pax6* and *Emx2* are expressed in opposing gradients across the cortical epithelium in E18 mouse cortex, and it has been suggested that these genes or their downstream effectors specify cortical areas directly by guiding TCAs to their proper targets (Bishop et al., 2000; Mallamaci et al., 2000). In the hindbrain (Keynes and Krumlauf, 1994) and spinal cord (Stoeckli and Landmesser, 1998), patterned gene expression directs the specification of axonal projection patterns, and the possibility that thalamocortical pathways would exhibit similar pre-specification is a compelling one.

We reasoned that if gradients of *Pax6* and *Emx2* provide targeting information for TCAs, then there should be differences in gene expression in different cortical areas at the time of TCA ingrowth. Evidence in favor of this hypothesis is that knockout of either of the genes causes respecification of cortical gene expression patterns, caudalizing cortex in the case of *Pax6* knockout, and anteriorizing it in the case of *Emx2* knockout. Unfortunately these knockout mice die before experimental identification of cortical areal boundaries is possible, necessitating reliance on downstream marker molecules to infer areal boundary location. Importantly, recent information from gain of function mutations (Leingartner et al., 2003; Hamasaki et al., 2004) has demonstrated that ectopically-expressed Emx2 leads to mistargeting of TCAs, supporting an important, though not necessarily direct, involvement of the gene in TCA targeting.

There are alternative interpretations of the results of these genetic manipulation studies, however. One issue that needs to be addressed is that in the cerebral cortex of *Pax6* and *Emx2* knockouts, TCAs actually become lost in the ventral telencephalon, not in the cortical subplate (López-Bendito and Molnár, 2003; Garel and Rubenstein, 2004). Thus the genes may direct the establishment of guidance cues there, and may not be involved in targeting within the neocortex itself (Caric et al., 1997; Jones et al., 2002; Garel et al., 2003). Furthermore, *Pax6* and *Emx2* are known to be involved in setting up gradients of neurogenesis (Estivill-Torrus et al., 2002; Hevner et al., 2002; Scardigli et al., 2003). It is possible that areal boundaries only appear to shift in the knockouts because of failure to generate neurons destined for that area. The brains in the Pax6 knockout mice are substantially reduced in size (Schmahl et al., 1993; Caric et al., 1997; Heins et al., 2001). Conversely, *Emx2* promotes proliferation of neural progenitors (Cecchi, 2002).

Ferret Sensory Cortex as a Model System

In order to investigate these alternative possibilities, we compared the time course of TCA ingrowth with gene expression patterns during postnatal development using ferrets as a model system. Although mice have the obvious benefits of a short generation time and tractable genetics, ferrets provide special advantages for studies of cortical development. They have a gyrencephalic cortex, meaning that there are reliable landmarks for future areal boundaries and that cortex is well differentiated functionally (Kelly et al., 1986; Law et al., 1988; Phillips et al., 1988; Jackson et al., 1989; Pallas et al., 1990; Pallas and Sur, 1993). Another advantage is that the physiology of ferret sensory cortex is similar to that of cats, which has been well described. In addition, ferrets are born early in the process of cortical development (Fig. 3). In both cats and ferrets, approximately 50 days pass between conception and thalamocortical invasion of the cortical plate. Ferret kits are born at embryonic day 42, however, three weeks before parturition in cats. Importantly, TCAs do not reach the cortical plate until approximately P14, and ferret

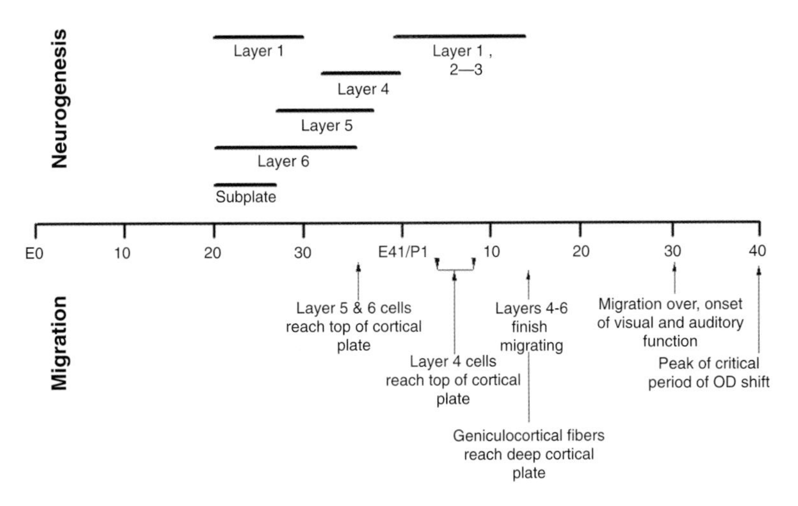

Figure 3 Ferrets have a protracted period of postnatal brain development. Unlike in rodents, cats or monkeys, much of the migration of neurons into the cortical plate occurs after birth, thalamic axons do not reach the cortical plate until approximately P14, and kits do not see or hear until one month of postnatal age. (Data taken from Shatz and Luskin, 1986; Jackson et al., 1989; Herrmann et al., 1994; Issa et al., 1999)

kits do not see or hear until after P30 (Cucchiaro and Guillery, 1984; Moore and Hine, 1992). Thus they provide the advantage of protracted development, which facilitates identification of causal relationships and postnatal access to important developmental events.

Timing of Thalamocortical Ingrowth to Sensory Cortex

We have been interested in the specification events that organize ferret sensory cortical areas, specifically primary visual (V1) and auditory (A1) cortex. The time course of thalamic innervation of ferret visual cortex in the second postnatal week has been described (Herrmann et al., 1994). We determined using retrograde and anterograde tracers placed in A1 and medial geniculate nucleus (MGN), respectively, that MGN is beginning to reach the cortical plate by P14, and has established large numbers of geniculocortical synapses by P20. We then determined when the putative patterning genes are expressed in relation to TCA ingrowth. We hypothesized that if gradients of *Pax6* or *Emx2* are directly involved in guidance of TCAs to their correct cortical target, then they would have to be present and differentially distributed in the cortical plate between P14 and P20. Alternatively, if the genes are providing positional information, specifying gradients of neurogenesis, or setting up expression of later-appearing targeting molecules, they would be present earlier but not necessarily at the time of TCA ingrowth. In addition to

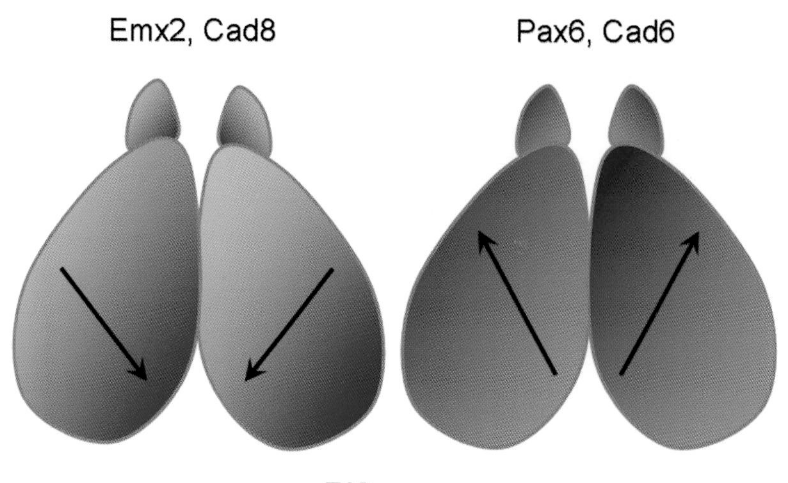

Figure 4 Genes differentially expressed in E18 mouse cortex and purported to be involved in patterning of cortical areas include the transcription factors *Emx2* and *Pax6* and the calcium-dependent, homophilic adhesion factors *Cad6* and *Cad8* (from Bishop et al., 2002). *Emx2* and *Cad8* are high caudomedially, where visual cortex is located; *Pax6* and *Cad6* are high rostrolaterally, and thus should be expressed more highly in auditory than visual cortex. These genes could in principle specify the location of cortical areal boundaries.

examining *Pax6* and *Emx2*, we examined the expression pattern of two other genes, *cadherin (Cad) 6* and *Cad8*, which are differentially expressed in embryonic mouse cortex (Fig. 4). The distribution of *Cad6* overlaps with the pattern of *Pax6* and *Cad8* expression overlaps that of *Emx2* expression (Suzuki et al., 1997; Nakagawa et al., 1999; Gil et al., 2002), although there is no evidence to date that there is a causal relationship. Cadherins are calcium-dependent, homophilic adhesion molecules that could serve to guide TCAs to targets that co-express them at the same time point.

Expression of Patterning Molecules During Thalamocortical Ingrowth

We examined the expression of *Pax6*, *Emx2*, *Cad6*, and *Cad8* in V1 and A1 during postnatal development, using a quantitative technique. Quantitative real-time (QRT-)PCR provides a measurement of the absolute amount of mRNA in a sample by comparison to a standard curve. We found that the expression gradient of *Pax6* and *Emx2* was sharply declining by P14, and that there was very little *Emx2* expressed in postnatal ferret cortex (Fig. 5). We interpret this data to mean that it is highly unlikely for *Pax6* and *Emx2* to be involved in guidance of TCAs as they

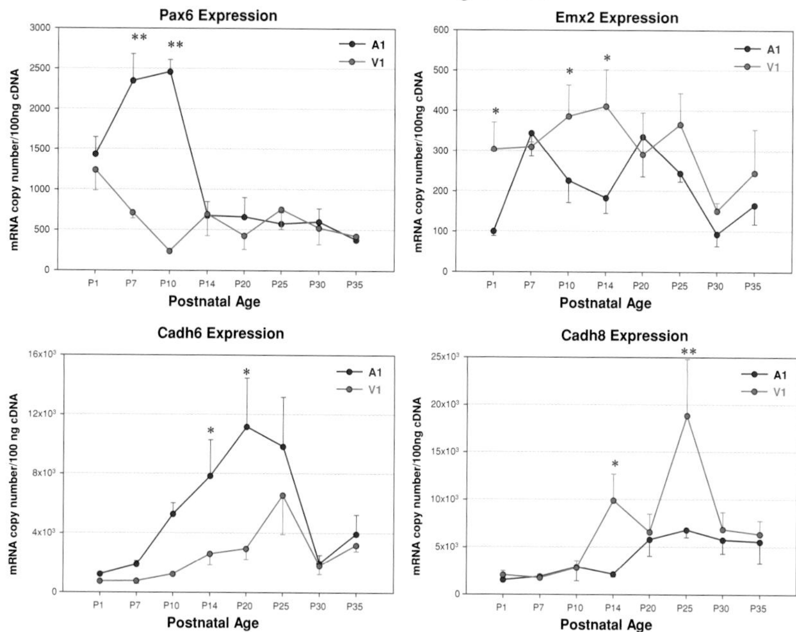

Figure 5 Quantitative real-time PCR was used to investigate the spatiotemporal relationship between thalamocortical axon ingrowth and patterning gene expression. Early in postnatal cortical development, *Pax6* and *Emx2* show complementary expression patterns. As the levels of these transcription factors decline, the complementary expression of *Cad6* and *Cad8* increases, reaching a peak as the thalamic axons arrive.

enter the cortical plate during the second postnatal week. On the other hand, *Cad6* and *Cad8* were differentially expressed in V1 and A1, respectively, during the time when TCAs choose a cortical target. Thus the cadherins are expressed at a place and time that would allow them to play an important role in thalamocortical targeting (Xu et al., 2003 and submitted).

The QRT-PCR technique requires that tissue be homogenized prior to measurement, and thus does not provide spatial resolution at a scale sufficient to assess its distribution within the cortical tissue. In addition, it measures mRNA, not protein. For a guidance molecule to be functional in guiding TCAs, the protein would have to be expressed below the cortical plate, where the axons make the decision to leave the white matter and enter the subplate. To examine the spatial distribution of the cadherin proteins, we employed Western blots as well as immunocytochemistry on coronal sections of A1 and V1. We found Cad6 protein expressed in migrating neurons in the deep layers of the cortical plate, in the subplate of both A1 and V1, and in the subventricular and ependymal zones of A1. Semi-quantitative Western blots of Cad6 protein showed that protein expression levels were higher in A1, whereas Cad8 levels were higher in V1, matching the pattern of gene expression. These data provide further support for a role of these particular cadherins in thalamocortical targeting. Evidence from other groups supports the involvement of the cadherin family in general in thalamocortical targeting (Inoue et al., 1998; Gil et al., 2002; Poskanzer et al., 2003; Redies et al., 2003). No doubt there are other guidance molecules involved as well, and the several labs currently doing the gene screening will certainly identify other candidates that can be tested in the near future.

Role of Thalamocortical Activity Patterns in Cortical Areal Specification

An alternative explanation for specific thalamocortical targeting patterns is that the TCAs have intrinsic information about where to project within cortex. They could then specify sharp areal borders in the cortical epithelium through differences in their own activity patterns. A previous study has shown that activity blockade can interfere with TCA targeting within the subplate (Catalano and Shatz, 1998), supporting this hypothesis. We have examined the role of TCA activity in cortical development using two different manipulations. In one case, the normal sensory input to a cortical area is eliminated by ablation of the peripheral sensory organ, resulting in miswiring of TCAs. In the other, rewiring at the periphery alters the modality of information carried by thalamocortical afferents, thus altering functional properties of cortex.

Peripheral Deafferentation Can Alter Thalamocortical Targeting

To address whether normal sensory activity is necessary for TCA targeting, we induced neonatal hearing loss in ferret kits by bilateral cochlear ablation prior to the onset of auditory function (Moore and Hine, 1992). We reasoned that if sensory activity is essential to guide TCA pathfinding, then a change in activity should affect targeting. On the other hand, if targeting is molecularly-specified, then manipulations of activity should have no effect. We found that loss of the peripheral hearing organs led to a redirection of axons from visual thalamus (LGN) to primary auditory cortex (Pallas et al., 2002, and in prep.) (Fig. 6). This is a startling finding, in part because many previous attempts to induce TCA mistargeting have failed (e.g. Miller et al., 1991; Miller et al., 1993; Croquelois et al., 2005), and in part because the cochleae are several synapses distant from the auditory cortex. These results support a model whereby targeting of thalamocortical axons is not irreversibly determined by intrinsic information, but is at least partially dependent on sensory activity.

Cross-Modal Plasticity Alters the Functional Identity of Sensory Cortex

In cross-modal plasticity, early loss of function in one sensory system is compensated for by another sensory system taking over the lost

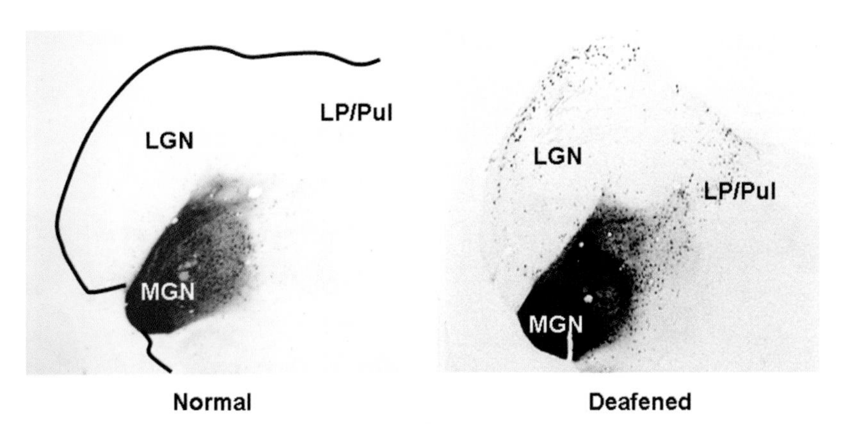

Figure 6 Early bilateral ablation of the cochlear hearing organs leads to ectopic innervation of primary auditory cortex by axons from visual thalamus (LGN and LP/Pul). Shown at left is a coronal section through thalamus following injection of HRP throughout primary auditory cortex. Label is restricted to auditory thalamus (MGN). At right is an equivalent section from an early-deafened ferret, showing that in addition to backfilled cells in auditory thalamus, injection of HRP in A1 backfills many cells in visual thalamus.

function. Sometimes referred to as sensory substitution (Rauschecker, 1995), this phenomenon has been demonstrated in clinical settings with deaf or blind individuals (Sadato et al., 1996; Finney et al., 2001; Bavelier and Neville, 2002). It is also familiar to evolutionary biologists who study fossorial species with reduced or absent visual systems (Heil et al., 1991; Bronchti et al., 2002).

Cross-modal retinal projections to auditory thalamus have been experimentally induced in ferrets by combining unilateral lesion of the superior colliculus (SC), which eliminates a retinal target, with bilateral lesion of the inferior colliculus, which eliminates the major source of input to MGN (Sur et al., 1988; Angelucci et al., 1997) (Fig. 7). Seeking target space, the retinal axons then innervate the ipsilateral, denervated MGN. In this procedure, unlike in the deafened ferrets described above, LGN is unaffected and does not form ectopic projections to auditory cortex. The retinal axons form arbors in MGN that are similar in morphology to those that would form in SC or LGN (Pallas et al., 1994; Pallas and Sur, 1994), conferring a retinotopic map of visual space onto the MGN (Roe et al., 1993). This cross-modal rewiring of retinal output thus alters the modality of patterned activity reaching primary auditory cortex, but without altering the source or identity of the TCAs. The manipulation thus effectively decouples the effects of patterned activity and the effects of patterned gene expression on cortical areal specification, because the activity pattern is changed without altering the thalamocortical pathway. This allows the effects of activity and gene expression patterns to be studied independently of each other.

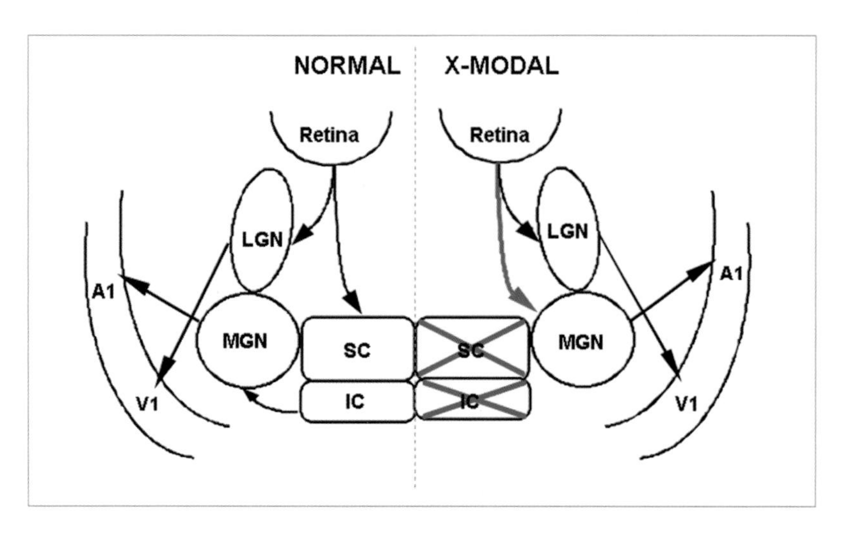

Figure 7 Cross-modal induction of retinal projections to auditory cortex is accomplished by midbrain ablations at birth. Unilateral ablation of the superior colliculus (SC) and bilateral ablation of the inferior colliculus (IC) promotes sprouting of retinal axons into the MGN. As a result, visual activity patterns reach A1, without changing the identity of the thalamocortical pathway from MGN.

The cross-modal pathway has been studied using both anatomical and physiological methods. With respect to anatomy, the TCA projection to A1 in cross-modal ferrets remains typically auditory in form. It retains its laminar, cochleotopic organization rather than adopting the point-to-point, two-dimensional, retinotopic type of pattern seen in the visual pathway (Pallas and Sur, 1993). Projections from cross-modal A1 to other cortical areas also remain unaffected; A1 in the cross-modal animals projects to other auditory cortical areas, and not to visual cortical areas (Pallas and Sur, 1993; see also Huffman et al., 2004). Functionally, however, cross-modal A1 becomes similar to visual cortex, with its retinotopic organization and its visual receptive field properties such as end-inhibition, orientation tuning, and simple or complex receptive field types (Roe et al., 1990; Roe et al., 1992). The emergence of retinotopy from the overlapped thalamocortical projections suggests that retinotopy is recreated within A1 itself, although the underlying mechanism remains unclear.

Further insight into the functional specification process came from our studies of modular organization in cross-modal A1. In sensory cortical areas, intracortical projections link clusters or laminae of neurons with similar response properties. In visual cortex, for example, neurons with similar orientation tuning and ocular dominance are interconnected (Katz and Callaway, 1992). In the auditory cortex, slabs of neurons tuned to the same sound frequency and binaural response type

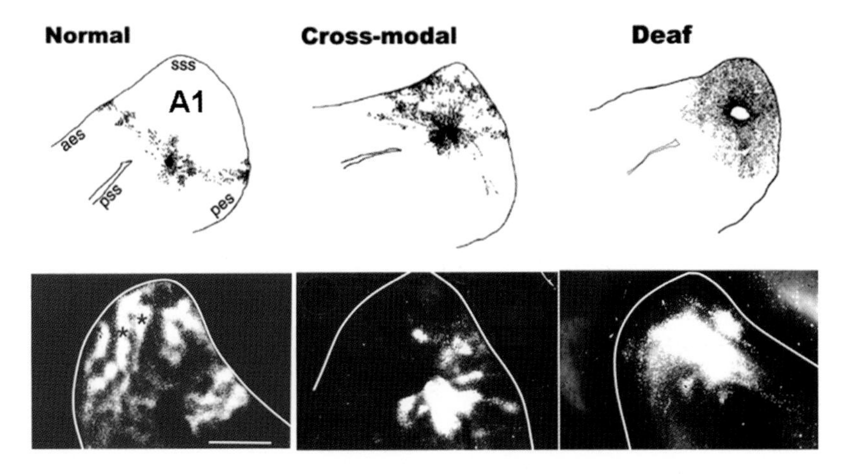

Figure 8 Intracortical horizontal connectivity and callosal connectivity of A1 in normal, deafened, and cross-modal animals. In normal ferrets (left), horizontal connections are revealed by restricted biotinylated dextran amine (BDA) injections of the left A1 to occupy isofrequency laminae. Callosal connections, as demonstrated by injection of HRP throughout the left A1, delineate the E–E binaural bands on the right side. In cross-modal cases (center panels), horizontal and callosal projections are more similar to those seen in visual cortex, and are shifted in location. This different pattern does not result from stabilization of an early, exuberant pattern, as shown by the more diffuse and extensive label produced by a loss of sensory input (right panels) (adapted from Gao and Pallas, 1999; Pallas et al, 1999, 2002)

are linked (Imig et al., 1982). The obvious question for cross-modal A1 was whether its visual inputs led to alterations in modular organization that could explain its visual response properties. We demonstrated that the intracortical connectivity of cross-modal A1 is altered from a slab-like, isofrequency band organization to resemble the clusters of terminals seen in visual cortex (Gao and Pallas, 1999) (FIG 8, top). Callosal projections are also modified (Pallas et al., 1999) (FIG 8, bottom). The callosal connectivity between cross-modal A1 and the contralateral, unmanipulated A1 are shifted in position, suggesting that A1 itself has been shifted. This suggestion is further supported by the observation that intracortical connections within cross-modal A1, which apparently support visual stimulus orientation tuning (Sharma et al., 2000), are arranged in an inverse pattern to the callosal projections. We would like to know whether the manipulated hemisphere has both 1) an auditory A1 that is laterally shifted and connected callosally with the contralateral, normal A1, and 2) a "visualized" A1, containing local projections that unite neurons with similar visual orientation tuning. If this is the case, the cross-modal rewiring manipulation creates a new cortical area, and thus a convenient model system for studying cortical evolution in a developmental framework. These results are consistent with extrinsic control of many important aspects of cortical identity in a functional sense. Our studies of cross-modal plasticity thus provide a concrete demonstration that changing only the modality and pattern of activity in TCAs can profoundly change cortical circuitry.

Does Cross-Modal Auditory Cortex have an Altered Genetic Identity?

The results of our gene expression studies in normal ferrets, combined with our results from the cross-modal animals, raised an interesting question. Does the expression of patterning genes influence the development of corticocortical connectivity patterns? If so, then we would expect the changes in connectivity that we saw in the experimental animals to be correlated with changes in expression. *Cadherins* are candidate molecules for this function (Morishita et al., 2004). There is evidence that they are involved in the establishment of normal corticocortical connectivity patterns. *Cadherin* expression is unaffected in *Mash-1* mutants lacking TCA projections (Nakagawa et al., 1999), indicating that the presence of TCAs is not required for expression. *Gbx2* mutants have no TCAs but have normal corticocortical connectivity (Huffman et al., 2004), indicating that TCAs are not required to establish normal intra- and intercortical projections. To test the hypothesis that rearrangements of cadherin expression underlie changes in corticortical connectivity patterns in cross-modal ferrets, we compared expression of *Cad6* and *Cad8* mRNA and protein in visual (V1) and auditory (A1) cortices during postnatal development in normal and cross-modal ferrets. As noted above, *Cad6* expression in normal animals was much greater in A1 than in V1, whereas *Cad8* expression was significantly higher in V1 than in A1. In contrast, in cross-modal ferrets expression of *Cad6* and *Cad8* was similar in A1 and V1 (Xu and Pallas, Soc Neurosci. Abstr. 2005), indicating a loss of molecular distinctions between these two cortical areas that correlates with A1's acquisition of visual processing circuitry. These data suggest that *Cad6* and *Cad8* may be involved in cross-modal re-specification through alterations in cortical features such as corticortical connectivity patterns.

Further Questions

The experimental approaches discussed here are providing important information about the relative roles of intrinsic genetic information and extrinsic neural activity in thalamocortical axon guidance and the areal specification of cortex. However much remains to be discovered. For example, given the broad expression patterns of genes known to be involved in cortical patterning, how are the sharp boundaries between areas established? Should we be looking for the putative factors that are restricted to cortical areas? Can functional identity be established independent of early genetic identity, as our cross-modal plasticity experiments suggest? If gene expression gradients define cortical areal identity, how is it possible to get entirely new cortical areas in evolution? The mechanisms for areal boundary formation must work in all species, for different degrees of subdivision of cortical areas representing particular sensory modalities. Does this require coordinated evolutionary changes in TCAs and cortical patterning, or can changes in TCA projection patterns drive formation of new boundaries? We now have the tools to answer these and other fascinating questions about the development of the most interesting part of the brain, the cerebral cortex.

Acknowledgments

Support contributed by: NIH EY-12696, NSF IBN-0078110, and the GSU Research Foundation to SLP, and the National Organization for Hearing Research to KAR.

References

Angelucci, A., Clascá, F., Bricolo, E., Cramer, K.S., and Sur, M. (1997) Experimentally induced retinal projections to the ferret auditory thalamus: Development of clustered eye-specific patterns in a novel target. *J. Neurosci.* 17:2040–2055.

Bavelier, D., and Neville, H.J. (2002) Cross-modal plasticity: where and how? *Nature Reviews Neurosci.* 3:443–452.

Bishop, K.M., Goudreau, G., and O'Leary, D.D.M. (2000) Regulation of area identity in the mammalian neocortex by *Emx2* and *Pax6*. *Science* 288:344–349.

Bronchti, G., Heil, P., Sadka, R., Hess, A., Scheich, H., and Wollberg, Z. (2002) Auditory activation of "visual" cortical areas in the blind mole rat (*Spalax ehrenbergi*). *J. Comp. Neurol.* 16:311–329.

Caric, D., Gooday, D., and Price, D.J. (1997) Determination of the migratory capacity of embryonic cortical cells lacking the transcription factor Pax-6. *Development* 124:5087.

Catalano, S.M., and Shatz, C.J. (1998) Activity-dependent cortical target selection by thalamic axons. *Science* 281:599–562.

Cecchi, C. (2002) *Emx2*: a gene responsible for cortical development, regionalization and area specification. *Gene* 291:1–9.

Chiu, C., and Weliky, M, (2001) Spontaneous activity in developing ferret visual cortex in vivo. *J. Neurosci.* 21:8906–8914.

Crandall, J.E., and Caviness, V.S. (1984) Thalamocortical connections in newborn mice. *J. Comp. Neurol.* 228:542–556.

Croquelois, A., Bronchti, G., and Welker, E. (2005) Cortical origin of functional recovery in the somatosensory cortex of the adult mouse after thalamic lesion. *Eur. J. Neurosci.* 21:1798–1806.

Cucchiaro, J., and Guillery, R.W. (1984) The development of the retinogeniculate pathways in normal and albino ferrets. *Proc. Roy. Soc. Lond.* 223:141–164.

Dufour, A., Seibt, J., Passante, L., Depaepe, V., Ciossek, T., Frisén, J., Kullander, K., Flanagan, J.G., Polleux, F., and Vanderhaeghen, P. (2003) Area specificity and topography of thalamocortical projections are controlled by *ephrin/Eph* Genes. *Neuron* 39:453–465.

Estivill-Torrus, G., Pearson, H., van Heyningen, V., Price, D.J., and Rashbass, P. (2002) *Pax6* is required to regulate the cell cycle and the rate of progression from symmetrical to asymmetrical division in mammalian cortical progenitors. *Development* 129:455–466.

Felleman, D.J., and Van Essen DC (1991) Distributed hierarchical processing in the primate cerebral cortex. *Cereb. Cortex* 1:1–47.

Finney, E.M., Fine, I., and Dobkins, K.R. (2001) Visual stimuli activate auditory cortex in the deaf. *Nat. Neurosci.* 4:1171–1173.

Gao, W.-J., and Pallas, S.L. (1999) Cross–modal reorganization of horizontal connectivity in auditory cortex without altering thalamocortical projections. *J. Neurosci.* 19:7940–7950.

Garel, S., and Rubenstein, J.L.R. (2004) Intermediate targets in formation of topographic projections: inputs from the thalamocortical system. *Trends Neurosci.* 27:533–539.

Garel, S., Huffman, K.J., and Rubenstein, J.L.R. (2003) Molecular regionalization of the neocortex is disrupted in *Fgf8* hypomorphic mutants. *Development* 130:1903–1914.

Ghosh, A., Antonini, A., McConnell, SK., and Shatz, C.J. (1990) Requirement for subplate neurons in the formation of thalamocortical connections. *Nature* 347:179–181.

Gil, O.D., Needleman, L., and Huntley, G.W. (2002) Developmental patterns of cadherin expression and localization in relation to compartmentalized thalamocortical terminations in rat barrel cortex. *J. Comp. Neurol.* 453:372–388.

Hamasaki, T., Leingartner, A., Ringstedt, T., and O'Leary, D.D.M. (2004) *Emx2* regulates size and positioning of the primary sensory and motor areas in neocortex by direct specification of cortical progenitors. *Neuron* 43:359–372.

Heil, P., Bronchti, G., Wollberg, Z., and Scheich, H. (1991) Invasion of visual cortex by the auditory system in the naturally blind mole rat. *Neuroreport* 2: 735–738.

Heins, N., Cremisi, F., Malatesta, P., Gangemi, R.M.R., Corte, G., Price, J., Goudreau, G., Gruss, P., and Götz, M. (2001) Emx2 promotes symmetric cell divisions and a multipotential fate in precursors from the cerebral cortex. *Molec. Cell. Neurosci.* 18:485–502.

Herrmann, K., Antonini, A., and Shatz, C.J. (1994) Ultrastructural evidence for synaptic interactions between thalamocortical axons and subplate neurons. *Eur. J. Neurosci.* 6:1729–1742.

Hevner, R.F., Miyashita-Lin, E., and Rubenstein, J.L.R. (2002) Cortical and thalamic axon pathfinding defects in *Tbr1*, *Gbx2*, and *Pax6* mutant mice: evidence that cortical and thalamic axons interact and guide each other. *J. Comp. Neurol.* 447:8–17.

Huffman, K.J., Garel, S., and Rubenstein, J.L.R. (2004) *Fgf8* regulates the development of intra-neocortical projections. *J. Neurosci.* 24:8917–8923.

Imig, T.J., Reale, R.A., and Brugge, J.F. (1982) The auditory cortex: Patterns of corticocortical projections related to physiological maps in the cat. In: Woolsey, C.N. (ed) *Cortical Sensory Organization, Volume 3, Multiple Auditory Areas.* Humana Press, Clifton, New Jersey, pp 1–41.

Innocenti, G.M., Fiore, L., and Caminiti, R. (1977) Exuberant projection into the corpus callosum from the visual cortex of newborn cats. *Neurosci. Lett.* 4:237–242.

Inoue, T., Tanaka, T., Suzuki, S.C., and Takeichi, M. (1998) Cadherin-6 in the developing mouse brain: expression along restricted connection systems and synaptic localization suggest a potential role in neuronal circuitry. *Dev. Dyn.* 211:338–351.

Issa, N.P., Trachtenberg, J.T., Chapman, B., Zahs, K.R., and Stryker, M.P. (1999) The critical period for ocular dominance plasticity in the ferret's visual cortex. *J. Neurosci.* 19:6965–6978.

Jackson, C.A., Peduzzi J.D., and Hickey, T.L. (1989) Visual cortex development in the ferret. I. Genesis and migration of visual cortical neurons. *J. Neurosci.* 9:1242–1253.

Jones, L., López-Bendito, G., Gruss, P., Stoykova, A., and Molnár, Z. (2002) *Pax6* is required for the normal development of the forebrain axonal connections. *Development* 129:5041–5052.

Kaas, J.H. (1993) Evolution of multiple areas and modules within neocortex. *Perspect. Dev. Neurobiol.* 1:101–107.

Katz, L.C., and Callaway, E.M. (1992) Development of local circuits in mammalian visual cortex. *Ann. Rev. Neurosci.* 15:31–56.

Kelly, J.B., Judge, P.W., and Phillips, D.P. (1986) Representation of the cochlea in primary auditory cortex of the ferret (Mustela putorius). *Hearing Res.* 24:111–115.

Keynes, R., and Krumlauf, R. (1994) *Hox* genes and regionalization of the nervous system. *Ann. Rev. Neurosci.* 17:109–132.

Law, M.I., Zahs, K.R., and Stryker, M.P. (1988) Organization of primary visual cortex (Area 17) in the ferret. *J. Comp. Neurol.* 278:157–180.

Leingartner, A., Richards, L.J., Dyck, R.H., Akazawa, C., and O'Leary, D.D.M. (2003) Cloning and cortical expression of rat *Emx2* and adenovirus-mediated overexpression to assess its regulation of area-specific targeting of thalamocortical axons. *Cereb. Cortex* 13:648–660.

López-Bendito, G., and Molnár, Z. (2003) Thalamocortical development: how are we going to get there? *Nature Rev. Neurosci.* 4:276–289.

Mallamaci, A., Muzio, L., Chan, C.-H., Parnavelas, J., and Boncinelli, E. (2000) Area identity shifts in the early cerebral cortex of *Emx2−/−* mutant mice. *Nature Neurosci.* 3:679–686.

Miller, B.M., Chou, L., and Finlay, B.L. (1993) The early development of thalamocortical and corticothalamic projections. *J. Comp. Neurol.* 335:16–41.

Miller, B.M., Windrem, M.S., and Finlay, B.L. (1991) Thalamic ablations and neocortical development: Alterations in thalamic and callosal connectivity. *Cereb. Cortex* 1:241–261.

Moore, D.R., and Hine, J.E. (1992) Rapid development of the auditory brainstem response threshold in individual ferrets. *Devel. Brain Res.* 66:229–235.

Morishita, H., Murata, Y., Esumi, S., Hamada, S., and Yagi, T. (2004) CNR/Pcdhalpha family in subplate neurons, and developing cortical connectivity. *Neuroreport* 15:2595–2599.

Nakagawa, Y., Johnson, J.E., and O'Leary, D.D.M. (1999) Graded and areal expression patterns of regulatory genes and cadherins in embryonic neocortex independent of thalamocortical input. *J. Neurosci.* 19:10877–10885.

O'Leary, D.D.M., and Stanfield, B.B. (1986) A transient pyramidal tract projection from the visual cortex in the hamster and its removal by selective collateral elimination. *Devel. Brain Res.* 27:87–99.

Olavarria, J., and Van Sluyters, R. (1985) Organization and postnatal development of callosal connections in the visual cortex of the rat. *J. Comp. Neurol.* 239:1–26.

Pallas, S.L. (2001) Intrinsic and extrinsic factors shaping cortical identity. *Trends Neurosci.* 24:417–423.

Pallas, S.L. (2002) Cross-modal plasticity as a tool for understanding ontogeny and phylogeny of cerebral cortex. In: Shuetz, A., and Miller, R., (eds): *Cortical Areas: Unity and Diversity*. Conceptual Advances in Brain Research 5: 245–272. Taylor and Francis, London.

Pallas, S.L. (2005) Pre- and postnatal sensory experience shapes brain connectivity. In: B. Hopkins and S.P. Johnson (eds.): *Advances in Infancy Research*. Praeger, Westport, CT, pp. 1–30.

Pallas, S.L. (in press) Compensatory innervation in development and evolution. In: G.F. Striedter and J.L.R. Rubenstein (eds.): *Evolution of the Nervous System*, Vol. 1: History of Ideas, Basic Concepts and Developmental Mechanisms.

Pallas, S.L., and Sur, M. (1993) Visual projections induced into the auditory pathway of ferrets. II. Corticocortical connections of primary auditory cortex with visual input. *J. Comp. Neurol.* 337:317–333.

Pallas, S.L., and Sur, M. (1994) Morphology of retinal axon arbors induced to arborize in a novel target, the medial geniculate nucleus. II. Comparison with axons from the inferior colliculus. *J. Comp. Neurol.* 349:363–376.

Pallas, S.L., Roe, A.W., and Sur, M. (1990) Visual projections induced into the auditory pathway of ferrets. I. Novel inputs to primary auditory cortex (AI) from the LP/Pulvinar complex and the topography of the MGN-AI projection. *J. Comp. Neurol.* 298:50–68.

Pallas, S.L., Hahm, J., and Sur, M. (1994) Morphology of retinal axons induced to arborize in a novel target, the medial geniculate nucleus. I. Comparison with arbors in normal targets. *J. Comp. Neurol.* 349:343–362.

Pallas, S.L., Littman, T., and Moore, D.R. (1999) Cross-modal reorganization of callosal connectivity in auditory cortex without altering thalamocortical projections. *Proc. Natl. Acad. Sci. USA* 96:8751–8756.

Pallas S.L., Razak, K.A., and Moore, D.R. (2002) Cross-modal projections from LGN to primary auditory cortex following perinatal cochlear ablation in ferrets. *Soc. Neurosci. Abstr.* 28:220.8.

Phillips, D.P., Judge, P.W., and Kelly, J.B. (1988) Primary auditory cortex in the ferret (Mustela putorius): neural response properties and topographic organization. *Brain Res.* 443:281–294.

Poskanzer, K., Needleman, L.A., Bozdagi, O., and Huntley, G.W. (2003) N-cadherin regulates ingrowth and laminar targeting of thalamocortical axons. *J. Neurosci.* 23:2294–2305.

Rauschecker, J.P. (1995) Compensatory plasticity and sensory substitution in the cerebral cortex. *Trends Neurosci.* 18:36–43.

Redies, C., Treubert-Zimmermann, U., and Luo, J. (2003) Cadherins as regulators for the emergence of neural nets from embryonic divisions. *J. Physiol. Paris* 97:5–15(11).

Roe, A.W., Pallas, S.L., Hahm, J., and Sur, M. (1990) A map of visual space induced in primary auditory cortex. *Science* 250:818–820.

Roe, A.W., Pallas, S.L., Kwon, Y., and Sur, M. (1992) Visual projections routed to the auditory pathway in ferrets: Receptive fields of visual neurons in primary auditory cortex. *J. Neurosci.* 12:3651–3664.

Roe, A.W., Garraghty, P.E., Esguerra, M., and Sur, M. (1993) Experimentally induced visual projections to the auditory thalamus in ferrets: Evidence for a W cell pathway. *J. Comp. Neurol.* 334:263–280.

Rosa, M.G.P., and Krubitzer, L.A. (1999) The evolution of visual cortex: where is V2? *Trends Neurosci.* 22:242.

Sadato, N., Pascual-Leone, A., Grafman, J., Ibañez, V., Deiber, M.P., Dold, G., and Hallett, M. (1996) Activation of the primary visual cortex by Braille reading in blind subjects. *Nature* 380:526–528.

Scardigli, R., Baumer, N., Gruss, P., Guillemot, F., and LeRoux, I., (2003) Direct and concentration-dependent regulation of the proneural gene *Neurogenin2* by *Pax6*. *Development* 130:3269–3281.

Schmahl, W., Knoedlseder, M., Favor, J., and Davidson, D. (1993) Defects of neuronal migration and the pathogenesis of cortical malformations are associated with Small eye (*Sey*) in the mouse, a point mutation at the *Pax-6*-locus. *Acta Neuropathol. (Berl.)* 86:126–135.

Sharma, J., Angelucci, A., and Sur, M. (2000) Induction of visual orientation modules in auditory cortex. *Nature* 404:841–847.

Shatz, C.J., and Luskin, M.B. (1986) The relationship between the geniculocortical afferents and their cortical target cells during development of the cat's primary visual cortex. *J. Neurosci.* 6:3655–3668.

Stoeckli, E.T., and Landmesser, L.T. (1998) Axon guidance at choice points. *Curr. Opin. Neurobiol.* 8:73.

Sur, M., Garraghty, P.E., and Roe, A.W. (1988) Experimentally induced visual projections into auditory thalamus and cortex. *Science* 242:1437–1441.

Suzuki, S.C., Inoue, T., Kimura, Y., Tanaka, T., and Takeichi, M. (1997) Neuronal circuits are subdivided by differential expression of Type-II classic cadherins in postnatal mouse brains. *Molec. Cell Neurosci.* 9:433–447.

Weliky, M., and Katz, L.C. (1999) Correlational structure of spontaneous neuronal activity in the developing lateral geniculate nucleus in vivo. *Science* 285:599–604.

Xu, M., Baro, D.J., and Pallas, S.L. (2003) A quantitative study of gene expression topography in visual and auditory cortex during thalamocortical development in postnatal ferrets. *Soc. Neurosci. Abstr*. 29: 673.7

Xu, M., and Pallas, S.L. (2005) Molecular patterning is altered in cross-modal plasticity. *Soc. Neurosci. Abstr.* 31: 140.19.

9

Pathways to Barrel Development

Mark W. Barnett*, Ruth F. Watson*, and Peter C. Kind

Abstract

Understanding the cellular mechanism by which glutamate receptors mediate changes in neuronal phenotype is key to understanding activity-dependent development of the nervous system. The primary somatosensory cortex (S1) of rodents offers a unique opportunity to identify key molecules that regulate glutamate-dependent cortical development because of its unique cytoarchitectonic structures in layer 4 termed "barrels". Analysis of knockout mice has revealed that both NMDA receptors, and metabotropic glutamate receptor 5 (mGluR5) activation of phospholipase C-β1 (PLC-β1), are necessary for normal barrel development (Erzurumlu and Kind, 2001). Over the last several years, we have been using analysis of barrel cortex development in knockout (KO) mice to identify the signalling pathways downstream of glutamate receptors that regulate cortical development. This approach has been greatly helped by the isolation and characterisation of proteins associated with the postsynaptic density (PSD; Walikonis et al., 2000; Husi et al., 2000). To date we have analysed more than 35 mice with selective deletion of key PSD components. Two of these mutants, those lacking *Syngap* (Barnett et al., 2006) and those lacking the RIIβ subunit of PKA (PKARIIβ$^{-/-}$, Watson et al., in press), also showed defects in barrel development. This chapter reviews the principle cellular processes involved in barrel development. It also reviews the current state of knowledge of the intracellular signalling pathways, initiated by glutamate neurotransmission, that regulate barrel development, with specific focus on SynGAP and mGluR5 activation of PLC-β1. Finally, we examine how the analysis of mutant mice has increased our knowledge about the cellular processes that underlie barrel development.

The role of activity in cortical development and plasticity has been the focus of intense research ever since Hubel and Wiesel demonstrated that the physiological and anatomical development of visual system

*These authors contributed equally to the writing of this chapter.

was dependent on the nature of the animal's early visual experience (Wiesel and Hubel, 1963a,b). The discovery of the N-methyl-D-aspartate receptor (NMDAR) and its control over visual plasticity during development greatly increased our understanding of the molecular basis of activity-dependent cortical plasticity (Kleinschmidt et al., 1987). Since that time, much has been learned of the NMDA receptor-dependent pathways and cellular mechanisms that control Long Term Potentiation (LTP) and Long Term Depression (LTD) in the hippocampus. However, far less is known of the cellular mechanisms by which NMDARs regulate neocortical development and plasticity. Using pharmacological blockade and more recently, transgenic animals, several NMDAR-dependent pathways have been shown to be involved in cortical development and plasticity including cyclic AMP/Protein kinase A (PKA; Abdel-Majid et al., 1998, Beaver et al., 2001; Lu et al., 2003; Fischer et al., 2004; Watson et al., in press; Kind and Neumann, 2001), Calcium/calmodulin-dependent kinase II (CaMKII) (Taha et al., 2002; Glasewski et al., 1996), Extracellular signal-regulated kinases (ERKs) (Di Cristo et al., 2001), Calcineurin (Yang et al., 2005) and phospholipase C-β1 (Hannan et al., 2001). All of these pathways have been shown to play crucial roles in various forms of LTP and LTD in hippocampus, suggesting a conservation of plasticity pathways in different brain structures. This chapter focuses on several PSD component mutants that we have identified as playing a crucial role in the differentiation of the primary somatosensory cortex.

In addition to their role in synaptic plasticity, NMDA receptors have been shown to regulate early developmental events that are likely to be distinct from synaptic plasticity. For example NMDARs regulate axon dynamics and dendritogenesis and hence map formation in a range of systems, including neocortex (Dickson and Kind, 2003). However, while NMDARs are known to play a crucial role in many of these early developmental events (see Erzurumlu and Iwasato, 2005, this volume), very little is known of the intracellular pathways through which they mediate their effects. One approach to identify the intracellular pathways by which NMDARs regulate development and plasticity is to screen mutants of NMDAR associated proteins for defects in cortical development. Such an approach has been greatly aided by the recent large-scale isolation and proteomic characterisation of the NMDAR complex (NRC) and postsynaptic density (PSD; Husi et al, 2000; Walikonis et al, 2000). These studies identified proteins that associate with NMDARs in adult forebrain and have provided a large list of candidate molecules that could be acting downstream of NMDAR signalling during cortical development. We adopted such an approach taking advantage of a large number of transgenic animals that have been generated previously, primarily for examining their role in hippocampal plasticity and learning and memory. We have focussed on the development of the somatosensory cortex because of the clear anatomical structures known as "barrels".

Trigeminal Pathway Organisation and Development

Brainstem and Thalamus

Each whisker follicle is innervated by bundles of axons from trigeminal ganglion cells (TGCs) that form individual branches of the infraorbital nerve. The TGCs project centrally to the caudal brainstem where they extend collaterals to four nuclei in the brainstem trigeminal complex (Hayashi et al., 1980). Both the incoming axons and the cell bodies in three of the four nuclei (principal nucleus (nVp), subnucleus interpolaris (nVi) and subnucleus caudalis (nVc) accurately recapitulate the ipsilateral whisker pattern (Bates and Killackey, 1985; Ma and Woolsey, 1983, 1984). These cytoarchitectonic arrangements are termed "barrelettes". Barrelettes can be first seen in the mouse on the day of birth (P0) using cytochrome oxidase histochemistry (Ma, 1993) and are well-segregated by P1. The cellular aggregation, as visualized by Nissl staining, begins on P1, however, complete segregation into five rows is not seen until P4 (Ma, 1993). Neurons in the nVp and nVi nuclei and to a lesser extent in nVo (subnucleus oralis) and nVc project axons across the midline to the ventral posterior nucleus (VpM) of the thalamus. Brainstem axon terminals and cell bodies within the contralateral thalamus replicate the whisker pattern in cytoarchitectonic arrangements termed "barreloids" (Van der Loos, 1976). Barreloids begin to form at P3 as visualized by cytochrome oxidase histochemistry (Yamakodo, 1985). Interestingly, during development the dendrites of VpM neurons project selectively towards the incoming brainstem axons to form a dense plexus of synapses in the centre of each barreloid (Brown et al., 1995; Zantua et al., 1996). This selective aggregation of dendrites is transient however and distal dendrites extend outside individual barreloids. By P18 the dendrites have grown to span several barrels a state that is maintained into adulthood (Brown et al., 1995; Zantua et al., 1996).

TCA Segregation and Barrel Formation

As is the case in brainstem and thalamus the primary somatosensory cortex (S1) of mammals is organized into a topographic map that reproduces the pattern of peripheral sensory receptors. Thalamocortical axons from VpM project into the barrel hollows (Killackey, 1973) and to a lesser extent to the interbarrel region or septa (Pierret et al., 2000). At P1, TCAs are uniformly distributed within layer 4 with each axon occupying 1 to 2 prospective barrel diameters (Rebsam et al., 2002: Agmon et al., 1995). At P7 TCAs arborize exclusively in a single barrel with few, if any, extending between barrels. Some fibres divide into two or three branches in layer 6a that converge on the same barrel in layer 4. The segregation of TCAs into whisker-related patches occurs between P2 and P5 and involves the retraction of inappropriately located arbors and elaboration of appropriately placed arbors (Rebsam et al., 2002). The developmental mechanisms mediating TCA patch formation, therefore, appear to be different from those that underlie eye specific segregation into ocular dominance bands in higher mammals since the latter process appears to involve specific targeting of TCAs rather than axon retraction (Katz and Crowley, 2002).

A unique feature of rodent somatosensory cortex that makes it particularly useful for identifying molecules involved in cortical development is that the layer 4 neurons also aggregate into functional units referred to as "barrels" (Woolsey and Van der Loos, 1970). Barrels are a highly stereotypical array of cellular aggregates that recapitulate the whisker array on the facepad. They consist of cell dense walls and cell sparse hollows; septa are the relatively acellular areas that separate individual barrels. The three dimensional structure of a barrel is that of a bowed cylinder and this shape, along with Hendrik Van der Loos' fascination with 17th century dutch art, led to the choice of "barrel" being used to illustrate the anatomy (the barrel used is depicted in Pieter Bruegel's *Fair of St. George's Day*; Woolsey, 1996). In the posteromedial barrel subfield (PMBSF), the region representing the large mystacial vibrissae, each barrel receives input from a single whisker.

The cell bodies of layer 4 neurons begin to aggregate to form cell-dense barrel walls subsequent to TCA segregation (Rice and Van der Loos, 1977). Evidence of this cellular segregation can first be seen on P4 and a full barrel field can be seen by P6, although barrel segregation is most distinct between P10–P14. The septa are clearly visible by P7. Another key feature of the layer 4 neurons is the selective orientation of their dendrites into the barrel hollows (Woolsey et al., 1975). The precise timing of this dendritic orientation is not clear, however, it appears to result from a selective pruning of inappropriate branches combined with an elaboration of appropriate branches (Greenough and Chang, 1988) and unlike in VpM, the dendrites remain selectively oriented in adults (Woolsey et al., 1975).

Though the cellular mechanisms responsible for barrel wall formation are not known, several possibilities exist (Figure 1). First there may be a selective cell death in the region of TCA innervation although no preferential distribution of apoptotic nuclei has been seen in developing S1 (Miller, 1995). Second, the massive elaboration of neuropil, both axons and dendrites, could cause the cortical neurons to be passively displaced to form a cell-dense barrel wall; analysis of mutant mice, however, indicates that this possibility is unlikely (see below, Hannan et al., 2001). Third, neurons of layer 4 may actively migrate away from the incoming thalamocortical afferents. Fourth, differential cell adhesion regulated by signals from TCAs combined with cortical growth may lead to a barrel pattern. Whether barrel development is controlled by one or some combination of these processes remains to be determined.

Synaptic Development in Layer 4 of S1

In order to understand the role of glutamate neurotransmission in barrel development, it is essential to briefly review the anatomical

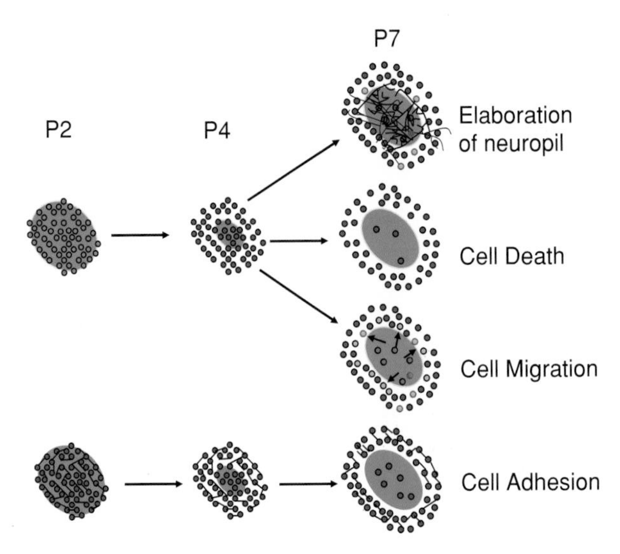

Figure 1 Schematic showing the various cellular mechanisms that could control layer 4 cell segregation. *Elaboration of Neuropil:* During barrel formation, TCAs segregate and elaborate terminal arbors to form whisker-related patches. Subsequent to TCA segregation, the dendrites of layer 4 neurons selectively elaborate branches within the TCA patches to form a dense plexus of synapses. This large increase in neuropil could passively displace the cell bodies of layer 4 neurons to form a cell-dense barrel wall. *Plc-β1*$^{-/-}$ mice develop TCA patches of normal size and distribution and layer 4 neuronal dendrites with normal complexity and orientation, but do not form barrels. Hence elaboration of neuropil is an unlikely mechanism for forming barrels. *Cell death* could also account for barrel formation if there were selective death of neurons within TCA patches. However Miller (1995) found no preferential distribution of dying cells within layer 4 (although differential distribution between emerging barrel walls and hollows were not examined). Also the decrease in cellular segregation in *Syngap*$^{+/-}$ cannot be explained by changes in cell death since the overall density of layer 4 neurons was unaltered. *Cell Migration:* It is possible that layer 4 neuronal soma migrate away from the TCA terminals. There is little evidence to speak to this possibility although glutamate can initiate cell migration. *Differential cell adhesion combined with cortical growth* could create a cell dense barrel wall. Glutamate receptor activation could cause a local decrease in cell adhesion molecules, such that only soma at the edge of TCA patches remained tightly bound. The increase in the tangential size of the cortex between P3 and P7 could result in the emergence of tightly bound, high-density regions around TCA patches (i.e. barrel walls), and low-density, loosely bound regions in the TCA patches (i.e. barrel hollows). We are currently addressing this possibility using computer models of patch and barrel formation.

and functional development of cortical, and more specifically, layer 4 synapses. In P4 rodent S1 cortex, the density of total synapses is about 15–25% that of the adult and postsynaptic densities can be clearly identified under the electron microscope. (Micheva and Beaulieu, 1996; Spires et al, 2005). In the mouse, the presence of asymmetrical (putatively excitatory) and symmetrical (putatively inhibitory) synapses has been demonstrated at P4 in all layers of the posteromedial barrel subfield

(De Felipe et al., 1997). In adult animals, 85% of synapses in the rodent barrel cortex are glutamatergic (Micheva and Beaulieu, 1995). From P4 to P8 there are 57% excitatory and 43% inhibitory synapses (De Felipe et al., 1997). Asymmetrical (excitatory) synapse density increases rapidly from P6 to P8 slowly from P9 to P12 and sharply between P13 and P14 along with the onset of patterned whisking (White et al., 1997). Twenty to twenty five percent of synapses in layer 4 come from the VpM and make asymmetrical contact on the stellate cells, and on all other neurons with processes in layers 4 and 6 (White, 1979). Importantly, even at early postnatal ages (P3) during TCA segregation, functional TCA synapses can be detected using electrophysiological methods (Lu et al., 2003). NMDAR-dependent synaptic plasticity (LTP/LTD) can only be induced during a narrow time window between P3 and P7 (Crair and Malenka, 1995; Lu et al., 2003). Similarly the ratio of NR2B to NR2A-containing NMDA receptors decreases over a similar, but not identical, time-course (Lu et al., 2001).

To identify the intracellular pathways by which these glutamate receptors initiate barrel formation and layer 4 synaptic plasticity, we have used biochemical approaches on isolated PSD fractions from barrel cortex to determine the protein constituents of the NRC during barrel formation. In agreement with the presence of clear PSDs and functional synapses in P4 mouse somatosensory cortex, we have shown that the main components of the NMDA receptor complex are present at these ages and can be isolated from biochemical postsynaptic density preparations. To address the role of PSD proteins in barrel development, we have examined barrel formation in transgenic mice with deletions of NRC components. The rest of the chapter reviews the data from these experiments and compares the results with proteins found to be involved in visual cortical development and plasticity.

The Cellular and Molecular Mechanisms of Barrel Formation

The development of orderly maps in sensory systems is a complex interplay between guidance molecules and activity (both spontaneous and sensory-evoked). In the rodent somatosensory cortex, Ephrins appear to play a key role in the tangential organisation of the whisker map (Vanderhaeghen et al., 2000), serotonin modulates TCA complexity and patch formation (Gaspar et al., 2003) and glutamate neurotransmission regulates TCA complexity and cellular aggregation in layer 4 (Erzurumlu and Kind, 2001, Datwani et al., 2002, Lee et al., 2005). The first demonstration that glutamate neurotransmission was necessary for segregation in the trigeminal system came from analysis of *Nr1*$^{-/-}$ (the essential subunit of NMDARs) mice that failed to develop barrelettes (Li et al. 1994). Iwasato et al. (2000) then generated conditional transgenic mice in which deletion of the *Nr1* gene was restricted to excitatory cortical neurones (Cx*nr1*$^{-/-}$ mice). These Cx*nr1*$^{-/-}$ mice developed normal barrelettes and barreloids but failed to form cellular aggregates in layer 4 even though TCAs segregated, albeit with decreased TCA patch size in the posteromedial barrel subfield (PMBSF). Hannan et al., (2001) later reported that mice with a deletion of metabotropic glutamate receptor 5 (*Mglur5*$^{-/-}$) also fail to form whisker-related cellular aggregates, despite partial TCA segregation into rows in the PMBSF. Analysis of both the Cx*nr1*$^{-/-}$ and *Mglur5*$^{-/-}$ mutants has convincingly demonstrated that postsynaptic neurotransmitter receptor activation is vital in communicating the peripherally related sensory patterns from TCAs to barrel cells (reviewed by Erzurumlu & Kind 2001; Kind & Neumann 2001).

These findings raise several questions that our laboratory has been addressing over the last 5 years. What are the signalling pathways downstream of these glutamate receptors that regulate cortical development? Do NMDARs and mGluR5 modulate the same intracellular pathways to regulate barrel development? And what are the cellular events underlying barrel formation? To elucidate the glutamate receptor dependent intracellular pathways underlying barrel differentiation, we have been examining mice with genetic deletion of genes encoding PSD proteins. To date we have completed a preliminary analysis of the barrel cortex phenotype (cellular aggregation and TCA segregation in layer 4) of over 30 mutants. From these mutants, animals with deletions in the genes encoding phospholipase C-beta1 (PLC-β1), synaptic ras GTPase Activating Protein (SynGAP) and Protein Kinase A type 2 regulatory beta subunit (PKAR2β) have indicated that each of these molecules play a role in barrel development. This review will now focus on the aberrant barrel phenotypes of the *Plc-β1*$^{-/-}$ and *Syngap*$^{-/-}$ mice and will examine what is known about their downstream targets.

mGluR5/PLC-β1 Pathway

PLC-β1 is one of four PLC-β family members that are part of a larger family of phosphoinositide (PI)-specific PLCs. The β subfamily are G-protein-coupled and, of the PLC-βs, β1 is the most highly expressed in the neocortex. The PI-specific PLCs hydrolyse phosphatidylinositol 4,5-bisphosphate (PIP2) into two second messengers, diacylglycerol (DAG) and 1,4,5-inositol triphosphate (IP3). Subsequently, DAG activates Protein Kinase C (PKC) and IP3 activates the IP3 receptor (IP3R) to release intracellular Ca^{2+} from the endoplasmic reticulum (ER; Figure 2A).

A role for G-protein coupled phospholipases in developing cortical neurons was first suggested when Dudek and Bear (1989) found that mGluR-mediated PI hydrolysis paralleled the sensitive period in visual cortex. PLC-β1 was subsequently isolated from a screen designed to identify molecules selectively expressed in the developing cat visual cortex (Kind et al., 1994, 1997) raising the possibility that PLC-β1 was a primary target of phosphoinositide-coupled (i.e. group 1) mGluR

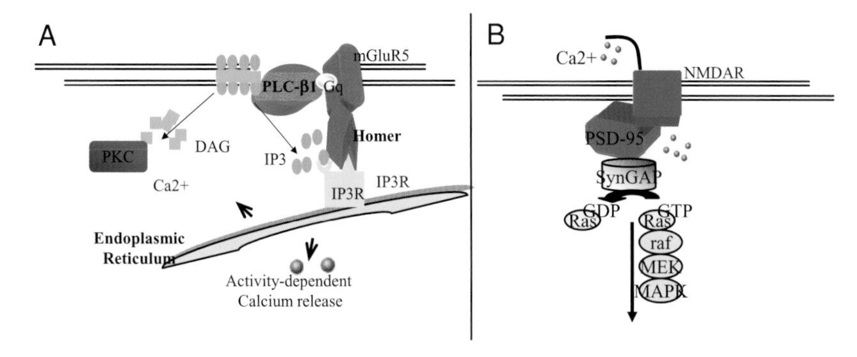

Figure 2 Schematic of the 3 major signalling pathways known to affect barrel development. a) mGluR5 activation of PLC-β1 is a key pathway to cortical cell segregation but not TCA segregation. b) SynGAP activation, likely resulting from NMDAR stimulation regulates both barreloid and barrel formation.

signalling during cat visual cortical development. Interestingly Kind et al (1997) found that PLC-β1 is highly expressed in intermediate-compartment-like organelles called botrysomes, located selectively near the roots of, and within, dendrites. These findings led to the hypothesis that PLC-β1 might regulate protein trafficking to dendrites in response to mGluR activation and hence may be a key regulator of activity-dependent dendritic development.

In support of a role for PLC-β1 in cortical development, Hannan et al. (1998, 2001) showed that PLC-β1 levels are high in layers 2–4 of rodent S1 in the first two postnatal weeks, corresponding spatially and temporally with barrel formation and dendritogenesis. Furthermore, genetic deletion of *Plc-β1* disrupted the cytoarchitectural differentiation of barrels, but did not affect the pattern, distribution or size of TCA patches (although the structure and size of individual TCA axons has not been examined). Group 1 mGluR-mediated PI-hydrolysis was dramatically reduced in neocortex of young *Plc-β1*$^{-/-}$ mice supporting the hypothesis that mGluR5 activation of PLC-β1 regulates cellular aggregation in layer 4. The normal pattern of TCAs in *Plc-β1*$^{-/-}$ mice compared with the disrupted pattern in *Mglur5*$^{-/-}$ mice, however, indicates that mGluR5 regulation of TCA segregation may be mediated by a PLC-β1-independent mechanism. mGluR5, but not PLC-β1 is expressed at high levels in the somatosensory thalamus (Munoz et al., 1999; Watanabe et al., 1998) suggesting that mGluR5 in VpM neurons may regulate TCA segregation. However, the defects in TCA segregation do not appear to result from a loss of segregation at lower levels in the trigeminal pathway since barreloid development, as revealed by cytochrome oxidase histochemistry, appears normal in both *Mglur5*$^{-/-}$ and *Plc-β1*$^{-/-}$ mice. These findings suggest that mGluR5 may regulate TCA segregation by controlling the release of retrograde signals from layer 4 in a similar manner to NMDARs (Iwasato et al., 2000) although a direct role in the TCAs cannot be ruled out.

To directly test the hypothesis that PLC-β1 regulates activity-dependent dendritic rearrangements during cortical development, we have recently examined dendritic complexity and orientation of layer 4 neurons in *Plc-β1*$^{-/-}$ mice using Rapid Golgi staining (Upton et al., in preparation). No significant difference in total dendrite length, dendrite number and number of branch points of layer 4 neurons was observed between *Plc-β1*$^{-/-}$ and wild type mice. In addition layer 4 neuronal dendrites were normally oriented toward TCA patches. This study indicates that PLC-β1 is not regulating dendritic complexity and orientation of layer 4 neurons and selective dendritic elaboration within TCA patches is not sufficient to form barrels. We have also analysed spine density in the barrel cortex of *Plc-β1*$^{-/-}$ mice. In contrast to the Cx*nr1*$^{-/-}$ (Datwani et al., 2002), spine density is also unaltered in *Plc-β1*$^{-/-}$ mice. These differences in dendritic phenotype between *Plc-β1*$^{-/-}$ and Cx*nr1*$^{-/-}$ mice indicate that multiple pathways activated by glutamate receptors are not simply converging to regulate cortical development. Instead different combinations of pathways are likely needed to differentially regulate aspects of cortical differentiation such as barrel development, dendritic complexity, dendritic orientation and spine density.

Interestingly, Spires et al. (2005) demonstrated a reduced symmetric/asymmetric synaptic ratio within the barrel cortex at P5 in *Plc-β1*$^{-/-}$ mice, possibly resulting in an imbalance in excitatory and inhibitory circuitry. Spine morphology of layer 5 pyramidal neurons passing through layer 4 also showed a reduction in mushroom type spines compared to age-matched wildtypes indicating a disruption in spine maturation. These observations correlate well with findings in the hippocampus that the stimulation of group 1 mGluRs causes spine elongation (Vanderklish & Edelman, 2002) and that calcium release from intracellular stores affects spine morphology (Harris, 1999). In conclusion, PLC-β1 signalling appears to be important in the development of cortical connectivity by regulating spine shape and synapse formation, but not spine number or dendritic complexity.

Although the cellular mechanisms regulated by PLC-β1 are beginning to be elucidated, the biochemical pathways through which mGluR5 activation of PLCβ1 regulates these cellular processes remains largely unknown. One hypothesis proposed by Spires et al. (2005) is that PLCβ1 could affect spine morphology through its interaction with Homer proteins. Homer proteins are encoded by 3 genes (Homers1-3) and are multidomain, scaffolding molecules that link mGluR5/PLCβ1 to the IP3R. They also link mGluR5 with the NMDAR complex (Xiao et al., 2000; Fagni et al., 2002) by binding to Shank, another scaffolding molecule that associates with the guanylate kinase-associated protein GKAP/PSD95 complex. Homers have previously been shown to be involved in regulating the morphology of dendritic spines (Sala et al., 2003) and one splice variant, Homerla, has been shown to regulate spine formation in an activity-dependent manner (Sala et al., 2003). In collaboration with Professor Paul Worley we have examined the barrel cortex of individual Homer null mutant mice and Homer triple mutants ($h1^{-/-}h2^{-/-}h3^{-/-}$).

Nissl staining showed normal cellular segregation of layer 4 neurons into barrel walls and barrel septa also appeared normal. Therefore the mechanisms by which PLC-β1 controls barrel development are Homer-independent. These findings suggest that IP3-stimulated release from the ER may not be a crucial step in barrel development. However, it may be that a limited release of Ca^{2+} from the ER, that is not dependent on a close association of mGluR5 with the IP3 receptor, may be sufficient to drive barrel formation. Alternatively, DAG activation of PKC or direct regulation of PIP2 levels by PLC-β1 may be the crucial step to barrel development. A role for PKC in spine plasticity via its interaction with Rac and Rho has been shown (Pilpel & Segal, 2004). It is important to note that we have not examined whether Homers are critical for PLC-β1 dependent spine and synapse development. Further work will be necessary to determine the cellular mechanism regulated by PLC-β1 that underlie barrel formation and spine maturation.

SynGAP

SynGAP is a Synaptic Ras-GTPase Activating Protein (Chen et al., 1998) that is highly enriched in excitatory synapses (Kim et al., 1998; Chen et al., 1998; Petralia et al., 2005). It associates with the NMDAR complex (Husi et al., 2000) via direct interactions with the PDZ domains of MAGUKs, namely SAP102 and PSD95 (Chen et al., 1998; Kim et al., 1998; Kim et al., 2005; Figure 2B). SynGAP regulates the level of phosphorylated ERK (Komiyama et al., 2002), by regulating levels of Ras-GTP (but see below; Chen et al., 1998; Oh et al., 2004). However, regulation of ERK levels is likely key to functional synapse maturation since following NMDAR activation, pERK levels rise and regulate AMPA receptor insertion into the PSD (Zhu et al, 2002).

Recently, we showed that *Syngap* mRNA and SynGAP protein are expressed in developing S1 with highest levels in layer 4 during barrel formation (Barnett et al., 2006). *Syngap* mRNA is also expressed at high levels in developing VpM but not in brainstem. Mice carrying targeted mutations in *Syngap* have defects in barrel cortex development (Barnett et al., submitted). Nissl staining revealed that P6/7 *Syngap*$^{-/-}$ mice have a complete loss of cellular segregation into barrels. 5-HT immunohistochemistry revealed a partial segregation into rows but no segregation into whisker-specific patches. Using cytochrome oxidase histochemistry, *Syngap*$^{-/-}$ mice also show only partial segregation of barreloids in the VpM with only a few of the largest barreloids visible at P6. In contrast *Syngap*$^{+/-}$ mice have significantly reduced barrel formation but show normal afferent (as identified by 5-HT staining) and barreloid segregation (as shown by CO) (Barnett et al., 2006). This deficit in cellular segregation in layer 4, despite normal TCA segregation indicates that SynGAP plays a pivotal role in barrel formation. SynGAP expression is low in the developing brainstem and barrelettes develop normally in both the PrV and nVi. Therefore although NMDARs are necessary for whisker patterning at all levels of the trigeminal axis, the intracellular pathways they utilise to achieve these whisker-related patterns are different.

Our data are in good agreement with research from Kennedy's laboratory showing a role for SynGAP in several developmental events (Vasquez et al., 2004; Kneusel et al., 2005). *Syngap*$^{-/-}$ neurons prematurely form spines and functional synapses in culture and develop much larger mature spines compared to wild type neurons (Vasquez et al., 2004). These larger spines demonstrate a precocious incorporation of PSD proteins relative to wild type spines indicating that SynGAP plays an important role in regulating spine formation and maturity. More recently, SynGAP has also been shown to regulate neuronal apoptosis (Kneusel et al., 2005). Measurements of caspase-3 activation in *Syngap*$^{-/-}$ brains show that significantly more neurons in the hippocampus and the cortex undergo apoptosis at P0/P1 and that this apoptosis only occurs in regions where the *Syngap* gene is lost (Kneusel et al., 2005). Also by examining heterozygote and homozygous deletion of *Syngap*, the level of neuronal apoptosis was shown to correlate with the level of SynGAP protein present. Cell death selectively in the presumptive barrel hollow could underlie barrel formation and the loss or decreased segregation of barrels in *Syngap*$^{-/-}$ and *Syngap*$^{+/-}$ animals, respectively, could reflect an altered distribution of apoptotic cells, however, this possibility seems unlikely for 2 reasons. First, Miller (1995) reported no pattern to the distribution of apoptotic cells in layer 4 during barrel formation. Second, *Syngap*$^{+/-}$ mice show a normal overall density of layer 4 neurons despite significantly reduced segregation of cortical cells (Barnett et al., 2006). Instead it appears that SynGAP is playing a critical role in barrel development by mediating active cellular aggregation, either through directed cell movement or differential cell adhesion.

SynGAP was previously shown to associate with the NMDARs via an interaction with PSD-95 a highly abundant scaffolding protein found in developing cortical PSDs (Chen et al., 1998; Kim et al., 1998). To determine whether its association with PSD-95 was critical for barrel formation we examined barrel formation in two different lines of *psd-95* null mutant mice. Both lines show normal cellular segregation in layer 4 producing clearly defined barrel fields. Furthermore using PSD preparations from barrel cortex of *psd-95*$^{-/-}$ animals we demonstrated that SynGAP can associate with the PSD during barrel development in a PSD-95 independent manner suggesting 1) that there is either compensation between scaffolding molecules or redundancy of these molecules with respect to barrel formation or 2) SynGAP uses other domains (i.e. its pleckstrin homology or C2 domain) or other PDZ-containing proteins (i.e. SAP-102, PSD-93, or MUPP1) to associate with the PSD (Barnett et al., 2006; Kim et al., 2005; Krapivinsky et al., 2004). The structure and *in vitro* activity of SynGAP suggested that H-Ras, a small G-protein, could be the substrate for SynGAP. However no barrel defects were detected in either adult or P7 *h-ras*$^{-/-}$ mutant mice (Barnett et al., submitted), so H-Ras does not appear to act as a key effector of cortical development. However other Ras isoforms could potentially provide the substrate for SynGAP during barrel development.

ERK and Barrel Development

Our findings of a role for SynGAP in barrel cortex development are also in good agreement with previous findings showing a role for the ERK pathway in many forms of NMDAR-dependent synaptic plasticity (Sweatt 2001), including the shift in ocular dominance that occurs in the visual cortex after monocular deprivation (MD; Di Cristo et al., 2001). As mentioned above, NMDA receptor activation causes an increase in pERK. One way that NMDA could increase pERK levels is by inhibiting SynGAP activity, causing a build-up of Ras-GTP and hence ERK phosphorylation (Chen et al., 1998; Kim et al., 2003). However, Oh et al. (2004) have shown that phosphorylation of SynGAP by CaMKII increases SynGAP's activity by 70–95%. Since CaMKII is activated by NMDA receptor stimulation, SynGAP activity would be expected to increase following NMDA receptor stimulation and pERK levels would decrease. In hippocampal slices, Komiyama et al. (2002), found an increase in basal pERK levels in *Syngap*$^{+/-}$ supporting the hypothesis that SynGAP regulates the ERK pathway. They also reported an increase in NMDA receptor-mediated pERK levels in *Syngap*$^{+/-}$ animals. These findings indicate that ERK can still be phosphorylated in an NMDAR-dependent manner in the absence of SynGAP (possibly via PKA or PKC-dependent pathways). Stimulation of NMDARs may simultaneously activate SynGAP-independent and SynGAP-dependent pathways to regulate the precise levels of Ras-GTP and pERK. The SynGAP-independent pathway could positively regulate ERK phosphorylation and the SynGAP/CaMKII pathway would negatively regulate ERK phosphorylation. The removal of SynGAP would therefore result in a release from SynGAP-mediated inhibition of NMDAR/CaMKII-activated ERK phosphorylation and an increase in NMDAR stimulated pERK levels.

It has previously been proposed that a primary role for ERK may be to integrate signals initiated from a variety of sources to produce coordinated cellular events (reviewed by Adams and Sweatt, 2002). Since multiple receptors (mGluR5 and NMDARs) signalling via numerous intracellular signalling pathways (PLC-β1, SynGAP, PKA) are necessary for barrel development, it is tempting to hypothesize that ERK may be acting to integrate these signals to mediate the various cellular events involved in barrel formation. In support of this notion, mGluR5 signalling has been shown to activate the ERK pathway (Gallagher et al., 2004; Berkeley & Levey 2003; Choe & Wang 2001) as can PKA (Cancedda et al., 2003) and PKC activity (Sweatt, 2004). The disruption of any single signalling pathway could alter ERK activity to such an extent that barrel development is disrupted. The degree to which ERK is altered in each mutant may dictate the severity of the phenotype. Deleting SynGAP has been shown to increase ERK activity and disrupt barrel formation, while blocking ERK activation using MEK inhibitors prevents the ocular dominance plasticity to MD and LTP in visual cortex and hippocampus (Di Cristo, 2001; Sweatt, 2001). Therefore there may be a critical band of ERK activity necessary for normal neuronal development.

Non-ERK Pathways

It is important to note that SynGAP could be regulating barrel development independently of ERK. For example, SynGAP could be regulating Ras-dependent, but ERK-independent pathways or Ras-independent pathways. Several ERK-independent signalling cascades have been described downstream of Ras in response to NMDA receptor activation (Cullen & Lockyer, 2002) including the Ras-PI3K pathway (Cullen & Lockyer, 2002). PI3K is a major effector of Ras and has been shown to be involved in cytoskeletal remodelling and is necessary for NMDAR-stimulated delivery of AMPARs to the neuronal surface, a mechanism of synaptic plasticity (Man et al., 2003; Opazo et al., 2003). Recent studies have also revealed that SynGAP can act as a Rab-GAP (Tomoda et al., 2004) and a Rap-GAP (Krapivinsky et al., 2004). While SynGAP is more closely related to other RasGAPs, Krapivinsky et al. (2004) found that for *in vitro* hippocampus, SynGAP has more efficient GAP activity for Rap than it does for Ras. Krapivinsky et al. (2004) also showed that dephosphorylated SynGAP that is dissociated from its complex with MUPP1 and CaMKII, causes an inactivation of Rap and subsequent increase in p38MAPK activity. p38MAPK is therefore another potential signalling cascade through which SynGAP might regulate barrel formation. Finally, Tomoda et al. (2004) showed that SynGAP regulates Rab5 activity, whose activity regulates the actin cytoskeleton during "circular ruffle" formation and cell migration (Lanzetti et al., 2004). It is possible therefore, that other small non-Ras G-proteins are crucial effectors of SynGAP during barrel development.

Conclusions

Cellular Processes of Barrel Formation

A central question in developmental neurobiology concerns the molecular mechanisms by which glutamate neurotransmission regulates cortical development. Over the last several years we have identified several key members of the postsynaptic density that are essential for the formation of "barrels", the prominent anatomical features of the rodent somatosensory cortex. Interestingly, no two mutants have identical phenotypes indicating that while different signalling pathways may regulate some of the same cellular processes, the cohort of processes regulated by each pathway must be unique. Furthermore, the finding that $Plc\text{-}\beta 1^{-/-}$ mice develop normal axon segregation and dendritic orientation and complexity indicates that barrel development is not simply a result of neuropil expansion within the barrel hollow. Instead a glutamate-dependent process of cell migration, cell adhesion or possibly cell death seems to be the driving force for barrel development.

Heterogeneity of PSDs

Our results also indicate that there is significant functional and structural heterogeneity between PSD complexes in different brain regions and within the same regions at different developmental stages. For example, while NMDA receptors are needed at all levels of the trigeminal axis (see Erzurumlu chapter), mGluR5 and PLC-β1 are crucial for cortical cellular differentiation to form barrels but not needed for segregation of the barrelettes or barreloids in the brainstem or thalamus, respectively. SynGAP is expressed in both thalamus and cortex where it plays a role in pattern formation, but is not expressed in the brainstem where whisker-related patterns form normally. These findings indicate that understanding the pathways that underlie cortical development will first require the characterisation of the spatio-temporal expression patterns of NRC components.

It is interesting to note that all of the molecules identified that regulate barrel development are either receptors or signalling enzymes. We have examined 5 scaffolding molecules (Homers 1-3, PSD-95 and AKAP79/150), none of which show an obvious defect in barrel formation (although roles in dendritogenesis have not been examined). These results suggest that the scaffolding molecules that tether each of these molecules to the PSD are not essential for barrel development. Alternatively, there may be a high degree of redundancy in scaffolding system such that the loss of an individual scaffolding molecule (i.e. PSD-95) does not disrupt the association of a particular enzyme from the protein complex. Indeed, genetic deletion of *psd-95* does not significantly alter NMDA receptor association with the PSD, although it does alter synaptic plasticity (Migaud et al., 1998). The issue of compensation will be addressed using double and triple knockout mice as well as conditional mutants in which a particular gene can be ablated at particular times during development, namely when barrels are developing.

Activity-dependent plasticity during neuronal development has been the subject of intense interest over the last several decades because of the hope that it could shed light on the mechanisms that underlie and possibly provide treatments for childhood learning disorders and neurodegenerative diseases. With the recent applications of advanced molecular and biochemical techniques, there is realistic optimism that these hopes will eventually become reality. For example, there is recent evidence suggesting that the primary defects in Fragile X mental retardation could result from alterations in mGluR5 signalling, a hypothesis that if correct, could lead to the development of powerful drug therapies for patients (Bear et al., 2004; Dolen and Bear, 2005). In addition, SAP-102, a key MAGUK present in the NRC during cortical development has been identified as the gene disrupted in a form of familial X-linked mental retardation (Tarpey et al., 2003). Similarly, several Rho or Rab GTPases also appear to be involved in certain forms of X-linked mental retardation (Renieri et al., 2005). Unfortunately, simply knowing the proteins necessary for normal development does not always lead to treatment for afflicted patients. However, knowing the biochemical pathways by which these genes regulate normal development may provide novel avenues for the development of treatments for a range of neurodevelopmental disorders.

References

Abdel-Majid RM, Leong WL, Schalkwyk LC, Smallman DS, Wong ST, Storm DR, Fine A, Dobson MJ, Guernsey DL, Neumann PE (1998) Loss of adenylyl cyclase I activity disrupts patterning of mouse somatosensory cortex. Nat Genet 19:289–291.

Adams JP, Sweatt JD (2002) Molecular psychology: roles for the ERK MAP kinase cascade in memory. Annu Rev Pharmacol Toxicol 42:135–163.

Agmon A, Yang LT, Jones EG, O'Dowd DK (1995) Topological precision in the thalamic projection to neonatal mouse barrel cortex. J Neurosci 15:549–561.

Barnett MW, Watson RF, Vitalis T, Porter K, Komiyama NH, Stoney PN, Gillingwater TH, Grant SG, Kind PC (2006) Synaptic Ras GTPase activating protein regulates pattern formation in the trigeminal system of mice. J Neurosci 26:1355–1365.

Bates CA, Killackey HP (1985) The organization of the neonatal rat's brainstem trigeminal complex and its role in the formation of central trigeminal patterns. J Comp Neurol 240:265–287.

Bear MF, Huber KM, Warren ST (2004) The mGluR theory of fragile X mental retardation. Trends Neurosci 27:370–377.

Beaver CJ, Ji Q, Fischer QS, Daw NW (2001) Cyclic AMP-dependent protein kinase mediates ocular dominance shifts in cat visual cortex. Nat Neurosci 4:159–163.

Berkeley JL, Levey AI (2003) Cell-specific extracellular signal-regulated kinase activation by multiple G protein-coupled receptor families in hippocampus. Mol Pharmacol 63:128–135.

Brown KE, Arends JJ, Wasserstrom SP, Zantua JB, Jacquin MF, Woolsey TA (1995) Developmental transformation of dendritic arbors in mouse whisker thalamus. Brain Res Dev Brain Res 86:335–339.

Cancedda L, Putignano E, Impey S, Maffei L, Ratto GM, Pizzorusso T (2003) Patterned vision causes CRE-mediated gene expression in the visual cortex through PKA and ERK. J Neurosci 23:7012–7020.

Chen HJ, Rojas-Soto M, Oguni A, Kennedy MB (1998) A synaptic Ras-GTPase activating protein (p135 SynGAP) inhibited by CaM kinase II. Neuron 20:895–904.

Choe ES, Wang JQ (2001) Group I metabotropic glutamate receptor activation increases phosphorylation of cAMP response element-binding protein, Elk-1, and extracellular signal-regulated kinases in rat dorsal striatum. Brain Res Mol Brain Res 94:75–84.

Crair MC, Malenka RC (1995) A critical period for long-term potentiation at thalamocortical synapses. Nature 375:325–328.

Cullen PJ, Lockyer PJ (2002) Integration of calcium and Ras signalling. Nat Rev Mol Cell Biol 3:339–348.

Datwani A, Iwasato T, Itohara S, Erzurumlu RS (2002) NMDA receptor-dependent pattern transfer from afferents to postsynaptic cells and dendritic differentiation in the barrel cortex. Mol Cell Neurosci 21:477–492.
De Felipe J, Marco P, Fairen A, Jones EG (1997) Inhibitory synaptogenesis in mouse somatosensory cortex. Cereb Cortex 7:619–634.
Di Cristo G, Berardi N, Cancedda L, Pizzorusso T, Putignano E, Ratto GM, Maffei L (2001) Requirement of ERK activation for visual cortical plasticity. Science 292:2337–2340.
Dickson KS, Kind PC (2003) NMDA receptors: neural map designers and refiners? Curr Biol 13:R920–922.
Dolen G, Bear MF (2005) Courting a cure for fragile X. Neuron 45:642–644.
Dudek SM, Bear MF (1989) A biochemical correlate of the critical period for synaptic modification in the visual cortex. Science 246:673–675.
Erzurumlu RS, Kind PC (2001) Neural activity: sculptor of 'barrels' in the neocortex. Trends Neurosci 24:589–595.
Fagni L, Worley PF, Ango F (2002) Homer as both a scaffold and transduction molecule. Sci STKE 2002:RE8.
Fischer QS, Beaver CJ, Yang Y, Rao Y, Jakobsdottir KB, Storm DR, McKnight GS, Daw NW (2004) Requirement for the RIIbeta isoform of PKA, but not calcium-stimulated adenylyl cyclase, in visual cortical plasticity. J Neurosci 24:9049–9058.
Gallagher SM, Daly CA, Bear MF, Huber KM (2004) Extracellular signal-regulated protein kinase activation is required for metabotropic glutamate receptor-dependent long-term depression in hippocampal area CA1. J Neurosci 24:4859–4864.
Gaspar P, Cases O, Maroteaux L (2003) The developmental role of serotonin: news from mouse molecular genetics. Nat Rev Neurosci 4:1002–1012.
Glazewski S, Chen CM, Silva A, Fox K (1996) Requirement for alpha-CaMKII in experience-dependent plasticity of the barrel cortex. Science 272:421–423.
Grant SG, O'Dell TJ (2001) Multiprotein complex signaling and the plasticity problem. Curr Opin Neurobiol 11:363–368.
Greenough WT, Chang FL (1988) Dendritic pattern formation involves both oriented regression and oriented growth in the barrels of mouse somatosensory cortex. Brain Res 471:148–152.
Hannan AJ, Kind PC, Blakemore C (1998) Phospholipase C-beta1 expression correlates with neuronal differentiation and synaptic plasticity in rat somatosensory cortex. Neuropharmacology 37:593–605.
Hannan AJ, Blakemore C, Katsnelson A, Vitalis T, Huber KM, Bear M, Roder J, Kim D, Shin HS, Kind PC (2001) PLC-beta1, activated via mGluRs, mediates activity-dependent differentiation in cerebral cortex. Nat Neurosci 4:282–288.
Harris KM (1999) Calcium from internal stores modifies dendritic spine shape. Proc Natl Acad Sci U S A 96:12213–12215.
Hayashi H (1980) Distributions of vibrissae afferent fiber collaterals in the trigeminal nuclei as revealed by intra-axonal injection of horseradish peroxidase. Brain Res 183:442–446.
Husi H, Ward MA, Choudhary JS, Blackstock WP, Grant SG (2000) Proteomic analysis of NMDA receptor-adhesion protein signaling complexes. Nat Neurosci 3:661–669.
Iwasato T, Datwani A, Wolf AM, Nishiyama H, Taguchi Y, Tonegawa S, Knopfel T, Erzurumlu RS, Itohara S (2000) Cortex-restricted disruption of NMDAR1 impairs neuronal patterns in the barrel cortex. Nature 406:726–731.

Katz LC, Crowley JC (2002) Development of cortical circuits: lessons from ocular dominance columns. Nat Rev Neurosci 3:34–42.

Killackey HP (1973) Anatomical evidence for cortical subdivisions based on vertically discrete thalamic projections from the ventral posterior nucleus to cortical barrels in the rat. Brain Res 51:326–331.

Kim JH, Liao D, Lau LF, Huganir RL (1998) SynGAP: a synaptic RasGAP that associates with the PSD-95/SAP90 protein family. Neuron 20:683–691.

Kim JH, Lee HK, Takamiya K, Huganir RL (2003) The role of synaptic GTPase-activating protein in neuronal development and synaptic plasticity. J Neurosci 23:1119–1124.

Kim MJ, Dunah AW, Wang YT, Sheng M (2005) Differential roles of NR2A- and NR2B-containing NMDA receptors in Ras-ERK signaling and AMPA receptor trafficking. Neuron 46:745–760.

Kind P, Blakemore C, Fryer H, Hockfield S (1994) Identification of proteins downregulated during the postnatal development of the cat visual cortex. Cereb Cortex 4:361–375.

Kind PC, Neumann PE (2001) Plasticity: downstream of glutamate. Trends Neurosci 24:553–555.

Kind PC, Kelly GM, Fryer HJ, Blakemore C, Hockfield S (1997) Phospholipase C-beta1 is present in the botrysome, an intermediate compartment-like organelle, and Is regulated by visual experience in cat visual cortex. J Neurosci 17:1471–1480.

Kleinschmidt A, Bear MF, Singer W (1987) Blockade of "NMDA" receptors disrupts experience-dependent plasticity of kitten striate cortex. Science 238:355–358.

Knuesel I, Elliott A, Chen HJ, Mansuy IM, Kennedy MB (2005) A role for synGAP in regulating neuronal apoptosis. Eur J Neurosci 21:611–621.

Komiyama NH, Watabe AM, Carlisle HJ, Porter K, Charlesworth P, Monti J, Strathdee DJ, O'Carroll CM, Martin SJ, Morris RG, O'Dell TJ, Grant SG (2002) SynGAP regulates ERK/MAPK signaling, synaptic plasticity, and learning in the complex with postsynaptic density 95 and NMDA receptor. J Neurosci 22:9721–9732.

Krapivinsky G, Medina I, Krapivinsky L, Gapon S, Clapham DE (2004) SynGAP-MUPP1-CaMKII synaptic complexes regulate p38 MAP kinase activity and NMDA receptor-dependent synaptic AMPA receptor potentiation. Neuron 43:563–574.

Lanzetti L, Palamidessi A, Areces L, Scita G, Di Fiore PP (2004) Rab5 is a signalling GTPase involved in actin remodelling by receptor tyrosine kinases. Nature 429:309–314.

Lee LJ, Iwasato T, Itohara S, Erzurumlu RS (2005) Exuberant thalamocortical axon arborization in cortex-specific NMDAR1 knockout mice. J Comp Neurol 485:280–292.

Li Y, Erzurumlu RS, Chen C, Jhaveri S, Tonegawa S (1994) Whisker-related neuronal patterns fail to develop in the trigeminal brainstem nuclei of NMDAR1 knockout mice. Cell 76:427–437.

Lu HC, Gonzalez E, Crair MC (2001) Barrel cortex critical period plasticity is independent of changes in NMDA receptor subunit composition. Neuron 32:619–634.

Lu HC, She WC, Plas DT, Neumann PE, Janz R, Crair MC (2003) Adenylyl cyclase I regulates AMPA receptor trafficking during mouse cortical 'barrel' map development. Nat Neurosci 6:939–947.

Ma PM (1993) Barrelettes–architectonic vibrissal representations in the brainstem trigeminal complex of the mouse. II. Normal post-natal development. J Comp Neurol 327:376–397.
Ma PM, Woolsey TA (1984) Cytoarchitectonic correlates of the vibrissae in the medullary trigeminal complex of the mouse. Brain Res 306:374–379.
Ma PM, Woolsey TA (1983) Cytoarchitectonic organization of the brainstem trigeminal nuclei in the mouse: Correlation with vibrissal inputs and postnatal development. Soc. Neurosci, Abstr. 9:922.
Man HY, Wang Q, Lu WY, Ju W, Ahmadian G, Liu L, D'Souza S, Wong TP, Taghibiglou C, Lu J, Becker LE, Pei L, Liu F, Wymann MP, MacDonald JF, Wang YT (2003) Activation of PI3-kinase is required for AMPA receptor insertion during LTP of mEPSCs in cultured hippocampal neurons. Neuron 38:611–624.
Micheva KD, Beaulieu C (1995) An anatomical substrate for experience-dependent plasticity of the rat barrel field cortex. Proc Natl Acad Sci U S A 92:11834–11838.
Micheva KD, Beaulieu C (1996) Quantitative aspects of synaptogenesis in the rat barrel field cortex with special reference to GABA circuitry. J Comp Neurol 373:340–354.
Migaud M, Charlesworth P, Dempster M, Webster LC, Watabe AM, Makhinson M, He Y, Ramsay MF, Morris RG, Morrison JH, O'Dell TJ, Grant SG (1998) Enhanced long-term potentiation and impaired learning in mice with mutant postsynaptic density-95 protein. Nature 396:433–439.
Miller MW (1995) Relationship of the time of origin and death of neurons in rat somatosensory cortex: barrel versus septal cortex and projection versus local circuit neurons. J Comp Neurol 355:6–14.
Munoz A, Liu XB, Jones EG (1999) Development of metabotropic glutamate receptors from trigeminal nuclei to barrel cortex in postnatal mouse. J Comp Neurol 409:549–566.
Oh JS, Manzerra P, Kennedy MB (2004) Regulation of the neuron-specific Ras GTPase-activating protein, synGAP, by Ca2+/calmodulin-dependent protein kinase II. J Biol Chem 279:17980–17988.
Opazo P, Watabe AM, Grant SG, O'Dell TJ (2003) Phosphatidylinositol 3-kinase regulates the induction of long-term potentiation through extracellular signal-related kinase-independent mechanisms. J Neurosci 23:3679–3688.
Petralia RS, Sans N, Wang YX, Wenthold RJ (2005) Ontogeny of postsynaptic density proteins at glutamatergic synapses. Mol Cell Neurosci.
Pierret T, Lavallee P, Deschenes M (2000) Parallel streams for the relay of vibrissal information through thalamic barreloids. J Neurosci 20:7455–7462.
Pilpel Y, Segal M (2004) Activation of PKC induces rapid morphological plasticity in dendrites of hippocampal neurons via Rac and Rho-dependent mechanisms. Eur J Neurosci 19:3151–3164.
Rebsam A, Seif I, Gaspar P (2002) Refinement of thalamocortical arbors and emergence of barrel domains in the primary somatosensory cortex: a study of normal and monoamine oxidase a knock-out mice. J Neurosci 22:8541–8552.
Renieri A, Pescucci C, Longo I, Ariani F, Mari F, Meloni I (2005) Non-syndromic X-linked mental retardation: From a molecular to a clinical point of view. J Cell Physiol 204:8–20.
Rice FL, Van der Loos H (1977) Development of the barrels and barrel field in the somatosensory cortex of the mouse. J Comp Neurol 171:545–560.

Sala C, Futai K, Yamamoto K, Worley PF, Hayashi Y, Sheng M (2003) Inhibition of dendritic spine morphogenesis and synaptic transmission by activity-inducible protein Homer1a. J Neurosci 23:6327–6337.

Spires TL, Molnar Z, Kind PC, Cordery PM, Upton AL, Blakemore C, Hannan AJ (2005) Activity-dependent regulation of synapse and dendritic spine morphology in developing barrel cortex requires phospholipase C-beta1 signalling. Cereb Cortex 15:385–393.

Sweatt JD (2001) The neuronal MAP kinase cascade: a biochemical signal integration system subserving synaptic plasticity and memory. J Neurochem 76:1–10.

Sweatt JD (2004) Mitogen-activated protein kinases in synaptic plasticity and memory. Curr Opin Neurobiol 14:311–317.

Taha S, Hanover JL, Silva AJ, Stryker MP (2002) Autophosphorylation of alphaCaMKII is required for ocular dominance plasticity. Neuron 36:483–491.

Tarpey P, Parnau J, Blow M, Woffendin H, Bignell G, Cox C, Cox J, Davies H, Edkins S, Holden S, Korny A, Mallya U, Moon J, O'Meara S, Parker A, Stephens P, Stevens C, Teague J, Donnelly A, Mangelsdorf M, Mulley J, Partington M, Turner G, Stevenson R, Schwartz C, Young I, Easton D, Bobrow M, Futreal PA, Stratton MR, Gecz J, Wooster R, Raymond FL (2004) Mutations in the DLG3 gene cause nonsyndromic X-linked mental retardation. Am J Hum Genet 75:318–324.

Tomoda T, Kim JH, Zhan C, Hatten ME (2004) Role of Unc51.1 and its binding partners in CNS axon outgrowth. Genes Dev 18:541–558.

Upton L, Katsnelson A, Blakemore C, Hannan AJ, Kind PC (in preparation 2006) Normal dendritic complexity and orientation are not sufficient to form barrels in mouse primary somatosensory cortex

Van der Loos, H (1976) Barreloids in mouse somatosensory thalamus. Neurosci. Lett. 2:1–6.

Vanderhaeghen P, Lu Q, Prakash N, Frisen J, Walsh CA, Frostig RD, Flanagan JG (2000) A mapping label required for normal scale of body representation in the cortex. Nat Neurosci 3:358–365.

Vanderklish PW, Edelman GM (2002) Dendritic spines elongate after stimulation of group 1 metabotropic glutamate receptors in cultured hippocampal neurons. Proc Natl Acad Sci U S A 99:1639–1644.

Vazquez LE, Chen HJ, Sokolova I, Knuesel I, Kennedy MB (2004) SynGAP regulates spine formation. J Neurosci 24:8862–8872.

Walikonis RS, Jensen ON, Mann M, Provance DW, Jr., Mercer JA, Kennedy MB (2000) Identification of proteins in the postsynaptic density fraction by mass spectrometry. J Neurosci 20:4069–4080.

Watanabe M, Nakamura M, Sato K, Kano M, Simon MI, Inoue Y (1998) Patterns of expression for the mRNA corresponding to the four isoforms of phospholipase C beta in mouse brain. Eur J Neurosci 10:2016–2025.

Watson RF, Abdel-Majid RM, Barnett MW, Willis BS, Katnelson A, Gillingwater TH, McKnight GS, Kind PC and Neumann PE (in press, 2006) Involvement of Protein Kinase A in patterning of the Mouse Somatosensory Cortex.

White EL (1979) Thalamocortical synaptic relations: a review with emphasis on the projections of specific thalamic nuclei to the primary sensory areas of the neocortex. Brain Res 180:275–311.

White EL, Weinfeld L, Lev DL (1997) A survey of morphogenesis during the early postnatal period in PMBSF barrels of mouse SmI cortex with emphasis on barrel D4. Somatosens Mot Res 14:34–55.

Wiesel TN, Hubel DH (1963) Effects Of Visual Deprivation On Morphology And Physiology Of Cells In The Cats Lateral Geniculate Body. J Neurophysiol 26:978–993.

Wiesel TN, Hubel DH (1963) Single-Cell Responses In Striate Cortex Of Kittens Deprived Of Vision In One Eye. J Neurophysiol 26:1003–1017.

Woolsey TA (1996) Barrels: 25 years later. Somatosens Mot Res 13:181–186.

Woolsey TA, Van der Loos H (1970) The structural organization of layer IV in the somatosensory region (SI) of mouse cerebral cortex. The description of a cortical field composed of discrete cytoarchitectonic units. Brain Res 17:205–242.

Woolsey TA, Dierker ML, Wann DF (1975) Mouse SmI cortex: qualitative and quantitative classification of golgi-impregnated barrel neurons. Proc Natl Acad Sci U S A 72:2165–2169.

Xiao B, Tu JC, Worley PF (2000) Homer: a link between neural activity and glutamate receptor function. Curr Opin Neurobiol 10:370–374.

Yamakodo, M (1985) Postnatal development of barreloid neuropils in the ventrobasal complex of mouse thalamus : a histochemical study for cytochrome oxidase. No To Shinkei 37:497–506.

Yang Y, Fischer QS, Zhang Y, Baumgartel K, Mansuy IM, Daw NW (2005) Reversible blockade of experience-dependent plasticity by calcineurin in mouse visual cortex. Nat Neurosci 8:791–796.

Zantua JB, Wasserstrom SP, Arends JJ, Jacquin MF, Woolsey TA (1996) Postnatal development of mouse "whisker" thalamus: ventroposterior medial nucleus (VPM), barreloids, and their thalamocortical relay neurons. Somatosens Mot Res 13:307–322.

Zhu JJ, Qin Y, Zhao M, Van Aelst L, Malinow R (2002) Ras and Rap control AMPA receptor trafficking during synaptic plasticity. Cell 110:443–455.

10

Patterning of the Somatosensory Maps with NMDA Receptors

Reha S. Erzurumlu and Takuji Iwasato

Abstract

Neural maps of the somatosensory periphery are characterized by their somatotopic organization, and whisker- and digit-specific patterning in rodents. While a variety of molecular guidance cues help set up the topographic axonal projections in the brain, activity-dependent interactions between pre- and postsynaptic elements play a key role in neural patterning. Here we review our and other groups' analyses of the phenotypes of mice with various types of NMDA receptor (NMDAR) subunit mutations as they relate to the development and patterning of somatosensory pathways. Our recent studies on axonal and dendritic development in region specific NMDAR subunit *NR1* knockout and transgenic rescue of global *NR1* knockout mice show that NMDAR signaling is necessary for dendritic and axonal pruning and patterning. Further development of region and cell type-specific gene targeting strategies in mice will undoubtedly reveal cellular and molecular mechanisms that underlie the formation of patterned somatotopic maps and their plasticity.

N-Methyl-D-Aspartate (NMDA) Receptors

Glutamate is the main excitatory neurotransmitter in the vertebrate central nervous system (CNS). The diverse functions of glutamate neurotransmission are mediated by glutamate receptors that are classified into two major groups, ionotropic receptors and metabotropic receptors. The ionotropic receptors are subdivided into N-methyl-D-aspartate (NMDA) receptors (NMDARs), α-amino-3-hydroxy-5-methyl-4-isoxazole propionic acid (AMPA) receptors and kainate receptors (Hollmann and Heinemann, 1994; Dingledine et al., 1999; Hollmann, 1999). Upon ligand binding, these receptors participate in a large variety of neural events from wiring of synaptic circuits to their plasticity. Of the three types of ionotropic glutamate receptors, NMDARs have distinctive characteristics (Mayer and Westbrook, 1987; Ascher and

Nowak, 1988; Bliss and Collingridge, 1993; Wisden and Seeburg, 1993; Spruston et al., 1995; Monyer et al., 1999; Wisden et al., 2000). First, a voltage-dependent Mg^{2+} block permits channel opening only when the patch of the membrane they are situated in is sufficiently depolarized at the time of ligand binding. Second, they show a slow response to L-glutamate by significantly longer channel opening time compared to other glutamate receptors. Finally they are highly permeable to Ca^{2+} ions linked to initiation of multiple intracellular signaling pathways from localized actions at the postsynaptic site to immediate early gene expression.

In both the developing and mature brain, NMDARs are thought to be coincidence detectors of pre- and postsynaptic activity because of their channel gating properties that require simultaneous postsynaptic membrane depolarization and presynaptic glutamate release (Bliss and Collingridge, 1993; Malenka and Nicoll, 1993). Such coincidence detectors of synaptic activity would then be poised to play a major role in selective consolidation of coactive synapses and elimination of others, a highly plausible mechanism underlying adult synaptic plasticity (e.g., learning and memory) and developmental sculpting and patterning of neural circuits (Stent, 1973; Constantine-Paton et al., 1990; Bliss and Collingridge, 1993; Fox and Daw, 1993.).

There are seven NMDAR subunits (NR1, NR2A-D, and NR3A, B) (Moriyoshi et al., 1991; Ikeda et al., 1992; Kutsuwada et al., 1992; Meguro et al., 1992; Monyer et al., 1992; Ciabarra et al., 1995; Sucher et al., 1995; Andersson et al., 2001; Nishi et al., 2001). Most NMDARs are thought to be tetramer complexes composed of two NR1 subunits and two NR2 subunits (Clements and Westbrook, 1991). NR1 subunits have binding sites for co-agonist, glycine and NR2 subunits have binding sites for L-glutamate (Kuryatov et al., 1994; Laube et al., 1997). The combination of NR1 with different NR2 subunits shows variability in electrophysiological and pharmacological properties (Monyer et al., 1994). Therefore, NR1 and NR2 subunits are essential and modulatory subunits, respectively.

Virtually all CNS neurons express NR1 subunit mRNA throughout development and in adulthood. While four NR2 subunit genes are differentially expressed temporally and spatially (Monyer et al., 1994; Watanabe et al., 1992; 1993; 1994a; 1994b). *NR2A* mRNA is hardly detected in the embryonic brain but increases in the entire brain during the first two weeks after birth. *NR2B* mRNA is detected in various brain regions at embryonic stages. Its expression in the cerebellum and brainstem is diminished by postnatal day (P)14, whereas that in the forebrain remains high. *NR2C* mRNA is mainly detected in cerebellar granule cells. *NR2D* expression is detected in the diencephalon and brainstem of embryonic and neonatal brains, but after P7 expression is faint. Protein distribution of NR1 and NR2 subunits was examined with several antibodies and similar conclusions were reached (Petralia et al., 1994; Laurie et al., 1997).

While most neurons in the CNS express both NR1 and NR2 (A, B, C and/or D) subunits, expression of NR3 subunits is very restricted.

NR3A is expressed primarily during brain development (Das et al., 1998; Sasaki et al., 2002). *NR3B* expression is restricted to motorneurons of the brainstem and spinal cord (Nishi et al., 2001). NR3A subunit forms a complex with NR1 and NR2 subunits and suppresses NMDA response *in vitro* and *in vivo*. NR3B subunit similarly suppresses glutamate-induced current *in vitro*, while its *in vivo* function is not known yet.

During the past decade, loss-of-function studies, particularly targeted gene disruption in vivo, have yielded important information on the role of NMDARs in adult neural plasticity and in patterning of neural connections during development of the sensory pathways. In this review, we describe and discuss the results of our and other groups' analyses of the phenotypes of mice with various types of NMDAR subunit mutations as they relate to the development and patterning of somatosensory pathways.

Sensory Periphery-Related Patterning in Rodent Somatosensory System

The somatosensory pathways of rodents have topographically ordered and patterned "somatotopic maps" in the primary somatosensory (SI) cortex and subcortical nuclei, which can be visualized by a number of histological stains (Figure 1). The trigeminal and dorsal column pathways are the major ascending systems to the SI cortex. The trigeminal pathway transmits somatosensory inputs from the face and oral structures to SI cortex via the brainstem trigeminal complex (BSTC) and the ventral posterior medial (VPM) nucleus of the thalamus. The dorsal column pathway carries the somatosensory inputs from the rest of the body and involves the dorsal column nuclei of the medulla and the ventral posterior lateral (VPL) nucleus of the thalamus (reviewed by Killackey et al., 1995). Along the trigeminal pathway, the afferent axons and target cells are organized in modules that replicate the patterned array of whiskers and sinus hairs on the animal's snout (Woolsey and Van der Loos, 1970; Van der Loos, 1976; Belford and Killackey, 1979; Ma and Woolsey, 1984). These modules are called 'barrels" in SI cortex, "barreloids" in VPM, and "barrelettes" in BSTC. A similar patterned organization is observed in the forepaw representation areas along the dorsal column pathway (Belford and Killackey, 1978; Dawson and Killackey, 1987). At each level of the somatosensory pathway histochemical stains for mitochondrial enzymes such as cytochrome oxidase or succinic dehydrogenase reveal the overall patterning of pre and postsynaptic elements, axonal markers distinguish patterning of presynaptic terminals, and Nissl stains and cellular labels allow visualization of postsynaptic elements; i.e., barrels, barreloids and barrelettes.

During embryonic development, cortical area map is initially determined by at least two types of molecular cues: gradient of locally

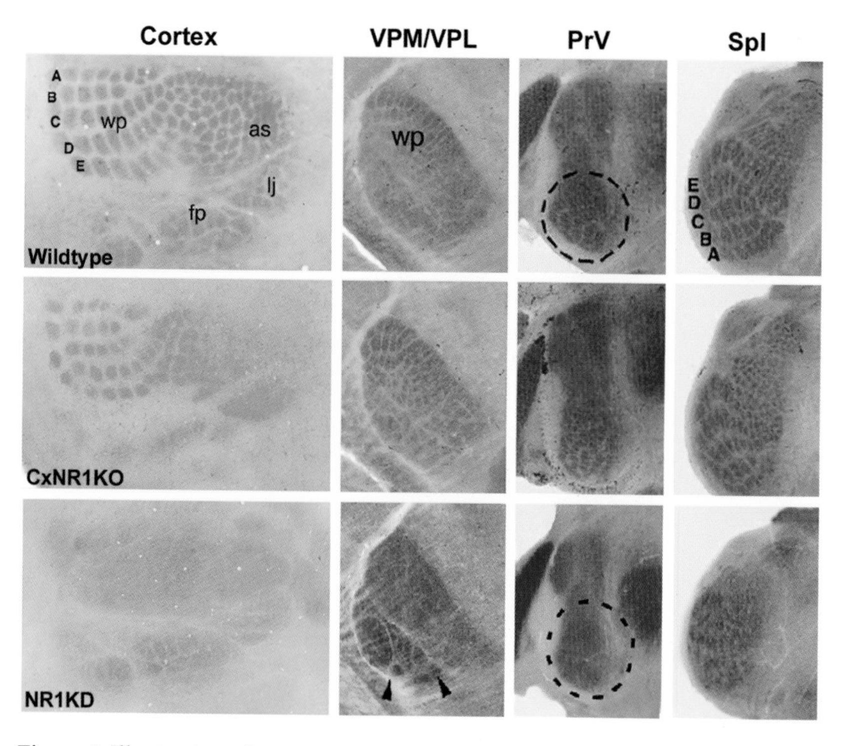

Figure 1 Illustration of somatosensory map phenotypes in the neocortex, thalamus, and trigeminal brainstem in wild type, Cx*NR1*KO, and *NR1*KD mice. Top panel shows cortical, thalamic and brainstem trigeminal (PrV and SpI) patterning in wild type mice as revealed with cytochrome oxidase (CO) histochemistry. Whisker rows A-E are indicated for the cortical map and in SpI. Dashed circles in the photomicrograph of the PrV indicate the whisker representation area. Wp: whiskerpad representation; as: anterior snout representation; lj: lower jaw representation; fp: forepaw representation. Note that in the middle panel of photomicrographs subcortical patterns in the thalamus, PrV and SpI are normal in Cx*NR1*KO mice, but in the SI cortex there is only patterning in the large whisker representation area, and these CO-patches are considerably smaller. In *NR1*KD mice, there is no patterning in the cortical face area, but digit-related patterns can be seen in the forepaw representation zone. In the thalamus, VPL has patterning but VPM does not. In these mice the PrV also lacks patterning (area indicated by dashed circle), but there is rudimentary patterning in the SpI.

secreted proteins such as FGF8, BMPs and Wnts, and graded expression of transcription factors such as Emx2 and Pax6 (Fukuchi-Shimogori and Grove, 2001; Grove and Fukuchi-Shimogori, 2003; Hamasaki et al., 2004, Polleux, 2004; Fukuchi-Shimogori and Grove, 2005, this volume). Within the primary somatosensory cortex, another molecular cue, *ephrin-A5* is expressed in a medial > lateral gradient, whereas within somatosensory thalamus (VPM and VPL), *EphA4* is expressed in a ventromedial > dorsolateral gradient during the perinatal stage. Studies using *Ephrin-A5* knockout mice indicate that cortical Ephrin-A5 is an important molecular cue for the topographic projection of EphA4 expressing thalamocortical axons (Vanderhaeghen et al., 2000). Interestingly ephrin-A5 also plays an important role in controlling the establishment of proper axonal projections between areas (Bolz et al., 2004). Consistent with this, the graded expression of the *ephrin-A5* is found not only in the cortex but also in the ventral thalamus during development (Dufour et al., 2003). Another molecule implicated in topographic projection of thalamocortical axons is phosphoprotein GAP-43 localized in thalamocortical axons and shows patterned expression during barrel formation (Erzurumlu et al., 1990). The *GAP-43* heterozygous knockout mice have enlarged barrels due to defects in thalamocortical path finding (McIlvain et al., 2003). The homozygous knockout mice of this molecule have normal barrelettes and barreloids in the brainstem and thalamus, respectively, but abnormal somatotopy of thalamocortical projection leading to impaired barrels. Furthermore, thalamocortical afferents often project to widely separated cortical targets (Maier et al., 1999). Current understanding is that, these topographically organized projections (somatotopic maps) along the somatosensory systems are established via neural activity-independent mechanisms (but see Catalano and Shatz, 1998; Molnár et al., 2005, this volume) and patterning of neural connections within "somatotopic" maps is controlled by neural activity-mediated mechanisms (Erzurumlu and Kind, 2001), particularly involving NMDARs. Aside from patterning of somatotopic maps, NMDARs also contribute to the areal parcellation of body map subdivisions in the mouse SI cortex. In transgenic mice with differential levels of *NR1* expression along the trigeminal and dorsal column pathways, the cortical face representation area, (which does not show any patterning) is diminished in size, while the paw representation areas with digit-related patterns expand (Lee and Erzurumlu, 2005). The shrinkage of cortical face representation in these mice can be attributed to diminished volume of principal sensory nucleus (PrV) of the trigeminal nerve and VPM. However, expansion of the paw representation areas indicates cortical activity-dependent competitive mechanisms, as there are no volumetric changes in the dorsal column nuclei and the VPL.

The rodent somatosensory pathway is a well-established model system for studies on mechanisms of sensory map formation and plasticity. The instructive role of the sensory periphery in sculpting central neural patterns have been demonstrated by lesion studies performed in perinatal rodents, or in mice selectively bred for aberrant numbers of whiskers (Welker and Van der Loos, 1986; Woolsey, 1990; O'Leary et al., 1994; Killackey et al., 1995; Ohsaki et al., 2002). Several lines of evidence also indicate that somatosensory periphery-related neural maps and patterns are conveyed to target cells by the afferents at each synaptic relay station (Erzurumlu and Jhaveri, 1990; 1992a, b; Senft and Woolsey, 1991). Recent studies have begun unveiling the mechanisms by which thalamocortical afferent terminals develop patterns, how their postsynaptic partners use these templates to pattern their dendritic trees, and how differential proportions of neural tissues are devoted to subcomponents isolated by septa within the body map.

Patterning of Somatosensory Pathways and NMDARs

First reports on the role of NMDARs in patterning of developing sensory projections were results of pharmacological blockade experiments in the vertebrate visual system. For example, NMDA antagonist APV application prevented eye specific segregation of retinal inputs in three-eyed tadpoles (Cline et al., 1987), ocular dominance plasticity in the cat visual cortex (Kleinschmidt et al., 1987; Bear et al., 1990), segregation of retinogeniculate sublaminae in the ferret (Hahm et al., 1991), and refinement of retinotectal projections in the rat (Simon et al., 1992). Pharmacological activity or NMDAR blockade studies along the whisker-barrel pathway of neonatal rats initially gave negative results (Chiaia et al., 1992a; 1994a, b; Henderson et al., 1992; Schlaggar et al., 1993), while later studies provided evidence for the disruption of barrel cortex organization and its functional attributes (Fox et al., 1996; Mitrovich et al., 1996). Such variations in the effects of pharmacological blockade of NMDARs in postnatal rats could be attributed to the selectivity and effectiveness of the blockade and the relatively late timing of drug application as whisker-specific patterns emerge in the brainstem of rats prenatally and shortly after birth in the SI cortex (Chiaia et al.,1992b; Schlaggar and O'Leary, 1994).

In contrast, targeted mutations of the NMDAR subunit genes in mice consistently demonstrated the importance of NMDAR-mediated activity in patterning of the somatosensory maps. At least, three laboratories developed knockout mice of the gene for *NR1* subunit, the critical NMDAR subunit (Forrest et al., 1994; Li et al., 1994, Tokita et al., 1996). Following complete deletion of *NR1*, topography of somatosensory brainstem projections were not altered but whisker related neural patterns failed to form in the first relay stations of the somatosensory pathway (Li et al., 1994). These mice followed a relatively normal developmental scheme until parturition, but after birth they could not feed, and died within 24 hrs due to respiratory failure. In an attempt to prolong their life span and to confirm the absence of brainstem somatosensory patterns at early postnatal times, Li et al., (1994) blocked the birth of mutant mice for an additional day by pharmacological interventions and after birth stimulation of mutant pups by CO_2. There were no whisker-specific patterns in the BSTC of *NR1* knockout mice at an age equivalent to postnatal day 2 (P2) in normal mice. Similar results were later reported for the *NR2B* subunit knockout mice (Kutsuwada et al., 1996). *NR2B*-deficient mice could not suckle and starved to death within 24 hrs after birth. When these mice were handfed to prolong their survival for two more days, they too failed to develop patterns in the BSTC. In contrast, knockout mice of *NR2D* subunit, the other NR2 subunit highly expressed in embryonic and perinatal brains, grew normally and had normal barrels, barreloids and barrelettes (Ikeda et al., 1995). Since both *NR1* knockout and *NR2B* knockout mice are postnatal lethal, pattern formation at higher trigeminal centers, including the barrel cortex could not be studied.

We then "rescued" the *NR1* knockout mice by transgenic expression of the *NR1-1a* splice variant (Iwasato et al., 1997). *NR1* knockout mice rescued with "high" levels of the transgene expression could survive to adulthood, though their body weight was slightly lower than that of wild-type mice. Their sensory periphery-related patterns were normal along both the trigeminal and dorsal column pathways. On the other hand, in *NR1* knockout mice rescued with "low" levels of transgene expression, pattern formation all along the trigeminal pathway, including the trigeminal recipient zone of the barrel cortex, was abolished. In these mice (*NR1KD* mice), there is approximately 70% reduction of *NR1* expression and NMDAR function in the PrV and rostral neuraxis. However, these mice had higher levels of *NR1* expression in dorsal column nuclei (DCN) than in PrV, and consequently developed digit-related patterning therein and in the VPL and the paw representation areas of the neocortex. As noted above, volumes of the subcortical trigeminal nuclei and the face representation area in the SI cortex were diminished in these mice (Lee and Erzurumlu, 2005). While there were no volumetric changes in subcortical dorsal column relay stations, the paw representation area of the SI cortex expanded, suggesting that NMDARs also play a role in allocation of cortical tissue to body map subdivisions.

A number of studies on recombinant NMDARs revealed that a single amino acid residue in the NR1 subunit aspargine 598 (N598) is critical for key properties of the NMDAR function such as Ca^{2+} influx and Mg^{2+} block (Burnashev et al., 1992; Sakurada et al., 1993; Kuner et al., 1996; Wollmuth et al., 1996; Kashiwagi et al., 1997; Schneggenburger and Ascher, 1997; Traynelis et al., 1998; Zheng et al., 1999). Mg^{2+} block and Ca^{2+} permeability of the NMDARs are abolished by the aspargine (N) to arginine (R) point mutation in the channel-lining region of the membrane domain (Burnashev et al., 1992; Sakurada et al., 1993). Introduction of the N598R mutation in the NR1 subunit impaired coincidence detection properties of NMDARs, and subsequent Ca^{2+} signaling, and *NR1* N598R mutant mice failed to develop barrelette patterns in the brainstem trigeminal nuclei (Rudhard et al., 2003). Previously, Single et al., (2000) also generated mice expressing mutant NMDARs by substituting aspargine (N) with glutamine (Q) or arginine (R). Animals with these point mutations in NR1 subunit displayed similar phenotypes to that of *NR1* knockout mice, and were postnatal lethal mainly due to respiratory failure. The phenotype was partially rescued in heterozygous mice that expressed both wild-type and mutant NR1 subunits at the same level. The authors noted that barrels in the somatosensory cortex formed in heterozygous mice in which 25% of the NMDARs are pure wild-type receptors and the remaining 75% are Ca^{2+}-impermeable receptors carrying one or two mutant NR1 subunits. This study and an earlier study by Iwasato et al., (1997) suggest that neural patterning within the somatosensory map regions requires a threshold level of NMDAR

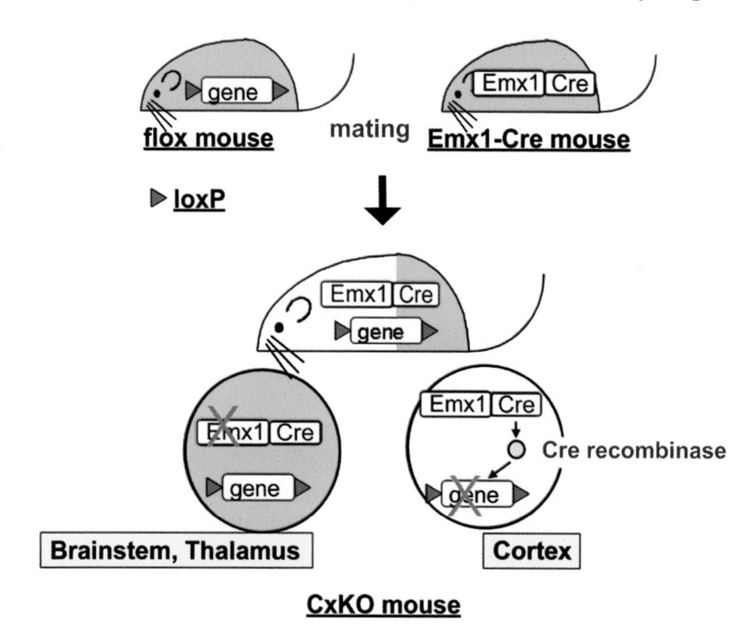

Figure 2 Cortical excitatory neuron-specific gene targeting by Cre/loxP recombination. In an *Emx1*-Cre mouse, *Cre recombinase* gene is expressed in the cortical excitatory neurons but not in cells in the thalamus or brainstem. In a flox mouse, a target gene is flanked by two loxP sequences but is functional. A cortex-specific gene knockout (CxKO) mouse carrying both *Emx1*-Cre and flox alleles is generated by mating between an *Emx1*-Cre and a flox mouse. In cortical excitatory neurons of CxKO mouse, the target gene is disrupted by Cre-mediated excision from the chromosome, while in thalamic and brainstem cells, target gene remains intact because Cre recombinase is not expressed.

function, the precise level of which is yet to be determined for different relay stations along the pathway.

To delineate the specific role of cortical NMDARs in patterning of the somatosensory cortex, we took advantage of the Cre/loxP conditional gene targeting approach (Figure 2). We expressed *Cre recombinase* gene under the control of the dorsal telencephalon-specific *Emx1* promoter by knock-in or BAC/PAC transgenic approaches (Iwasato et al., 2000, 2004). By crossing each *Emx1*-Cre mouse line with floxed *NR1* mice, we generated cortex-restricted *NR1* knockout (Cx*NR1*KO) mice (Iwasato et al., 2000; T.I. and S. Itohara, unpublished data). In these mice, *NR1* gene is deleted in virtually all of the excitatory neurons of the cerebral cortex, hippocampus and olfactory bulb, but remained intact in the thalamus, brainstem, striatum and cerebellum during the period of patterning of the somatosensory pathways. In this mouse model, inhibitory interneurons that migrate to the neocortex from Emx1 negative regions of the pallidum also escaped the NR1 deletion (Iwasato et al., 2000). Unlike *NR1* global knockout mice, Cx*NR1*KO mice survived through postnatal stages. In Cx*NR1* KO mice, whisker-specific patterns in the brainstem trigeminal nuclei and in the thalamus, where the *NR1* expression remains intact, developed normally. In the SI cortex, however, thalamocortical afferents (TCAs) formed only small, rudimentary patterns (as visualized with histochemical and immunohistochemical markers) in regions corresponding to the representation of larger whiskers. Furthermore, layer IV granule cells failed to develop barrels even in regions where there were rudimentary patterning of TCAs (Iwasato et al., 2000). These results demonstrated critical roles of cortical NMDAR-mediated activity in the patterning of both presynaptic component (TC axonal termini) and postsynaptic component (layer IV granule cell bodies and dendrites). Rudimentary patterning of TCAs in Cx*NR1*KO mice might be due to NMDAR activation in cortical GABAergic cells, which escaped the *NR1* deletion. This is a possibility that remains to be experimentally tested. Another interesting observation in these mice was that neonatal whisker-induced structural plasticity followed the same time course as in wild type mice, and rudimentary cortical patterns could be altered up to postnatal day 3 but not thereafter (Datwani et al., 2002a).

An important caveat to all these reports underscoring the involvement of NMDARs in patterning of developing somatosensory pathways (as well as the vast majority of other studies documenting barrel cortex phenotypes in other lines of mutant mice) is that the morphological assays have been done at a gross microscopic level using a variety of histological and immunohistochemical markers for barrel patterns (Erzurumlu and Kind, 2001; López-Bendito and Molnár, 2003). If NMDARs serve as coincidence detectors between pre and postsynaptic elements, how is presynaptic terminal and dendritic differentiation of pattern forming elements in somatosensory centers affected in mice with various types of genetic alterations of the NMDAR function? What are the downstream signaling mechanisms that allow detection of patterning of presynaptic inputs by postsynaptic cells, and consolidation of patterns at both sites? To date there is very little documentation of fine structural defects in mice with reported barrel pattern defects. Below we review our recent results on axonal and dendritic differentiation in mice with *NR1* mutations.

NMDAR-Mediated Differentiation of Trigeminal Sensory and Thalamocortical Axon Terminals

Whisker-specific information is carried to the CNS by the axonal processes of the infraorbital (IO) branch of the maxillary division of the trigeminal nerve. Damage to this nerve or to whisker follicles up to 3 days after birth (critical period) irreversibly alters all neural patterning at each level of the trigeminal neuraxis (Woolsey 1990; O'Leary et al. 1994). During normal development, whisker-related brainstem patterns (barrelettes) appear around P2-3 and are consolidated by P5(Ma 1993). Both in *NR1* knockout mice and in *NR1*KD mice these patterns fail to develop in the PrV (the nucleus which is solely responsible from transmitting whisker-patterns to the contralateral ventrobasal thalamus and subsequently to the barrel cortex), but rudimentary patterns can be seen in portions of the spinal trigeminal nucleus, subnucleus interpolaris (SpI) only in *NR1*KD mice. Peripheral trigeminal (IO) axons invade the developing whisker fields around embryonic day (E) 10 in the mouse, and their central counterparts lay down the trigeminal tract in the brainstem by E13 (Stainier and Gilbert 1990; 1991). Once the tract extends caudally to the level of the cervical spinal cord, axon extension is halted, and these single axons begin emitting radial collaterals into the brainstem trigeminal nuclei, where they eventually start to form whisker-specific patchy terminals by E17.

Carbocyanine dye labeling of single trigeminal axons from individual whisker follicles during development of the central trigeminal pathway in NMDAR mutant and control animals revealed striking differences (Lee et al., 2005a; Figure 3). In control, *NR1* knockout and *NR1*KD mice initial arborization patterns in the PrV are simple and similar at E15, but by E17 terminal fields show notable differences. Trigeminal arbors in control cases show patchiness and elaboration of small terminal branches, while much larger and highly branched terminal arbor field is emergent in *NR1*KD and more so in *NR1* knockout cases. At the time of birth, the whisker afferent arbors are the largest and most complex in *NR1* knockout mice and conspicuously larger in *NR1*KD animals in comparison to controls. By P5, after the end of the critical period for whisker-lesion induced morphological plasticity (Woolsey 1990), trigeminal terminal arbors in the *NR1*KD PrV occupy five-fold larger area than those in control cases. Clearly wide spread terminal arbors, increased branch tips and overlapping distribution of whisker afferents within the PrV of *NR1* knockout and *NR1*KD mice are major defects that contribute to the absence of barrelette patterns. Most whisker afferents bifurcate upon entry into the brainstem, one branch extends rostrally to form the ascending component of the central trigeminal tract and the other elongates caudally to contribute to the descending trigeminal tract (Jacquin et al., 1993). In *NR1*KD mice, there are differential levels of expression of *NR1* between the PrV and SpI, the latter having more expression (Iwasato et al., 1997). Consequently in the middle portion of the SpI in *NR1*KD mice, there is rudimentary whisker-specific patterning. Comparison of single axons terminating in the PrV and SpI in *NR1* knockout and *NR1*KD mice revealed that the same axon can form restricted terminal patches in the SpI of the *NR1*KD mice in comparison to wide terminal fields in the PrV, whereas both branches formed extensive terminals in the SpI and PrV of the *NR1* knockout mice (Lee and Erzurumlu, unpublished observations). These findings, along with those from Cx*NR1*KO mice (see below), provide a strong argument for the involvement of postsynaptic NMDARs in restricting terminal arbor fields of whisker afferents and formation of whisker-specific patches.

Cx*NR1*KO mice develop normal barrelette and barreloid patterns in the brainstem and thalamus, respectively, but in the barrel cortex, cellular aggregates (barrels) fail to form (Figure 1). A rudimentary patterning corresponding to the large whiskers can be visualized with serotonin

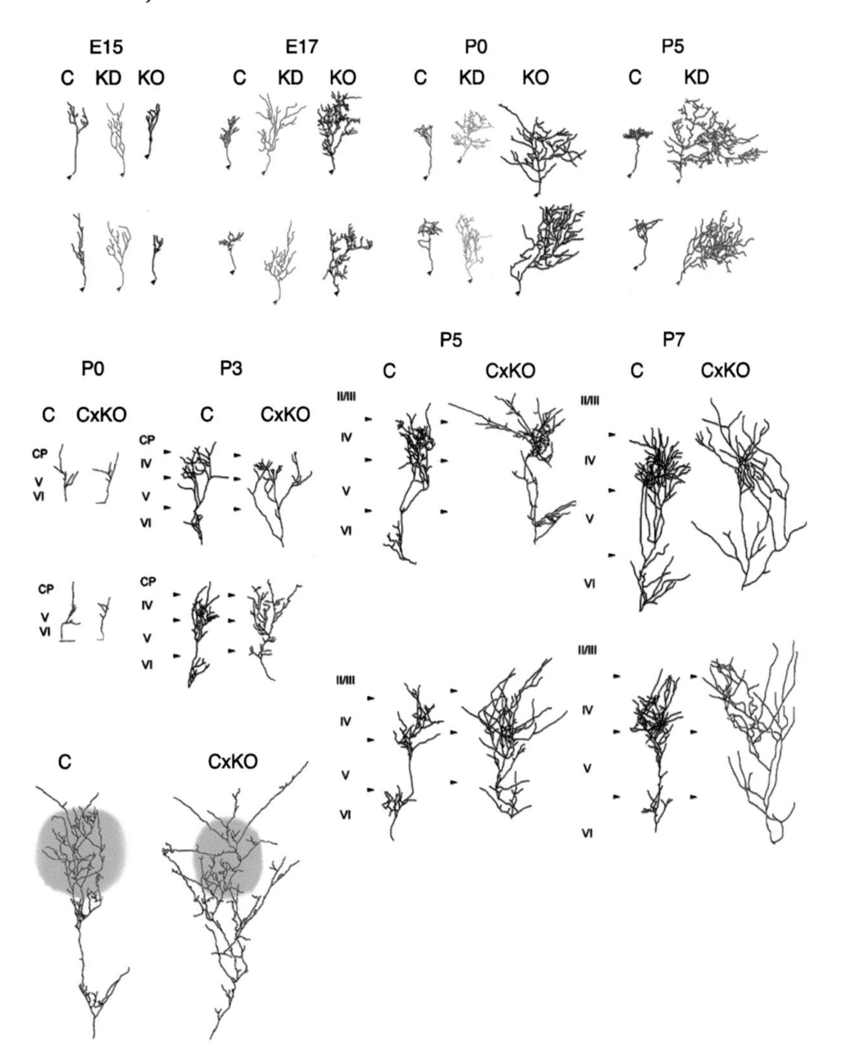

Figure 3 Illustration of trigeminal afferent terminals in the PrV of control, *NR1*KD, and global *NR1* knockout, mice and TCA terminals in control, and Cx*NR1*KO mice at different developmental time points. Top two rows show examples of single whisker afferent terminals in the PrV at E15, E17, and P0 for control (C), NR1*KD* (KD), and *NR1* global knockout (KO) cases (from left to right) and control and NR1*KD* cases at P5. Bottom two rows show examples of single TCA terminals labeled from the VPM for control (C) and Cx*NR1*KO (CxKO) cases at P0, P3, P5, and P7 (from left to right). Cortical laminae are indicated to the left of each pair. Note that both in the PrV and SI cortex these afferent terminals start branching in a similar fashion at early stages of development, but when *NR1* gene is disrupted afferent terminals grow extensive branches. Bottom panel shows comparison of single TCA terminals with respect to CO-dense patches seen in the control and CxNR1 cortex. Figure adapted from Lee et al., 2005a and 2005b).

transporter (5-HTT) immunohistochemistry and cytochrome oxidase (CO) histochemistry (Iwasato et al., 2000; Datwani et al., 2002b). Because developing TCAs transiently express 5-HTT (Lebrand et al., 1996; 1998), it has been used as a reliable marker for developing somatosensory, visual and auditory TCAs. CO histochemistry is also a common barrel pattern marker for pre and postsynaptic zones rich in mitochondria (Wong-Riley and Welt, 1980; Wong-Riley, 1989). These individual patches in the Cx*NR1*KO mice are much smaller and inter-patch distances are wider than those in the wild-type mice. The emergence of whisker specific patterning in the wild-type mice barrel cortex was visualized with 5-HTT immunohistochemistry as early as P3. This marker for TCAs and a CO histochemistry show the emergence and consolidation of whisker-specific TCA patterns in the wild-type mouse barrel cortex between P3-7. In Cx*NR1*KO mice, these patterns consolidate during the same period as in wild-type mice (Lee et al., 2005b).

Detailed analyses of single TCA development in the barrel cortex of Cx*NR1*KO mice between P1-7 revealed that while whisker-specific TCAs target proper cortical layers at first and begin arborization similar to that seen in control cases, their growth is not confined to layer IV (Lee et al., 2005b). At P1, TCAs invade the cortical plate as simple axons with few small branches, and their morphological appearance is similar in both control and Cx*NR1*KO mice (Figure 3). By P3, TCAs of the control animals display focalized branches in layer IV and to a lesser extent in layer VI. In contrast, in Cx*NR1*KO cortex, TCAs display a wider terminal territory and more branches in other layers. At P5 and later on, as the normal TCAs consolidate their focal terminal arbors in layer IV and fewer terminal arbors in layer VI, TCAs of Cx*NR1*KO mice continue their expansion and branching in other layers. In this study (Lee et al., 2005b) it was calculated that in control animals, bifurcation points and terminal tips were mostly distributed in layer IV (about 75-80% of the total number), with some in layer VI (10–15%). In Cx*NR1*KO cases from P5 and on, significantly reduced numbers of both bifurcation points and terminal tips were counted in layer IV. Greatly increased bifurcations of TCAs, as well as their terminals, were found in layers II/III and V in Cx*NR1*KO cortices at P3 and older ages, also suggesting that when cortical excitatory neurons lack functional NMDARs, TCAs fail to recognize any putative layer-specific "stop" signals (Molnár and Blakemore, 1995; see Yamamoto et al., 2005, this volume). Interestingly, the total numbers of bifurcation points and terminal tips in all cortical layers for each age did not show any significant differences, however, the total axonal branch length within the terminal field of all reconstructed single axons for each age was significantly higher in Cx*NR1*KO cases beginning on P5. This increase indicates that as the terminal arbors begin shaping, terminal branch segments get longer, thereby contributing to the wider span of terminal arbors seen in Cx*NR1*KO animals. Despite these large arbors in the Cx*NR1*KO cortex, zones of TCA terminal condensations were seen in layer IV. These terminal condensations correspond to the rudimentary patterning seen with histochemical and immunohistochemical markers.

In addition to terminal spreading in several cortical layers, the mediolateral span of Cx*NR1*KO TCAs was also double that of control TCAs by P7 (Lee et al., 2005b). Since in control and knockout phenotypes

TCA arbor mediolateral extent is similar during initial phases of cortical target invasion, exuberant growth of TCA terminals in Cx*NR1*KO SI cortex indicate that postsynaptic NMDARs might act as "stop and elaborate" signals for their presynaptic partners. Both studies (Lee et al., 2005a, b) at the level of the first (brainstem) and third (SI cortex layer IV) relay stations of the trigeminal pathway clearly show that NMDAR deficiency leads to exuberant presynaptic axon terminal branching, suggesting the presence of retrograde signals released through NMDAR activation of cortical cells that control pruning and patterning of presynaptic terminals.

NMDAR Function on Dendritic Differentiation of Barrelette and Barrel Cells

In the rodent PrV, there are three main cell types: small GABAergic interneurons, small trigeminothalamic projection cells (barrelette cells), and very large internuclear projection cells (interbarrelette cells) with dendrites that span multiple barrelettes. All three cell types have distinguishing electrophysiological properties in addition to their morphological characteristics (Lo et al., 1999; Lo and Erzurumlu 2001). Afferent patterning is detected only by the trigeminothalamic projection or "barrelette" neurons. These cells orient their dendrites toward discrete patches of trigeminal afferent terminals. As a result, whisker-specific barrelette units are formed (Ma and Woolsey 1984; Bates and Killackey 1985; Ma 1993). PrV barrelette cells convey these patterns to the thalamic barreloids and subsequently to the somatosensory barrel cortex (Woolsey and Van der Loos 1970; Van der Loos 1976; Killackey and Fleming 1985; Erzurumlu and Jhaveri 1990; Senft and Woolsey 1991). In *NR1*KD mice, electrophysiological properties of barrelette neurons are not altered, and synaptic communication is intact, except for 80% reduction in NMDAR currents (Lee et al., 2005a). While membrane properties of barrelette neurons remain unchanged in these transgenic mice, their dendritic differentiation is dramatically altered. Normally barrelette neurons have three primary dendrites emanating from the soma with restricted dendritic fields that are oriented towards the barrelette centers. *NR1*KD barrelette neurons were found to have four primary dendrites that radiated in all directions from the soma (Lee et al., 2005a). Overall, *NR1*KD barrelette neuron dendrites showed little or no orientation preference, had longer segments and fewer high order branches, indicating that NMDAR-mediated mechanisms play a major role in dendritic sculpting, complexity, and orientation (Figure 4).

Similar dendritic defects were reported for the layer IV spiny stellate (or barrel) cells in the barrel cortex of the Cx*NR1*KO mice (Datwani et al., 2002b). Normally, barrel cells in layer IV orient their dendrites towards tufts of TCA terminals, and in mice form cellular rings around them. In Cx*NR1*KO mice both the dendritic orientation and

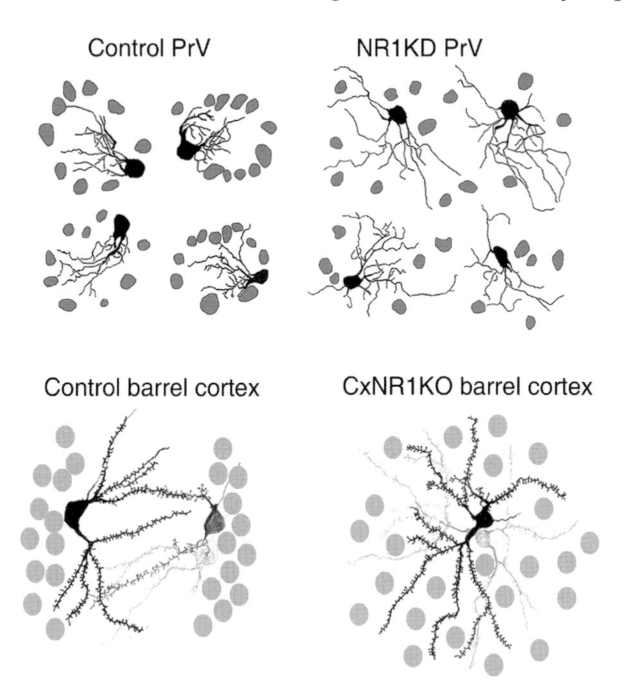

Figure 4 Illustration of barrelette cells in *NR1*KD (top row) and barrel cells in Cx*NR1*KO (bottom row) cases. In normal mice barrelette and barrel cells orient their dendrites toward the center of whisker-related cytoarchitectonic units (indicated by purple cellular profiles). In *NR1*KD mice there are no barrelettes in the PrV and barrelette cells (as identified by their electrophysiological properties and morphologies) fail to orient their dendrites and show exuberant dendritic growth. In the barrel cortex of Cx*NR1*KO mice, spiny stellate cells also fail to develop dendritic orientation bias and grow extensive dendritic trees. Figure adapted from Lee et al., 2005a; Datwani et al., 2002b).

cellular rings were absent (Figure 4). Additionally, increased dendritic spines were noted in second order dendritic branches. Thus, it is likely that a major consequence of NMDAR impairment is overgrowth of pre and postsynaptic neuronal processes. During the process of concurrent addition and pruning of presynaptic terminal and postsynaptic dendritic branches NMDAR-mediated activity could act as a stop/stabilization signal, thereby contributing to their focalization and patterning. Whether these structural changes in pre- and postsynaptic elements occur independently and concurrently or one follows the other remains to be determined. Comparison of presynaptic arbor differentiation between control, *NR1* knockout and *NR1*KD animals suggest the presence of a threshold level of NMDAR function below which morphological differentiation is affected while synaptic transmission is not. Presently this threshold and the signaling pathways downstream from NMDARs utilized in clustering of whisker afferent terminals and dendrites of their postsynaptic partners are not known.

Molecules Involved in Patterning of the Somatosensory System

Besides NMDAR subunit mutants, there are other genetically altered or spontaneous mutant mice, which show deficiencies in patterning of the SI cortex (see Gaspar and Rebsam, 2005, this volume and Molnár et al., 2005; this volume). In a transgenic mouse line (*MAOA* knockout) in which insertion of a transgene disrupted a monoamine oxidase (MAO)-A gene, patterning in the SI cortex was completely impaired, though barreloids and barrellettes formed (Cases et al., 1996). Subsequently, it was found that in *MAOA* knockout mice, barrel impairment is caused by the excessive activation of serotonin (5-HT) 1B receptors on thalamocortical axons by excess levels of extracellular 5-HT (Salichon et al., 2001). Consistently, 5-HTT knockout mice revealed a nearly complete absence of barrels and barreloids and less organized barrelettes (Persico et al., 2001). In 1996, a spontaneous mutant mouse line, *barrelless*, was reported. In this mouse, barrels are completely impaired and barreloids in the thalamus are partially impaired (Welker et al., 1996). By a linkage analysis, the *barrelless* gene was found to be adenylyl cyclase Type 1 (AC1), the neuron-specific calcium/calmodulin-stimulated adenylyl cyclase. *AC1* knockout mice generated by a regular gene targeting method also showed similar phenotypes as *barrelless* mice (Abdel-Majid et al., 1998).

These results demonstrate that serotoninergic system and cAMP pathway play critical roles in somatosensory cortical patterning. However, it is important to note that these studies used global knockout mice in which gene product is lost in all cells. Therefore except for a few serendipitous cases, it is not clear where along the somatosensory system a given molecule plays a role in development. For example, it is not clear whether AC1 plays an important role at the pre or post synaptic site (or both) during TCA patterning and barrel formation. *AC1* is highly expressed in the somatosensory brainstem and thalamus in addition to the neocortex during early postnatal stages (Matsuoka et al., 1997). In hippocampus, cerebellum and other systems, AC1 is reported to play a role in neuronal plasticity at the pre and postsynaptic sites (Wu et al., 1995, Villacres, et al., 1998, Storm et al., 1998, Ferguson and Storm, 2004; Wang et al., 2004). In the SI cortex, *barrelless* mice show impairment of both TCA terminal arbors and layer IV neurons. *Barrelless* mice contain few functional AMPARs at the postsynaptic site and LTP and LTD at thalamocortical synapses are difficult to induce (Lu et al., 2003). At the presynaptic side, neurotransmitter release is reduced (Lu et al., 2002). By global knockout studies, only PLC-β1 is clearly shown to play a role in the postsynaptic side of thalamocortical synapses (Hannan et al., 2001), because in wild-type brain, *PLC-β1* is predominantly expressed in the telencephalon during early postnatal stages (Watanabe et al., 1998). Hannan et al. (2001) further showed that *mGluR5* knockout mice also have impairment of barrel formation and they suggested that mGluR5 works an upstream molecule of PLC-β1 at the postsynaptic side. However, because mGluR5 protein is found in the somatosensory thalamus and brainstem in addition to the cortex during the barrel formation (Munoz et al., 1999), again global knockout mice cannot help pinpoint where in the somatosensory system, mGluR5 plays a role in patterning. Region-specific knockout models using the Cre/loxP system is a powerful tool to overcome such problems and can help dissociate between pre and postsynaptic mechanisms and the specific role of each molecule in barrel formation (Iwasato et al., 2000, 2004).

In the case of cortex-restricted *NR1* knockout (Cx*NR1*KO) mice, it is clear that NMDARs in cortical cells play a major role in development and patterning of both pre and postsynaptic components. Morphological defects at both sites indicate active communication via anterograde and retrograde signaling mechanisms. The nature of these signals is poorly understood. In other systems, nitric oxide (NO), brain-derived neurotrophic factor (BDNF), and arachidonic acid (AA) have been implicated as potential retrograde signals that might affect structural differentiation of presynaptic terminals (see Schmidt, 2004 for a review). In addition, a number of molecules downstream from NMDAR-initiated Ca^{2+} influx and those that act cooperatively with NMDARs at the postsynaptic density have been noted in modulating dendritic cytoskeletal dynamics, spine morphology, and presynaptic terminal sculpting (Carroll and Zukin, 2002; Scheiffele, 2003; Wenthold et al., 2003). Calcium/calmodulin dependent protein kinase II (CaMKII) (Wu and Cline 1998; Zou and Cline 1999), neuroligins and neurexins (Nguyen and Südhof, 1997; Scheiffele et al., 2000) and Eph proteins (Dalva et al., 2000) are among these candidates.

At present, molecular cascades downstream of NMDAR in pattern formation are unclear. They should be gradually revealed by studies with new lines of genetically altered mice. Especially it is important to use and further develop conditional knockout systems, which allow gene targeting with distinct region and cell type-specificities, such as cortical excitatory neuron-, all cortical neuron- and thalamic cell-specific ones.

Acknowledgments

We are grateful to our colleagues; P.T. Huerta, D.F. Chen, T. Sasaoka, S. Tonegawa, A.M. Wolf, H. Nishiyama, Y. Taguchi, T. Knöpfel, R. Ando, S. Itohara, and F.-S. Lo, and former graduate students; E. Ulupinar, A. Datwani and L.-J. Lee for their contributions to the studies reviewed here. Research in the authors' laboratories is supported by NIH (NS 039050 and NS37070, R.S.E) LSUHSC School of Medicine Research Enhancement Fund (R.S.E.) and Grant-in-Aid for Scientific Research on Priority Areas-A from the Ministry of Education, Culture, Sports, Science, and Technology of Japan (T.I.).

References

Abdel-Majid, R.M., Leong, W.L., Schalkwyk, L.C., Smallman, D.S., Wong, S.T., Storm, D.R., Fine, A., Dobson, M.J., Guernsey, D.L., and Neumann, P.E. (1998). Loss of adenylyl cyclase I activity disrupts patterning of mouse somatosensory cortex. *Nat. Genet.* 19:289–291.

Andersson, O., Stenqvist, A., Attersand, A., and von Euler, G. (2001). Nucleotide sequence, genomic organization, and chromosomal localization of genes encoding the human NMDA receptor subunits NR3A and NR3B. *Genomics* 78:178–184.

Ascher, P., and Nowak, L. (1988). The role of divalent cations in the *N*-Methyl-D-aspartate responses of mouse central neurons in culture. *J. Physiol. (Lond.)* 399:247–266.

Bates, C.A., and Killackey, H.P. (1985). The organization of the neonatal rat's brainstem trigeminal complex and its role in the formation of central trigeminal patterns. *J. Comp. Neurol.* 240:265–87.

Bear. M.F., Kleinschmidt, A., Gu, Q., and Singer, W. (1990). Disruption of experience-dependent synaptic modifications in striate cortex by infusion of an NMDA receptor antagonist. *J. Neurosci.* 10:909–925.

Belford, G.R., Killackey, H.P. (1978). Anatomical correlates of the forelimb in the ventrobasal complex and the cuneate nucleus of the neonatal rat. *Brain Res.* 158:450–455.

Belford, G.R., Killackey, H.P. (1979). The development of vibrissae representation in subcortical trigeminal centers of the neonatal rat. *J. Comp. Neurol.* 188:63–74.

Bliss, T.V.P., and Collingridge, G.L. (1993). A synaptic model of memory: long-term potentiation in the hippocampus. *Nature* 361:31–39.

Bolz, J., Uziel, D., Muhlfriedel, S., Gullmar, A., Peuckert, C., Zarbalis, K., Wurst, W., Torii, M., and Levitt, P. (2004). Multiple roles of ephrins during the formation of thalamocortical projections: maps and more. *J. Neurobiol.* 59:82–94.

Burnashev, N., Schoepfer, R., Monyer, H., Ruppensberg, J.P., Günther, W., Seeburg, P.H., and Sakmann, B. (1992). Control by aspargine residues of calcium permeability and magnesium blockade in the NMDA receptor. *Science* 257:1415–1419.

Carroll, R.C., and Zukin, R.S. (2002) NMDA-receptor trafficking and targeting: implications for synaptic transmission and plasticity. *Trends. Neurosci.* 25:571–577.

Cases, O., Vitalis, T., Seif, I., De Maeyer, E., Sotelo, C., and Gaspar, P. (1996). Lack of barrels in the somatosensory cortex of monoamine oxidase A-deficient mice: Role of a serotonin excess during the critical period. *Neuron* 16:297–307.

Catalano, SM., and Shatz, C.J. (1998). Activity-dependent cortical target selection by thalamic axons. *Science* 281:559–562.

Chiaia, N.L., Fish, S.E., Bauer, W.R., Bennett-Clarke, C.A., and Rhoades, R.W. (1992a). Postnatal blockade of cortical activity by tetrodotoxin does not disrupt the formation of vibrissa-related patterns in the rat's somatosensory cortex. *Dev. Brain Res.* 66:244–250.

Chiaia, N.L., Bennett-Clarke, C.A., Eck, M., White, F.A., Crissman, R.S., and Rhoades, R.W. (1992b). Evidence for prenatal competition among the central arbors of trigeminal primary afferent neurons. *J. Neurosci.* 12:62–76.

Chiaia, N.L., Fish, S.E., Bauer, W.R., Figley, B.A., Eck, M., Bennett-Clarke, C.A., and Rhoades, R.W. (1994a). Effects of postnatal blockage of cortical activity with tetrodotoxin upon lesion-induced reorganization of vibrissae-related patterns in the somatosensory cortex of rat. *Dev. Brain Res.* 79:301–306.

Chiaia, N.L., Fish, S.E., Bauer, W.R., Figley, B.A., Eck, M., Bennett-Clarke, C.A., and Rhoades, R.W. (1994b). Effects of postnatal blockade of cortical activity with tetrodotoxin upon the development and plasticity of vibrissa-related patterns in the somatosensory cortex of hamsters. *Somatosens. Mot. Res.* 11:219–228.

Ciabarra, A.M., Sullivan, J.M., Gahn, L.G., Pecht, G., Heinemann, S., and Sevarino, K.A. (1995). Cloning and characterization of chi-1: a developmentally regulated member of a novel class of the ionotropic glutamate receptor family. *J. Neurosci.* 15:6498–4508.

Clements, J.D. and Westbrook, G.L. (1991). Activation kinetics reveal the number of glutamate and glycine binding sites on the N-methyl-D-aspartate receptor. *Neuron* 7: 605–613.

Cline, H.T., Debski, E.A., and Constantine-Paton, M. (1987). N-methyl-D-aspartate receptor antagonist desegregates eye-specific stripes. *Proc. Natl. Acad. Sci. USA* 84:4325–4342.

Constantine-Paton, M., Cline, H.T., and Debski, E. (1990). Patterned activity, synaptic convergence, and the NMDA receptor in developing visual pathways. *Annu. Rev. Neurosci.* 13:129–54.

Dalva, M.B., Takasu, M.A., Lin, M.Z., Shamah, S.M., Hu, L., Gale, N.W., and Greenberg, M.E. (2000). EphB receptors interact with NMDA receptors and regulate excitatory synapse formation. *Cell* 103:945–956.

Das, S., Sasaki, Y.F., Rothe, T., Premkumar, L.S., Takasu, M., Crandall, J.E., Dikkes, P., Conner, D.A., Rayudu, P.V., Cheung, W., Chen, H.S., Lipton, S.A., and Nakanishi, N. (1998). Increased NMDA current and spine density in mice lacking the NMDA receptor subunit NR3A. *Nature* 393:377–381.

Datwani, A., Iwasato, T., Itohara, S., and Erzurumlu, R.S. (2002a). Lesion-induced thalamocortical axonal plasticity in the S1 cortex is independent of NMDA receptor function in excitatory cortical neurons. *J. Neurosci.* 22:9171–9175.

Datwani, A., Iwasato, T., Itohara, S., and Erzurumlu, R.S. (2002b). NMDA receptor-dependent pattern transfer from afferents to postsynaptic cells and dendritic differentiation in the barrel cortex. *Mol. Cell. Neurosci.* 21:477–492.

Dawson, D.R., and Killackey, H.P. (1987). The organization and mutability of the forepaw and hindpaw representations in the somatosensory cortex of the neonatal rat. *J. Comp. Neurol.* 256:246–256.

Dingledine, R., Borges, K., Bowie, D., and Traynelis, S.F. (1999). The glutamate receptor ion channels. *Pharmacol. Rev.* 51:7–61.

Dufour, A., Seibt, J., Passante, L., Depaepe, V., Ciossek, T., Frisen, J., Kullander, K., Flanagan, J.G., Polleux, F., and Vanderhaeghen, P. (2003). Area specificity and topography of thalamocortical projections are controlled by ephrin/Eph genes. *Neuron* 39:453–465.

Erzurumlu, R.S., and Jhaveri, S. (1990). Thalamic axons confer a blueprint of the sensory periphery onto the developing rat somatosensory cortex. *Dev. Brain. Res.* 56:229–234.

Erzurumlu, R.S., and Jhaveri, S. (1992a). Trigeminal ganglion cell processes are spatially ordered prior to the differentiation of the vibrissa pad. *J. Neurosci.* 12:3946–3955.

Erzurumlu, R.S., and Jhaveri, S. (1992b). Emergence of connectivity in the embryonic rat parietal cortex. *Cereb. Cortex* 2:336–352.

Erzurumlu, R.S., and Kind, P.C. (2001). Neural activity: sculptor of 'barrels' in the neocortex. *Trends. Neurosci.* 24:589–595.

Erzurumlu, R.S., Jhaveri, S., and Benowitz, L.I. (1990). Transient patterns of GAP-43 expression during the formation of barrels in the rat somatosensory cortex. *J. Comp. Neurol.* 292:443–456.

Ferguson, G.D., and Storm, D.R. (2004), Why calcium-stimulated adenylyl cyclases? *Physiology* 19: 271–276.

Forrest, D., Yuzaki, M., Soares, H.D., Ng, L., Luk, D.C., Sheng, M., Stewart, C.L., Morgan, J.I., Connor, J.A., and Curran, T. (1994). Targeted disruption of NMDA receptor 1 gene abolishes NMDA response and results in neonatal death. *Neuron* 13:325–338.

Fox, K., and Daw, N.W. (1993). Do NMDA receptors have a critical function in visual cortical plasticity? *Trends. Neurosci.* 16:116–122.

Fox, K., Schlaggar, B.L., Glazewski, S., and O'Leary, D.D. (1996). Glutamate receptor blockade at cortical synapses disrupts development of thalamocortical and columnar organization in somatosensory cortex. *Proc. Natl. Acad. Sci. USA* 93:5584–9.

Fukuchi-Shimogori, T., and Grove, E.A. (2001). Neocortex patterning by the secreted signaling molecule FGF8. *Science* 294:1071–1074.

Grove, E.A., and Fukuchi-Shimohoro, T. (2003). Generating the cerebral cortical area map. *Annu. Rev. Neurosci.* 26:355–380.

Hahm, J.O., Langdon, R.B., and Sur, M. (1991). Disruption of retinogeniculate afferent segregation by antagonists to NMDA receptors. *Nature* 351:568–570.

Hamasaki, T., Leingartner, A., Ringstedt, T., and O'Leary, D.D. (2004). EMX2 regulates sizes and positioning of the primary sensory and motor areas in neocortex by direct specification of cortical progenitors. *Neuron* 43:359–372.

Hannan, A.J., Blakemore, C., Katsnelson, A., Vitalis, T., Huber, K.M., Near, M., Roder, J., Kim, D., Shin, H.S., and Kind, P.C. (2001). PLC-B1, activated via mGluRs, mediates activity-dependent differentiation in cerebral cortex. *Nat. Neurosci.* 4:282–328.

Henderson, T.A., Woolsey, T.A., and Jacquin, M.F. (1992). Infraorbital nerve blockade from birth does not disrupt central trigeminal pattern formation in the rat. *Dev. Brain Res.* 66:146–52.

Hollmann, M. (1999). Structure of ionotropic glutamate receptors. In: Jonas, P., and Monyer, H. (eds.), *Ionotropic Glutamate Receptors in the CNS*. Springer, Berlin, pp 3–98.

Hollmann, M., and Heinemann, S. (1994). Cloned glutamate receptors. *Annu. Rev. Neurosci.* 17:31–108.

Ikeda, K., Nagasawa, M., Mori, H., Araki, K., Sakimura, K., Watanabe, M., Inoue, Y., and Mishina, M. (1992). Cloning and expression of the epsilon 4 subunit of the NMDA receptor channel. *FEBS Lett.* 313:34–38.

Ikeda, K., Araki, K., Takayama, C., Inoue, Y., Yagi, T., Aizawa, S., and Mishina, M. (1995). Reduced spontaneous activity of mice defective in the epsilon 4 subunit of the NMDA receptor channel. *Mol. Brain Res.* 33: 61–71.

Iwasato, T., Erzurumlu, R.S., Huerte, P.T., Chen, D.F., Sasaoka, T., Ulupinar, E., and Tonegawa, S.T. (1997). NMDA receptor-dependent refinement of somatotopic maps. *Neuron* 19:1201–1210.

Iwasato, T., Datwani, A., Wolf, A.M., Nishiyama, H., Taguchi, Y., Tonegawa, S., Knöpfel,T., Erzurumlu, R.S., and Itohara, S. (2000). Cortex-restricted disruption of NMDAR1 impairs neuronal patterns in the barrel cortex. *Nature* 406: 726–731.

Iwasato, T., Nomura, R., Ando, R., Ikeda, T., Tanaka, M., and Itohara, S. (2004). Dorsal telencephalon-specific expression of Cre recombinase in PAC transgenic mice. *Genesis* 38: 130–138.

Jacquin, M.F., Renehan, W.E., Rhoades, R.W., and Panneton, W.M. (1993). Morphology and topography of identified primary afferents in trigeminal subnuclei principalis and oralis. *J. Neurophysiol.* 70:1911–1936.

Kashiwagi, K., Pahk, A.J., Masuko, T., Igarashi, K., and Williams, K. (1997). Block and modulation of *N*-methyl-D-aspartate receptors by polyamines and protons: role of amino acid residues in the transmembrane and pore forming regions of NR1 and NR2 subunits. *Mol. Pharmacol.* 52:701–713.

Killackey, H.P., and Fleming, K. (1985). The role of the principal sensory nucleus in central trigeminal pattern formation. *Dev. Brain Res.* 22:141–145.

Killackey, H.P., Rhoades, R.W., and Bennett-Clarke, C.A. (1995). The formation of a cortical somatotopic map. *Trends. Neurosci.* 18:402–407.

Kleinschmidt, A., Bear, M.F., and Singer, W. (1987). Blockade of "NMDA" receptors disrupts experience-dependent plasticity of kitten striate cortex. *Science* 238:355–358.

Kuner, T., Wollmuth, L.P., Karlin, A., Seeburg, P.H., and Sakmann, B. (1996). Structure of the NMDA receptor M2 segment inferred from the accessibility of substituted cysteines. *Neuron* 17:343–352.

Kuryatov, A., Laube, B., Betz, H., and Kuhse, J. (1994). Mutational analysis of the glycine-binding site of the NMDA receptor: structural similarity with bacterial amino acid-binding proteins. *Neuron* 12:1291–1300.

Kutsuwada, T., Kashiwabuchi, N., Mori, H., Sakimura, K., Kushiya, E., Araki, K., Meguro, H., Masaki, H., Kumanishi, T., Arakawa, M., et al. (1992). Molecular diversity of the NMDA receptor channel. *Nature* 358:36–41.

Kutsuwada, T., Sakimura, K., Manabe, T., Takayama, C., Katakura, N., Kushiya, E., Natsume, R., Watanabe, M., Inoue, Y., Yagi, T., Aizawa, S., Arakawa, M., Takahashi, T., Nakamura, Y., Mori, H., and Mishina ,M. (1996). Impairment of suckling response, trigeminal neuronal pattern formation, and hippocampal LTD in NMDA receptor epsilon 2 subunit mutant mice. *Neuron* 16: 333–344.

Laube, B., Hirai, H., Sturgess, M., Betz, H., and Kuhse, J. (1997). Molecular determinants of agonist discrimination by NMDA receptor subunits: analysis of the glutamate binding site on the NR2B subunit. *Neuron* 18:493–503.

Laurie, D.J., Bartke, I., Schoepfer, R., Naujoks, K., and Seeburg, P.H. (1997). Regional, developmental and interspecies expression of the four NMDAR2 subunits, examined using monoclonal antibodies. *Mol. Brain Res.* 51:23–32.

Lebrand, C., Cases, O., Adelbrecht, C., Doye, A., Alvarez, C., El Mestikawy, S., Seif, I., and Gaspar, P. (1996). Transient uptake and storage of serotonin in developing thalamic neurons. *Neuron* 17:823–835.

Lebrand, C., Cases, O., Wehrle, R., Blakely, R.D., Edwards, R.H., and Gaspar, P. (1998). Transient developmental expression of monoamine transporters in the rodent forebrain. *J. Comp. Neurol.* 401:506–524.

Lee, L.J., and Erzurumlu, R.S. (2005). Altered parcellation of neocortical somatosensory maps in N-Methyl-D-Aspartate receptor-deficient mice. *J. Comp. Neurol.* 485:57–63.

Lee, L.J., Lo, F.S., and Erzurumlu, R.S. (2005a). NMDA receptor-dependent regulation of axonal and dendritic branching. *J. Neurosci.* 25:2304–2311.

Lee, L.J., Iwasato, T., Itohara, S., and Erzurumlu, R.S. (2005b). Exuberant thalamocortical axon arborization in cortex-specific NMDAR1 knockout mice. *J. Comp. Neurol.* 485:280–292.

Li, Y., Erzurumlu, R.S., Chen, C., Jhaveri, S., and Tonegawa, S. (1994). Whisker-related neuronal patterns fail to develop in the brainstem trigeminal nuclei of NMDAR1 knockout mice. *Cell* 76: 427-437.

Lo, F.S., and Erzurumlu, R.S. (2001). Neonatal deafferentation does not alter membrane properties of trigeminal nucleus principalis neurons. *J. Neurophysiol.* 85:1088–1096.

Lo, F.S., Guido, W., and Erzurumlu, R.S. (1999). Electrophysiological properties and synaptic responses of cells in the trigeminal principal sensory nucleus of postnatal rats. *J. Neurophysiol.* 82:2765–2775.

López-Bendito, G., and Molnár, Z. (2003) Thalamocortical development: how are we going to get there? *Nat. Rev. Neurosci.* 4:276–289.

Lu, H.C., Shi, W-C., Neumann, PE., Janz, R., and Crair, M.C. (2002). Altered presynaptic function in adenylyl cyclase I "barrelless" mutant mice. *Soc. Neuro. Sci. Abst.* No. 629.12.

Lu, H.C., She, W.C., Plas, D.T., Neumann, P.E., Janz, R., and Crair M.C. (2003). Adenylyl cyclase I regulates AMPA receptor trafficking during mouse cortical 'barrel' map development. *Nat. Neurosci.* 6:939–947.

Ma, P.M. (1993). Barrellettes: Architectonic vibrissal representations in the brainstem trigeminal complex of the mouse. II. Normal postnatal development. *J. Comp. Neurol.* 327:376–397.

Ma, P.M., and Woolsey, T.A. (1984). Cytoarchitectonic correlates of the vibrissae in the medullary trigeminal complex of the mouse. *Brain Res.* 306:374–379.

Maier, D.L., Mani, S., Donovan, S.L., Soppet, D., Tessarollo, L., McCasland, J.S., and Meiri, K.F. (1999). Disrupted cortical map and absence of cortical barrels in growth-associated protein (GAP)-43 knockout mice. *Proc. Natl. Acad. Sci. USA* 96:9397–9402.

Malenka, R.C., and Nicoll, R.A. (1993). NMDA receptor-dependent synaptic plasticity: multiple forms and mechanisms. *Trends. Neurosci.* 16:521–527.

Matsuoka, I., Suzuki, Y., Defer, N., Nakanishi, H., and Hanoune, J. (1997). Differential expression of type I, II, and V adenylyl cyclase gene in the postnatal developing rat brain. *J. Nueochem.* 68: 498–506.

Mayer, M.L., and Westbrook, G.L. (1987). The physiology of excitatory amino acids in the vertebrate central nervous system. *Prog. Neurobiol.* 28:197–276.

McIlvain, V.A., Robertson, D.R., Maimone, M.M., and McCasland, J.S. (2003). Abnormal thalamocortical pathfinding and terminal arbors lead to enlarged barrels in neonatal GAP-43 heterozygous mice. *J. Comp. Neurol.* 462:252–264.

Meguro, H., Mori, H., Araki, K., Kushiya, E., Kutsuwada, T., Yamazaki, M., Kumanishi, T., Arakawa, M., Sakimura, K., and Mishina, M. (1992). Functional characterization of a heteromeric NMDA receptor channel expressed from cloned cDNAs. *Nature* 357:70–74.

Mitrovic, N., Mohajeri, H., and Schachner, M. (1996). Effects of NMDA receptor blockade in the developing rat somatosensory cortex on the expression of the glia-derived extracellular matrix glycoprotein tenascin-C. *Eur. J. Neurosci.* 8:1793–802.

Molnár, Z., and Blakemore,C. (1995) How do thalamic axons find their way to the cortex? Trends. Neurosci. 18:389–397.

Monyer, H., Sprengel, R., Schoepfer, R., Herb, A., Higuchi, M., Lomeli, H., Burnashev, N., Sakmann, B., and Seeburg, P.H. (1992). Heteromeric NMDA receptors: molecular and functional distinction of subtypes. *Science* 256:1217–1221.

Monyer, H., Burnashev, N., Laurie, D.J., Sakmann, B., and Seeburg, P.H. (1994). Developmental and regional expression in the rat brain and functional properties of four NMDA receptors. *Neuron* 12:529–540.

Monyer, H., Jonas, P., and Rossier, J. (1999). Molecular determinants controlling functional properties of AMPARs and NMDARs in the mammalian CNS. In: Jonas, P., and Monyer, H. (eds.), *Ionotropic Glutamate Receptors in the CNS.* Springer, Berlin, pp. 309–339.

Moriyoshi, K., Masu, M., Ishii, T., Shigemoto, R., Mizuno, N., and Nakanishi, S. (1991). Molecular cloning and characterization of the rat NMDA receptor. *Nature* 354:31–37.

Munoz, A., Liu, X-B., and Jones, E.G. (1999). Development of metabotropic glutamate receptors from trigeminal nuclei to barrel cortex in postnatal mouse. *J. Comp. Neurol.* 409: 549–566.

Nguyen, T., and Südhof, T.C. (1997). Binding properties of neuroligin 1 and neurexin 1beta reveal function as heterophilic cell adhesion molecules. *J. Biol. Chem.* 272:26032–26039.

Nishi, M., Hinds, H., Lu, H.P., Kawata, M., and Hayashi, Y. (2001). Motoneuron-specific expression of NR3B, a novel NMDA-type glutamate receptor subunit that works in a dominant-negative manner. *J. Neurosci.* 21:RC185.

Ohsaki, K., Osumi, N., and Nakamura, S. (2002). Altered whisker patterns induced by ectopic expression of Shh are topographically represented by barrels. *Dev. Brain Res.* 137:159–170.

O'Leary, D.D.M., Ruff, N.L., and Dyck, R.H. (1994). Development, critical period plasticity, and adult reorganizations of mammalian somatosensory systems. *Curr. Opin. Neurobiol.* 4: 535–544.

Persico, A.M., Mengual, E., Moessner, R., Hall, F.S., Revay, R.S., Sora, I., Arellano, J., DeFelipe, J., Gimenez-Amaya, J.M., Conciatori, M., Marino, R., Balde, A., Cabib, S., Pascucci, T., Uhl, G.R., Murphy, D.L., Lesch, K.P., and Keller, F. (2001). Barrel pattern formation requires serotonin uptake by thalamocortical afferents, and not vesicular monoamine release. *J. Neurosci.* 21:6862–6873.

Petralia, R.S., Wang, Y.X., and Wenthold, R.J. (1994). The NMDA receptor subunits NR2A and NR2B show histological and ultrastructural localization patterns similar to those of NR1. *J. Neurosci.* 14:6102–6120.

Polleux, F. (2004). Generation of the cortical area map; emx2 strikes back. *Neuron* 43:295–297.

Rudhard, Y., Kneussel, M., Nassar, M.A., Rast, G.F., Annala, A.J., Chen, P.E., Tigaret, C.M., Dean, I., Roes, J., Gibb, A.J., Hunt, S.P., and Schoepfer, R. (2003). Absence of Whisker-related pattern formation in mice with NMDA receptors lacking coincidence detection properties and calcium signaling. *J. Neurosci.* 23:2323–2332.

Sakurada, K., Masu, M., and Nakanishi, S. (1993). Alteration of Ca2+ permeability and sensitivity to Mg2+ and channel blocker by a single amino acid substitution in the N-methyl-D-aspartate receptor. *J. Biol. Chem.* 268: 410–415.

Salichon, N., Gaspar, P., Upton, A.L., Picaud, S., Hanoun, N., Hamon, M., De Maeyer, E., Murphy, D.L., Mossner, R., Lesch, K.P., Hen, R., and Seif, I. (2001). Excessive activation of serotonin (5-HT) 1B receptors disrupts the formation of sensory maps in monoamine oxidase a and 5-ht transporter knock-out mice. *J. Neurosci.* 21: 884–896.

Sasaki, Y.F., Rothe, T., Premkumar, L.S., Das, S., Cui, J., Talantova, M.V., Wong, H.K., Gong, X., Chan, S.F., Zhang, D., Nakanishi, N., Sucher, N.J., and Lipton, S.A. (2002). Characterization and comparison of the NR3A subunit of the NMDA receptor in recombinant systems and primary cortical neurons. *J. Neurophysiol.* 87:2052–2063.

Scheiffele, P. (2003). Cell-cell signaling during synapse formation in the CNS. *Annu. Rev. Neurosci.* 26:485–508.

Scheiffele, P., Fan, J., Choih, J., Fetter, R., and Serafini, T. (2000). Neuroligin expressed in nonneuronal cells triggers presynaptic development in contacting axons. *Cell* 101:657–669.

Schlaggar, B.L., and O'Leary, D.D. (1994). Early development of the somatotopic map and barrel patterning in rat somatosensory cortex. *J. Comp. Neurol.* 346:80–96.

Schlaggar, B.L., Fox, K., and O'Leary, D.D. (1993). Postsynaptic control of plasticity in developing somatosensory cortex. *Nature* 364:623–626.

Schmidt, J.T. (2004). Activity-driven sharpening of the retinotectal projection: the search for retrograde synaptic signaling pathways. *J. Neurobiol.* 59:114–133.

Scneggenburger, R., and Ascher, P. (1997). Coupling of permeation and gating in an NMDA-channel pore mutant. *Neuron* 18:167–177.

Senft, S.L., and Woolsey, T.A. (1991). Growth of thalamic afferents into mouse barrel cortex. *Cereb. Cortex* 1:308–335.

Simon, D.K., Prusky, G.T., O'Leary, D.D., and Constantine-Paton, M. (1992). N-methyl-D-aspartate receptor antagonists disrupt the formation of a mammalian neural map. *Proc. Natl. Acad. Sci. USA* 89:10593–10597.

Single, F.N., Rozov, A., Burnashev, N., Zimmermann, F., Hanley, D.F., Forrest, D., Curan, T., Jensen, V., Hvalby, O., Sprengel, R., and Seeburg, P.H. (2000). Dysfunctions in mice by NMDA receptor point mutations NR1(N598Q) and NR1(N598R). *J. Neurosci.* 20:2558–2566.

Spruston, N., Jonas P., and Sakmann, B. (1995). Dendritic glutamate receptor channels in rat hippocampal CA3 and CA1 pyramidal neurons. *J. Physiol. (Lond.)* 482:325–352.

Stainier, D.Y., and Gilbert, W. (1990). Pioneer neurons in the mouse trigeminal sensory system. *Proc. Natl. Acad. Sci. USA* 87: 923–927.

Stainier, D.Y., and Gilbert, W. (1991). Neuronal differentiation and maturation in the mouse trigeminal sensory system, in vivo and in vitro. *J. Comp. Neurol.* 311:300–312.

Stent, G.S. (1973). A physiological mechanism for Hebb's postulate of learning. *Proc. Natl. Acad. Sci. USA* 70:997–1001.

Storm, D.R., Hansel, C. Hacker, B. Parent, A. and Linden, D. (1998). Impaired cerebellar long-term potentiation in type I adenylyl cyclase mutant mice. *Neuron* 20: 1199–1210.

Sucher, N.J., Akbarian, S., Ch,i C.L., Leclerc, C.L., Awobuluyi, M., Deitcher, D.L., Wu, M.K., Yuan, J.P., Jones, E.G., and Lipton, S.A. (1995). Developmental and regional expression pattern of a novel NMDA receptor-like subunit (NMDAR-L) in the rodent brain. *J. Neurosci.* 15:6509–6520.

Tokita, Y., Bessho, Y., Masu, M., Nakamura, K., Nakao, K., Katsuki, M., and Nakanishi, S. (1996). Characterization of excitatory amino acid neurotoxicity in N-methyl-D-aspartate receptor-deficient mouse cortical neuronal cells. *Eur. J. Neurosci.* 8:69–78.

Traynelis, S.F., Burgess, M.F., Zheng, F., Lyuboslavsky, P., and Powers, J.L. (1998). Control of voltage-dependent zinc inhibition of NMDA receptors by the NR1 subunit. *J. Neurosci.* 18:6163–6175.

Vanderhaeghen, P., Lu, Q., Prakash, N., Frisen, J., Walsh, C.A., Frostig, R.D., and Flanagan, J.G. (2000). A mapping label required for normal scale of body representation in the cortex. *Nat. Neurosci.* 3:358–365.

Van der Loos, H. (1976). Barreloids in mouse somatosensory thalamus. *Neurosci. Lett.* 2:1–6.

Villacres, E.C. Wong, S.T. Chavkin, C. and Storm, D.R. (1998). Type I adenylyl cyclase mutant mice have impaired mossy fiber long-term potentiation. *J. Neurosci.* 18: 3186–3194.
Wang, H., Ferguson, G.D., Pineda, V.V., Cundiff, P.E., and Storm, D.R. (2004). Overexpression of type-1 adenylyl cyclase in mouse forebrain enhances recognition memory and LTP. *Nat. Neurosci.* 2004 7:635–642.
Watanabe, M., Inoue, Y., Sakimura, K., and Mishina, M. (1992). Developmental changes in distribution of NMDA receptor channel subunit mRNAs. *Neuroreport* 3:1138–1140.
Watanabe, M., Inoue, Y., Sakimura, K., and Mishina, M. (1993). Distinct distributions of five N-methyl-D-aspartate receptor channel subunit mRNAs in the forebrain. *J. Comp. Neurol.* 338:377–390.
Watanabe, M., Mishina, M., and Inoue, Y. (1994a). Distinct spatiotemporal expressions of five NMDA receptor channel subunit mRNAs in the cerebellum. *J. Comp. Neurol.* 343:513–519.
Watanabe, M., Mishina, M., and Inoue, Y. (1994b). Distinct distributions of five NMDA receptor channel subunit mRNAs in the brainstem. *J. Comp. Neurol.* 343:520–531.
Watanabe, M., Nakamura, M., Sato, K., Kano, M., Simon, M.I., and Inoue, Y. (1998). Patterns of expression for the mRNA corresponding to the four isoforms of phospholipase Cb in mouse brain. *Eur. J. Neurosci.* 10:2016–2025.
Welker, E., and Van der Loos, H. (1986). Quantitative correlation between barrelfield size and the sensory innervation of the whiskerpad: a comparative study in six strains of mice bred for different patterns of mystacial vibrissae. *J. Neurosci.* 6:3355–3373.
Welker, E., Armstrong-James, M., Bronchti, G., Ourednik, W., Gheorghita-Baechler, F., Dubois, R., Guernsey, D.L., Van der Loos, H., and Neumann, P.E. (1996). Altered sensory processing in the somatosensory cortex of the mouse mutant barrelless. *Science* 27:1864–1867.
Wenthold, R.J., Prybylowski, K., Standley, S., Sans, N., and Petralia, R.S. (2003). Trafficking of NMDA receptors. *Annu. Rev. Pharmacol. Toxicol.* 43:335–358.
Wisden, W., and Seeburg, P.H. (1993). Mammalian ionotropic glutamate receptors. *Curr. Opin. Neurobiol.* 3:291–298.
Wisden, W., Seeburg, P.H., and Monyer, H. (2000). AMPA, kainite and NMDA ionotropic glutamate receptor expression-an in situ hybridization atlas. In: Björklund, A., and Hökfelt, T. (eds), *Handbook of Chemical Neuroanatomy.* Elsevier, Amsterdam, pp. 99–143.
Wollmuth, L.P., Kuner, T., Seeburg, P.H., and Sakmann, B. (1996). Differential contribution of the NR1- and NR2A-subunits to the selectivity filter of recombinant NMDA receptor channels. *J. Physiol. (Lond.)* 491:779–797.
Wong-Riley, M.T. (1989). Cytochrome oxidase: an endogenous metabolic marker for neuronal activity. *Trends. Neurosci.* 12:94–101.
Wong-Riley, M.T., and Welt, C. (1980). Histochemical changes in cytochrome oxidase of cortical barrels after vibrissal removal in neonatal and adult mice. *Proc. Natl. Acad. Sci. USA* 77:2333–2337.
Woolsey, T.A. (1990). Peripheral alteration and somatosensory development. In: E.J. Coleman (ed.) *Development of Sensory Systems in Mammals*, Wiley, New York, pp. 461–516.
Woolsey, T.A., and Van der Loos, H. (1970). The structural organization of layer IV in the somatosensory region (SI) of mouse cerebral cortex. The descrip-

tion of a cortical field composed of discrete cytoarchitectonic units. *Brain Res.* 17:205–242.
Wu, G.Y., and Cline, H.T. (1998). Stabilization of dendritic arbor structure in vivo by CaMKII. *Science* 279: 222–226.
Wu, Z-L., Thomas, S.A. Villacres, E.C. Xia, Z. Simmons, M.L. Chavkin, C. Palmiter, R.D. and Storm, D.R. (1995). Altered behavior and long-term potentiation in type I adenylyl cyclase mutant mice. *Proc. Natl. Acad. Sci. USA.* 92: 220–224.
Zheng, X., Zhang, L., Wang, A.P., Araneda, R.C., Lin, Y., Zukin, R.S., and Bennett, M.V.L. (1999). Mutation of structural determinants lining the *N*-methyl-D-aspartate receptor channel differentially affects phencyclidine block and spermine potentiation and block. *Neuroscience* 93:125–134.
Zou D.J., and Cline, H.T. (1999). Postsynaptic calcium/calmodulin-dependent protein kinase II is required to limit elaboration of presynaptic and postsynaptic neuronal arbors. *J. Neurosci.* 19: 8909–8918.

11

Presynaptic Mechanisms Controlling Axon Terminal Remodeling in the Thalamocortical and Retinogeniculate Systems

Alexandra Rebsam and Patricia Gaspar

Abstract

The establishment of point to point sensory maps requires that afferent inputs restrict their connections to a limited number of target neurons. This targeting involves axon terminal and synaptic remodeling, as clearly shown in the mammalian visual system. Retinal ganglion cells (RGCs) axon terminals from each eye segregate into separate territories after selective branching and pruning. In the primary somatosensory cortex (S1), the remodeling of the thalamocortical axons (TCAs) remained controversial but was recently shown by the use of specific markers and by single axon reconstructions. Moreover, molecular genetic studies in mice demonstrated that similar presynaptic mechanisms control the segregation of the retinogeniculate projections and the emergence of TCA barrels in S1. The thalamic neurons and the RGCs both express the serotonin transporter (5-HTT), the vesicular monoamine transporter, the 5-HT1B receptors, and the calcium-stimulated adenylate cyclase 1 (AC1) at the height of the plasticity period for these systems, during the first postnatal week. Mutations that affect the levels of serotonin (monoamine oxidase A and 5-HTT-null mice) prevent the segregation of eye-specific inputs in the lateral geniculate nucleus and the emergence of barrels in S1. Double knockout strategies indicated that an abnormal activation of the presynaptic 5-HT1B receptor plays a key role in this developmental abnormality. The 5-HT1B receptors can modulate the growth and branching of TCAs. Downstream events could involve the control of glutamate release and/or the control of cAMP levels. Observations in the AC1 defective mice, showing abnormal axon branching of the TCAs and RGCs suggest that the latter mechanism is critical.

Thus, current evidences indicate that presynaptic 5-HT receptors and cAMP mediated signaling are important modulators of axon terminal remodeling in the barrelfield and the retinogeniculate system.

1. Introduction

Establishing topographic point to point projections in the brain requires from axons to have a high degree of selectivity in the choice of the neurons with which they establish synaptic contacts. A large part of this selectivity is ensured by molecular guidance of the axons to their targets by attractive and repulsive factors. However, once the axons reach these targets, the precision of the connection is acquired only gradually and involves activity-dependent mechanisms and competitive interactions among axon terminals for target sites (Purves and Lichtman, 1980; Goodman and Shatz, 1993). This construction scheme has been most clearly validated in the visual system, where developing RGC axons initially extend broadly in their target areas, the lateral geniculate nucleus (LGN) and the superior colliculus (SC), and subsequently restrict their terminal arbors and the number of neurons with which they establish synaptic contacts (Shatz, 1996, Toborg and Feller, 2005; Guido, this volume). The secondary refinement of the map involves axon terminal branching, stabilisation of branches, as well as axon terminal elimination, in coordination with activity-dependent synaptic stabilisation and synaptic elimination (Sanes and Lichtman, 1999).

In the rodent somatosensory cortex, the barrel patterning is determined by the organisation of the thalamocortical axons (TCAs) from the ventrobasal thalamic nucleus (VB) into axon clusters that replicate the organisation of sensory receptors in the periphery. However, the mechanisms involved in the clustering of TCAs into barrel domains has remained controversial. After a period when a tight parallelism was thought to exist between the visual and the somatosensory systems, conflicting evidence suggested that the development of the two systems was opposed (O'Leary et al., 1994). Thus, the importance of activity-dependent refinement mechanisms was generally accepted in the visual system, but was considered to be marginal in the whisker to barrel pathway. This was based on studies that suggested that TCAs arrived into their target layer IV in the cerebral cortex more or less "pre-assembled" into periphery related patterns (Agmon et al., 1993). Furthermore, pharmacological experiments interfering with neural transmission did not alter the development of barrels in the primary somatosensory cortex (Chiaia et al., 1992; Schlaggar et al., 1993b; Henderson et al., 1992).

As shown by a number of chapter contributors of this book, this view has now largely evolved owing to a body of genetic evidence for shared molecular mechanisms during the development of the barrelfield and the retinogeniculate projections. First, it now appears clearly that axon terminal remodeling is required for the emergence of periphery-related patterns (Erzurumlu and Iwasato, this volume). Second, the role of glutamatergic neurotransmission and NMDA-dependent transduction pathways has now been demonstrated in both the visual system and the barrelfield (Erzurumlu and Kind, 2001, and see Erzurumlu and Iwasato, this volume). Finally, as discussed in the present chapter, there is mounting evidence that similar presynaptic mechanisms are required for the emergence of periphery-related patterns in the retinal visual system and in the barrelfield. This view results largely from molecular genetic studies demonstrating the role of serotonin neurotransmission and adenylate cyclase signaling for the formation of barrel patterns, and the emergence of eye-specific segregation of the retinogeniculate projections.

2. Axon Terminal Remodeling in the Retinogeniculate and Thalamocortical Projections

2.1. Retinal Projections to the Lateral Geniculate Nucleus

Numerous anatomical studies, single axon reconstructions, and cellular electrophysiological studies, have demonstrated that the sorting of eye-specific retinal afferents involves extensive axon remodeling (reviewed in Shatz 1996, Guido, this volume). During embryonic life and early postnatal period, retinal afferents from each eye largely overlap in the LGN, and converge onto single neurones, which can receive inputs from as much as 20 RGC neurons (Chen and Regehr, 2000). Over the first two postnatal weeks, the RGC axons segregate into separate territories and the convergence onto a single neuron becomes minimal (Figure 1A). This segregation coincides with the removal of retinal axons from the inappropriate eye and with extensive terminal branching of the retinal axons from the appropriate eye (Shatz and Stryker, 1988, but see Snider et al., 1999). In amphibians and fish, although the topographic connection appears to be more specific from the outset, *in vivo* visualisation of single axons at different developmental times showed that axons reaching out for their targets are highly dynamic with a constant process of branch addition and retraction (Witte et al., 1996; Cohen-Cory, 1999; Niell et al., 2004).

2.2. Thalamocortical Projections to the Somatosensory Cortex

A similar progressive activity-dependent remodeling of the somatosensory TCAs was initially proposed in the barrelfield (Senft and Woolsey, 1991) but was subsequently dismissed. Anatomical studies showed that the topography of the trigeminal projection is established early during embryonic life. Trigeminal axons reach the whisker pad by E14 and thalamocortical axons reach the cortex by E14.5 in mice (E15.5 in rats) and both have a clear topographic organisation from the outset (Erzurumlu and Killackey, 1983; Erzurumlu and Jhaveri, 1990). At these embryonic stages, each component of the sensory circuit develops independently: trigeminal axons are guided to the whisker pad by

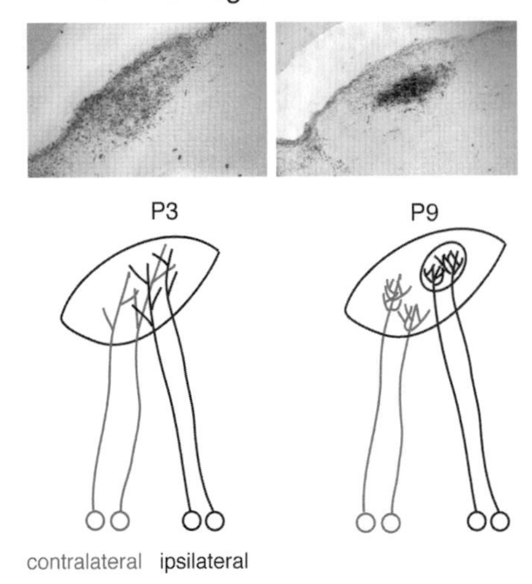

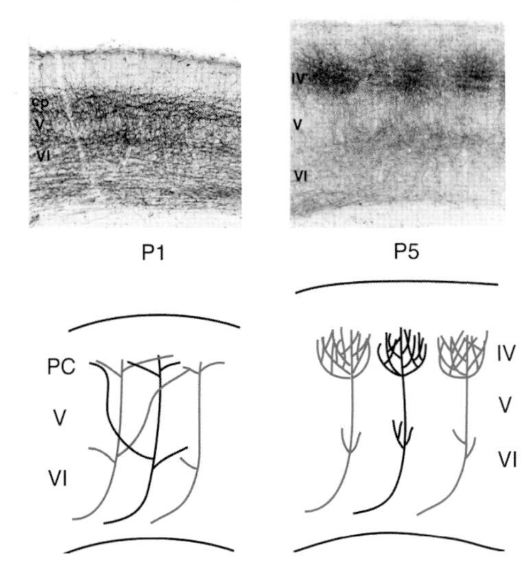

Figure 1 Axon terminal remodeling of retinogeniculate projections and of the whisker to barrel pathway. (A) Progressive separation of the ipsi and contralateral retinal afferents in the mouse lateral geniculate nucleus. As shown on the photomicrographs (upper panel), the HRP-labeled ipsilateral retinal axons, cover a large portion of the lateral geniculate nucleus (LGN) at P3, and become restricted to a central core of the LGN at P9. The diagram (lower panel) illustrates the intermix of the ipsilateral and contralateral retinal axons at P3; at this stage the retinal axons display very few axon collateral branches. At P9, the ipsi- and contralateral fibers are segregated and occupy distinctive terminal fields; the axon terminals have then produced abundant collateral branches. (B) The photomicrographs illustrate the somatosensory thalamocortical axons (TCAs), labeled with 5-HT transporter immunoreactivity. At P1, TCAs are distributed uniformly in the cortical plate (CP) and in layer VI. This uniform pattern is related to the intermix of individual TCAs extending in the tangential plane. At P5, sensory TCAs become segregated into barrel domains. This segregation is correlated with a pronounced remodeling of the TCAs, involving the addition of collateral branches towards the centre of the barrel and the retraction of collaterals from neighbouring domains.

attractive and branching guidance cues (Ulupinar et al., 1999; Ozdinler and Erzurumlu, 2002) whereas the thalamic axons are guided to the somatosensory cortex by a complex interplay of repulsive and attractive guidance cues that are posted along their way to the cortex (semaphorins, ephrins, and netrin); they also use descending cortical output axons and transient subplate neurons as guides (Molnár et al., 1998; López-Bendito and Molnár, 2003). Once the axons reach the cortex they immediately start growing into the cortex, and they are topologically organised with a degree of order that has been estimated to be either high (Dawson and Killackey, 1985; Agmon et al., 1993; Agmon et al., 1995) or more divergent than in adults (Naegele et al., 1988; Krug et al., 1998). The emergence of barrel patterning occurs only secondarily, approximately seven days after the TCA projections have reached the cortex. This delay can vary among species, it may be slightly shorter in rats; six days if one refers to the data of Schlaggar et al. (1993a) but it can be much longer: in the wallaby for instance, thalamic axons remain in their target for about 60 days before forming clusters (Marotte et al., 1997). The formation of the whisker related patterns is initiated by the segregation of the TCAs which conduct the sensory information from one whisker. This is shortly followed by a response of the target neurons which organise their cell bodies and dendrites around these clusters (Rice and Van der Loos, 1977; Jhaveri et al., 1991). The postnatal emergence of the whisker-related patterning follows a peripheral to central sequence: it is first visible in the brainstem (by E19 in rats and P1 in mice), (Belford and Killackey, 1979; Ma, 1993) and becomes apparent in the thalamus and the cortex two days later, by P3 in mice (Erzurumlu and Jhaveri, 1990). Within the cerebral cortex, as in the lower sensory relays (Ma, 1993), the emergence of the periphery-related pattern is gradual. The first segregation step is the distribution of the VB somatosensory axons along broad unsegmented rows that correspond to the principal whisker rows of the muzzle. This organisation emerges at P2 in mice (Rebsam et al., 2002) but it is already visible at P0–P1 in rats (Rhoades et al., 1990; Schlaggar et al., 1993a; Auso et al., 2001). The next step, at P3–P4, is the separation of the rows into clusters that correspond to the large whiskers. Finally, at P5, the smaller barrels that correspond to the smaller face whiskers and to the digits appear (Rhoades et al., 1990; Rebsam et al., 2002). As shown by lesions of the sensory receptors or of the trigeminal nerve, the patterning of the afferent axons requires the transmission of a peripheral signal that is provided by the whisker pad during a critical period (P0–P4) in mice (Van der Loos and Woolsey, 1973; Jeanmonod et al., 1981; Jensen and Killackey, 1987; Rebsam et al., 2005).

The way periphery-related TCA barrel patterns emerge in the cerebral cortex has been controversial. Are they sculpted within the cortex by activity-dependent remodeling of the thalamic axon arbors, or do they arrive pre-formed in the cortex? The first anterograde tracing studies using DiI showed that TCAs were uniformly distributed in layer IV before they segregated into separate clusters (Senft and Woolsey, 1991).

However, subsequent studies that used an elegant combination of anterograde and retrograde tracers in mice (Agmon et al., 1993; Agmon et al., 1995) or acetylcholinesterase labelling of the sensory thalamic axons in the rat (Schlaggar et al., 1993a) concluded that the barrel patterns emerged in the deep cortical layers and were then "projected" into layer IV. Subsequent analyses of single TCAs in the developing rat cerebral cortex supported this interpretation, by showing no signs of TCA exuberance or axon overgrowth within the cortex, or cortical plate (Catalano et al., 1991). Thus, the general scheme that was retained was that TCAs branched off in the deep cortical layers/subplate and then grew radially into the cortex to their appropriate radial and laminar destination. Our recent observations in this system, using a new marker of the somatosensory TCAs, give credit to the first interpretation of Woolsey and colleagues (Figure 1B). We benefited from the fact that the 5-HT transporter (5-HTT) is selectively expressed in all the VB axons from E14.5 to about P10, in mice and rats (Lebrand et al., 1998; Boylan et al., 2000). Thus, 5-HT or 5-HTT immunocytochemistry, allow a clear visualization of the entire TCA projection during barrel development (Figure 2). This approach removed a potential source of confusion that exists when using tracers such as DiI which can either label too many axons (e.g. from neighbouring thalamic nuclei such as the posteromedial thalamic nucleus) or too few axons (small ramifications of the TCAs may be lost when using small amounts of fluorescent tracers). With 5-HTT immunolabeling, TCAs are seen to invade the cortical plate at E16.5 (Figure 2B) they are uniformly distributed within layers IV and VI during the first 3 postnatal days (Figure 1B), and barrel separations emerge within layer IV itself over the next two days (Figure 1B, 2C). Clustering of the TC axons was also visible in layer VI but this feature was much clearer in rats (Figure 2D) than in mice (Figure 2C). No periphery related patterns were revealed at the level of the cortical subplate.

Single axon reconstructions showed that at P1, TCAs begin by extending beyond their prospective terminal field. In the tangential dimension, they can cover distances equivalent to 2 barrel domains (Rebsam et al., 2002) (Figure 3); in the radial dimension they frequently overshoot layer IV extending in layers 1–2. A more confined distribution of the TCAS in the tangential and radial dimension is visible at P7 (Figure 2C, 3). Similar observations, suggestive of a pruning of the TCAs between P1 and P4, were made by Welker and collaborators (Hage, 2003). This is also coherent with electron microscopic observations, that showed the presence of degenerated axons in the mouse barrel field during the first postnatal week (White et al., 1997) although the nature of the degenerated profiles was not identified. Thus, although the TCA overgrowth is less marked in the somatosensory cortex than in the retinal visual system, some degree of TCA pruning coincides with the emergence of barrel domains in S1. This retraction of branches coincides also with an extensive and focused axon branching toward the barrel centers, an observation that is highly consistent across studies (Agmon et al., 1993; Catalano et al. 1996; Rebsam et al., 2002; Hage, 2003 ; Lee et al., 2005).

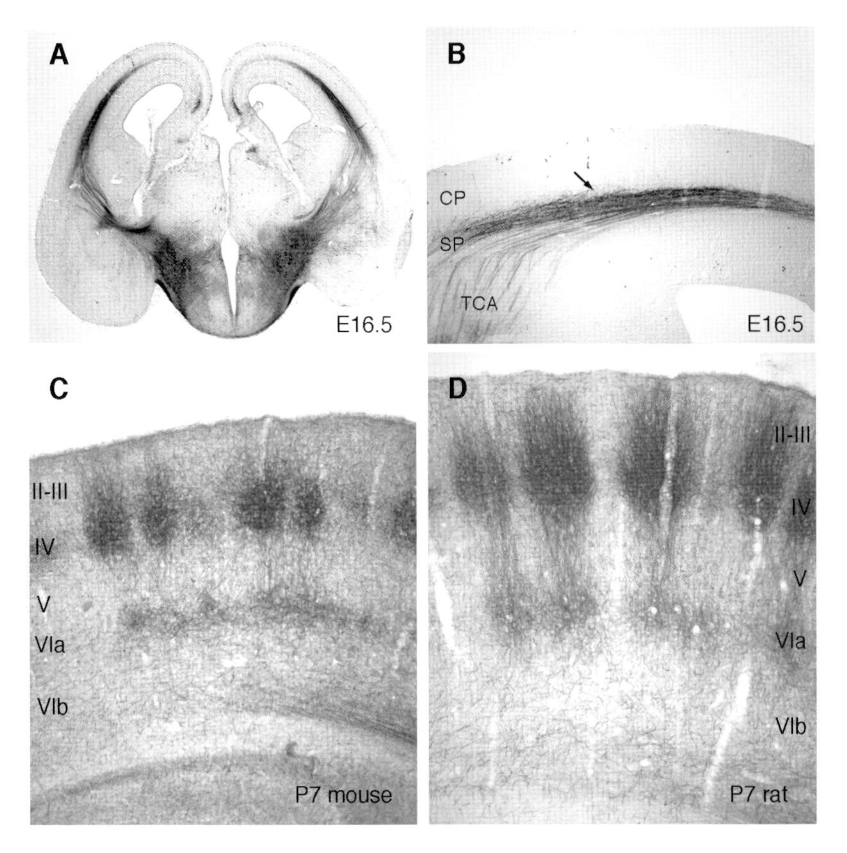

Figure 2 5-HT and 5-HTT immunolabeling of the developing sensory thalamocortical axons. At E16.5, the 5-HT labeled thalamocortical fibers are visible in the internal capsule and course up to the cortex. Dense staining in the basal forebrain corresponds to the ascending serotoninergic fibers coursing in the medial forebrain bundle (A). TCAs course within the subplate and invade the cortical plate (arrow) (B). At P7 in mice (C) and rats (D) the 5-HTT-labeled thalamocortical fibers are most dense within layers IV and VIa. The segregation of terminal fields into barrel domains is most clear in layer IV. Coronal segregation of the TCAs in layer VI is often difficult to visualise in mice (C) and is somewhat clearer in rats (D).

In mice, a 10 fold increase in the number of terminal branches per axon was found between P1 and P7 (Rebsam et al., 2002; Lee et al. 2005).

An interesting point that has not yet been investigated to our knowledge is the relationship that exists between axon terminal remodeling and synaptic refinement in the barrel field. In the retinogeniculate system (Shatz, 1996; Chen and Regehr, 2000), the climbing fibers of the cerebellum (Hashimoto and Kano, 2003), and the neuromuscular junction (Sanes and Lichtman, 1999), the reshaping of the axon terminal arbors has been correlated to the acquisition of an increased synaptic specificity: a reduced level of convergence of afferents on a single neuron coincides with the retraction of exuberant branches. In the retinogeniculate system, the synaptic refinement persists beyond the visible

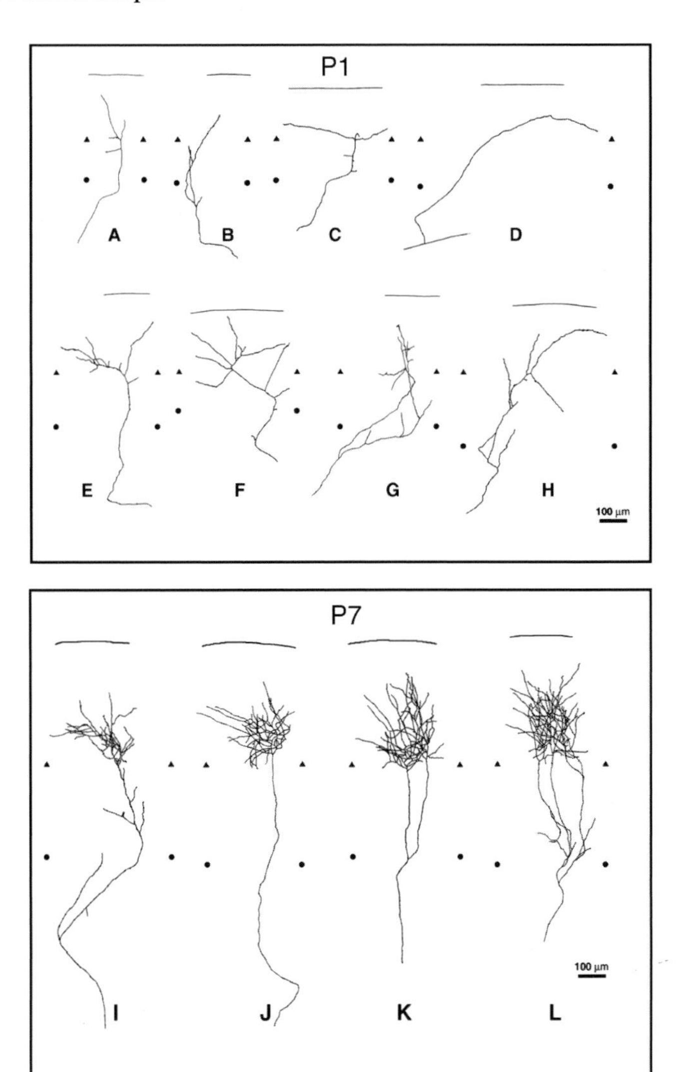

Figure 3 Resconstructions of individual axon arbors in the somatosensory cortex. At P1, the TCAs have very heterogeneous phenotypes. Some axons run tangentially in the cortical plane and emit few branches (A, B); others have a wide lateral extension (C, D, F, H) and a greater number of branches (E, F, G, H). The mean number of branches is of 2, 9 $\pm$ 2 in layer IV and 1.4 $\pm$ 1.1 in layer VIa (n = 42). At P7, the TCAs have a stereotyped phenotype, with the main arborization confined to a barrel domain, in layer IV. The mean number of branches is of 32 $\pm$ 10.9 in layer IV and 2 $\pm$ 3.7 in layer VIa (n = 13). These quantifications show the strong elaboration of branches in layer IV between P1 and P7.

anatomical changes, until P21 (Chen and Regehr, 2000). Electrophysiological studies in the sensory cortex analysing the receptive field of single barrels have indicated that there is a focusing of the receptive fields and that responses to the stimulation of neighbouring barrels decreases over the first postnatal weeks (Armstrong-James and Fox, 1987).

However, no physiological studies have, to our knowledge, analysed the convergence of different TCA inputs on single layer IV neurons in the somatosensory cortex. Synaptic responses can be recorded at the TC synapses (Agmon and Connors, 1992; Feldman et al., 1999; Laurent et al., 2002), and an age-dependent LTP has been demonstrated at the thalamocortical synapse between P3 and P10 (Feldman et al., 1999). However this synaptic plasticity coincides only partially with the major phase of axon arbor remodeling which occurs between P2 and P5. The number of morphologically identified synapses is very low at this time and the peak of synaptic production occurs a few days later around the second postnatal week (White et al., 1997), that is after the emergence of barrels. Analysing the degree of convergence of TCA inputs on single layer IV spiny neurons would be of interest as it may well reveal a more extensive period of synaptic remodeling than what is currently detectable with the anatomical methods.

3. Transient Expression of Genes in the Thalamus and Retina During the Period of Axon Terminal Remodeling

Changes in the intrinsic growth properties of afferent axons are likely to underlie the dynamic activity of the retinal and thalamic axon growth cones as they explore their target zones to select their preferred partners. In addition to the neurotrophins, a growing number of neurotransmitter receptors and their signaling pathways have been involved in axon remodeling. This was principally indicated by gene expression studies that showed transient expression of candidate genes in the VB and the RGCs. Studies of the null mutations of these genes, further showed alterations in the patterning and refinement of afferents.

3.1. Growth Associated Proteins

Genes encoding growth associated proteins such as *GAP43* and *L1* are transiently expressed in the retinal ganglion cells and the thalamus during the development of their projections (Erzurumlu et al., 1990; Jung et al., 1997; Fukuda et al., 1997). These proteins have been implicated in neurite growth, branching, and/or axon terminal sprouting (Aigner and Caroni, 1993). The loss of function of these genes however results in an abnormal fasciculation and growth of the retinal and thalamic axons (Zhu and Julien, 1999; Maier et al., 1999; Demyanenko and Maness, 2003; Wiencken-Barger et al., 2004), making it difficult to interpret the altered patterning of the axon terminals. Interestingly, the afferent axons are still able to segregate into periphery-related patterns, even though their shape is somewhat distorted. This is reminiscent of a number of mouse mutants that have thalamocortical axon guidance defects but in which the TCAs are still able to form barrel clusters once they have reached the cerebral cortex (López-Bendito and Molnár, 2003).

3.2. Neurotrophins

Neurotrophins and their receptors are another class of candidate molecules that could play a major role to regulate axon patterning since the expression of neurotrophins is modulated by neural activity and in turn, neurotrophins control the growth and branching of axons. In the retinal ganglion cells, BDNF and its high affinity tyrosine kinase receptor TrkB are expressed at high levels during late embryonic and early postnatal life (Masana et al., 1993; Ugolini et al., 1995). Inhibition of BDNF activity by siRNA in the RGCs causes the retraction of the corresponding RGCs in the lateral geniculate nucleus (Menna et al., 2003) and an altered patterning of the ipsi-contralateral projection. However, genetic invalidation of the TrkB gene failed to show clear segregation abnormalities in the ipsi/contralateral retinal axons (Pollock et al., 2003). A similar situation has been described in the barrelfield. Both BDNF and TrkB are highly expressed in the thalamic neurons during the critical period of TCA segregation. Blockade of function experiments indicated that TrkB could be a major player in the segregation of the TCAs in the visual cortex (Berardi et al., 2000). However in the TrkB KO, an almost normal barrel development was found although TCAs were abnormally outspread in the upper cortical layers, in layers II to III (Vitalis et al., 2002). This could indicate that TrkB signaling is required for the appropriate branching within layer IV. Alternatively, TrkB could be required for the retraction of exuberant collaterals from layers III and II. In normal mice, a fair amount of TCAs are visible in these upper layers by P3 that are no longer observed by P7, possibly as a consequence of axon retraction from the upper layers (Rebsam et al., 2002). This retraction could be compromised in the TrkB KO mice (Vitalis et al., 2002).

3.3. Serotonin

3.3.1. Transient Expression of the Serotonin Transporter

This is probably one of the most intriguing changes of gene expression pattern that occurs in the retina and the thalamus during this period (reviewed in Gaspar et al., 2003). It appears as though at the same period when the axons are still undecided on the choice of their final target they are also flickering about which neurotransmitter and receptors they should use. The thalamic neurons and the RGCs are glutamatergic and they establish functional glutamatergic synapses early during development (Agmon and Connors, 1992 ; Mooney et al., 1996). Nevertheless, both the thalamic and retinal neurons express a fairly large amount of genes that are related to serotonin neurotransmission. These genes are normally found only in the raphe neurons. However, as already mentioned, the serotonin plasma membrane transporter (5-HTT) is expressed in the somatosensory thalamus at the height of its plasticity period (E15-P10). This finding explains early descriptions of transient 5-HT innervation in the primary sensory areas (D'Amato et al., 1987; Rhoades et al., 1990). This was initially interpreted as a transient projection from the raphe (Killackey et al., 1995), however the raphe neurons, establish in fact a relatively loose axon terminal network in the cortex at that age (Lidov and Molliver, 1982 see also Fig. 1B in Rebsam et al. 2002). It was demonstrated that the 5-HT produced and released by raphe terminals is captured with nanomolar affinity by the thalamic axon terminals. Since the vesicular transporter for monoamines, VMAT2, is also expressed by the thalamic neurons, 5-HT could be accumulated in synaptic vesicles (Lebrand et al., 1996). The term "borrowed neurotransmitter" was coined, to describe this phenomenon since 5-HT is not produced by the thalamic neurons themselves. A release of 5-HT by thalamic terminals has not yet been directly demonstrated, although indirect support has recently been brought by electrophysiological studies (Binshtok et al., 2004). Another intriguing and as yet unresolved issue in this system is the possibility of a co-release of 5-HT and glutamate. Glutamate was shown to be released at the thalamocortical synapse during barrel development (Laurent, et al., 2002) and the TCAs express the vesicular transporters of glutamate, VGlut2 (Boulland et al., 2004). Interestingly, a co-existence between serotonin and glutamate as transmitters could exist also in the adult raphe neurons, in which the vesicular transporter of glutamate Vglut3 is expressed (Gras et al., 2002; Herzog et al., 2004).

A similar combination of transporter expression (VMAT2 and 5-HTT) was found in the developing retina (Figure 4). In the retina, 5-HTT

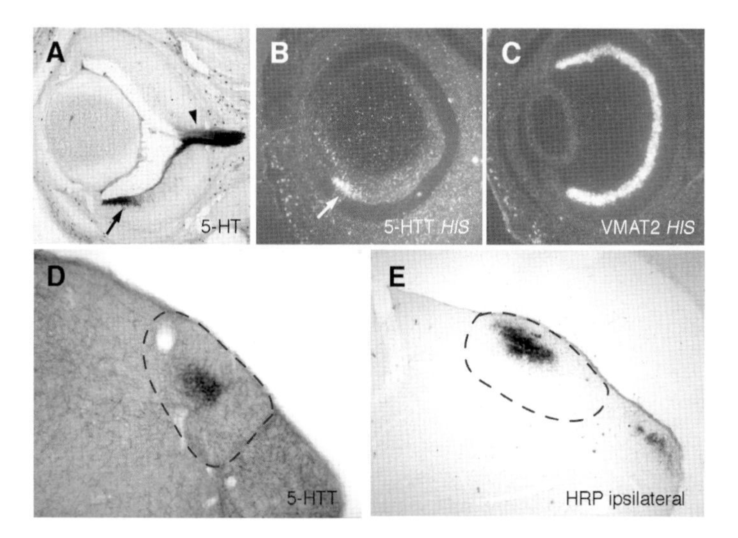

Figure 4 Transient expression of serotoninergic markers in the retinogeniculate pathway. 5-HT is found in a subset of retinal ganglion cells (RGCs) that are localized in the ventrotemporal part of the retina (E16 mouse retina). Labeled axons can be followed in the optic tract (arrowhead) (A). This localization of 5-HT is due to the uptake of extracellular 5-HT by the 5-HT transporter (5-HTT) which is expressed in the ventrotemporal RGCs (arrowhead in B). The vesicular transporter for monoamines, VMAT2, is expressed in all the RGCs (C). Labeled 5-HTT retinal axons can be followed up to the dLGN, where they form a central patch (D, immunolabeling at P9), which resembles the central patch of HRP labeled ipsilateral fibers at the same age (E).

expression and 5-HT high affinity capture was restricted to a peripheral ventral crescent of the retina (Figure 4B) which coincides in part with the localization of the ipsilateral retinal ganglion cells. The distribution of 5-HTT axon in the lateral geniculate nucleus and the superior colliculus also partially overlapped with the ipsilateral retinal terminals (Figure 4D, E), although a complete overlap could not be demonstrated (Upton et al., 1999). It is possible that the unequal capture of 5-HT among different contingents of afferent retinal axons could confer different growth properties to the retinal axons. Interestingly, transient expression of 5-HTT and VMAT2 is also observed in the primate retinal axons during embryogenesis in humans (Verney et al., 2002) and in monkeys (Lebrand et al., in press).

3.3.2. *Role of 5-HT for Emergence of Patterns in the Barrelfield and Retinogeniculate Projections*

The role of 5-HT in these systems was initially indicated by neurotoxins or pharmacological treatments that reduced 5-HT levels. These experiments delayed the separation of the ipsi-contralateral projections in the lateral geniculate nucleus (Rhoades et al., 1993) and caused a delay in barrel development and a reduction in barrel size (Osterheld-Haas et al., 1994; Bennett-Clarke et al., 1994). Similar observations were made in the VMAT2-KO mice, a model in which the brain levels of monoamines are drastically reduced (Persico et al., 2001 ; Alvarez et al., 2002). However the most conclusive demonstration of the 5-HT effects in these 2 systems came from genetic manipulations that increased 5-HT brain levels, rather than decreasing them. The monoamine oxidase A (MAOA)-KO mice proved a most valuable model in this regard, since they displayed a clear disruption of the thalamocortical and retinogeniculate segregation (Cases et al., 1996; Upton et al., 1999). MAOA-KO mice are unable to degrade 5-HT and accumulate it in considerable amounts in the brain during the first postnatal weeks. In P3 MAOA-KO mice, 5-HT brain levels are increased ten fold in comparison with control mice (Cases et al., 1995).

In the MAOA-KO mice, the development of the TCAs appear to be normal until P2, but thereafter, the process of thalamocortical axon segregation does not occur and the TCAs retain a uniform distribution within layers IV and VI. Similarly, retinal axons develop normally until P3 , but the later separation of the ipsi and contralateral retinal axons does not occur in the LGN or the superior colliculus (Upton et al., 1999). These phenotypes can be reproduced by the pharmacological inhibition of MAOs in normal mice during the first postnatal week (Vitalis et al., 1998) and they can be rescued in the MAOA-KO mice and 5-HTT-KO mice by lowering the levels of 5-HT with parachlorophenylanine (PCPA), a drug that inhibits, the synthetic enzyme of 5-HT, tryptophane hydroxylase (Cases et al., 1996). The optimal time period for a rescue of the phenotype matches the time of segregation of the TCAs and of the RGCs. After this period, the capacity for the RGCs and TCAs to remodel from a diffuse to a segregated pattern is gradually decreased (Persico et al., 2001; Upton, et al. 2002, Rebsam et al., 2005). In the MAOA-KO mice, barrels can be induced to form beyond the normal developmental period, between P6 and P10, but thereafter, the capacity of TCAs to remodel is lost. Interestingly, when barrels form at these relatively late stages they do not appear to require a signal from the whiskers (Boylan et al., 2001; Rebsam et al., 2005). This indicates that the periphery-related patterns are probably fixed in the subcortical sensory relays by P3.

Similar phenotypes were demonstrated in the 5-HTT-KO mice. An interesting difference of the 5-HTT-KO model relatively to the MAOA-KO mice is that the 5-HTT-KO mice are unable to take up 5-HT in the RGCs or the TCAs (Cases et al., 1998; Persico et al., 2001). This indicated that the accumulation of 5-HT in the extracellular space and not within the thalamic and retinal neurons causes the abnormal segregation of the axon terminals (Persico et al., 2001; Salichon et al., 2001). In the 5-HTT-KO, the increase in 5-HT levels in the brain is not as extreme as in the MAOA-KO mice, because of feedback mechanisms that reduce serotonin synthesis (Ravary et al., 2001). This might explain why the phenotypic changes are less marked than in the MAOA-KO. For instance, in the 5-HTT-KO, the larger caudal barrels are still formed and the enlargement of the ipsilateral retinal projections is less marked than in the MAOA-KO mice (Figure 5).

3.3.3. *Role of the 5-HT1B Receptor*

The presynaptic receptor 5-HT1B, is another serotonin-related gene that is transiently expressed in the developing sensory thalamocortical neurons (Bennett-Clarke et al., 1993). In the retinal ganglion cells, the 5-HT1B receptor expression begins by E15 (Upton et al., 1999) but persists throughout adult life (Mooney et al., 1994). The 5-HT1B receptor is presynaptic (Boschert et al., 1994), it is localized on axon terminals and has been shown to modulate neurotransmitter release. In particular, it negatively regulates glutamate release at the developing retinotectal and thalamocortical synapses (Rhoades et al., 1994; Laurent et al., 2002). The crucial role of this receptor in the developmental abnormalities of the MAOA-KO mice was shown with a double knockout strategy. The MAOA/5-HT1B DKO, the 5-HTT/5-HT1B DKO as well as the MAOA/5-HTT/5-HT1B triple KO mice develop an almost normal barrelfield and display a normal segregation of the retinal axons, despite the increased levels of brain serotonin (Salichon et al., 2001) (Figure 5). The lack of the 5-HT1B receptor by itself is insufficient to cause a visible alteration in the segregation of the thalamocortical axons in the somatosensory cortex (Figure 5), although a subtle phenotype is observed in the superior colliculus where the ipsilateral retinal axons are more broadly distributed than in control mice (Upton et al., 2002).

3.4. Adenylate Cyclase 1

Given the role of cAMP modulation for the growth of axons and their response to chemical guidance cues (Lohof et al., 1992; Song and Poo, 1999), the developmental expression of the calcium calmodulin

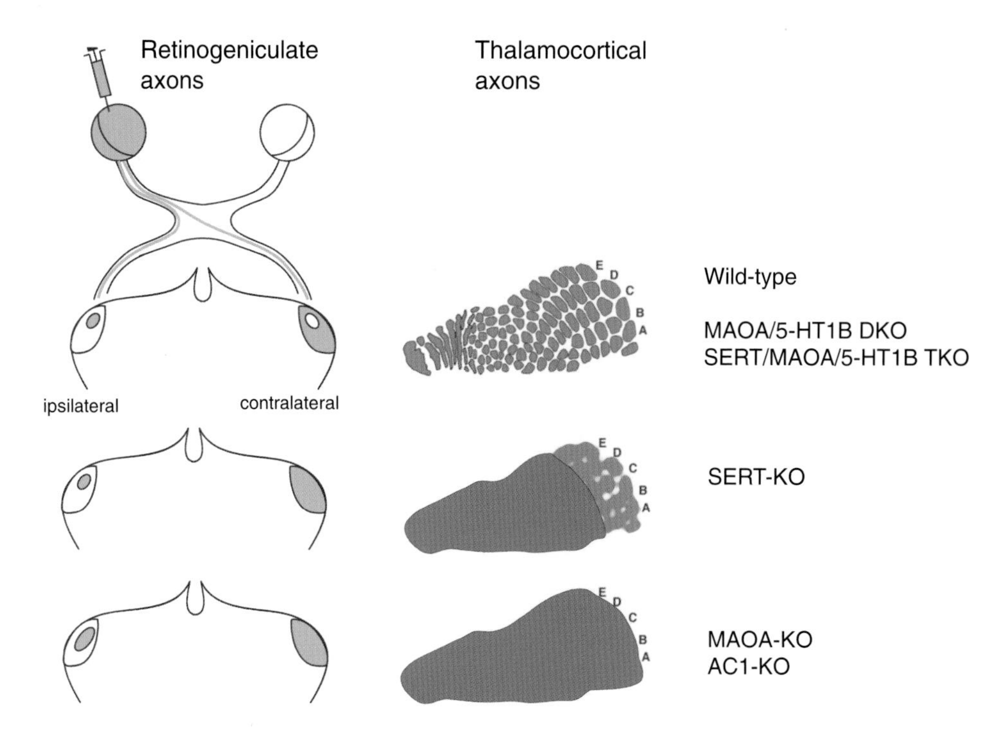

Figure 5 Summary diagram of the altered patterning of retinogeniculate and of the barrelfield in different mutant mouse strains. In wild type mice (WT), retinogeniculate axons from each eye are segregated: in the ipsilateral dLGN the retinal axons are grouped in a central patch; in the contralateral dLGN retinal projection are widely distributed leaving an empty central core. In the primary somatosensory cortex, the VB thalamocortical axons are segregated in clusters that define the barrel domains. The largest TCA clusters correspond to large vibrissae of the whisker pad, whereas smaller rostral barrels correspond to the small vibrissae of the anterior snout. In the 5-HT1B-KO, the MAOA/5-HT1B-DKO, the 5-HTT/5-HT1B-DKO and the 5-HTT/MAOA/5-HT1B-TKO mice the patterning of these inputs is undistinguishable from normal. In the 5-HTT-KO mice, the contralateral retinogeniculate projections cover the entire dLGN, and do not leave an empty gap, but the size of the ipsilateral retinal projection is not modified. In the somatosensory cortex, poorly outlined barrel domain can be distinguished in the caudal whisker pad, but TCAs retain a diffuse, non-segregated, distribution in the rest of the somatosensory representation. In the *brl* and the MAOA-KO mice, the ipsilateral projection domain is enlarged and the contralateral fibers occupy the entire dLGN. Thalamocortical axons are distributed uniformly in the whisker representation of the somatosensory cortex, and no segregation into barrel domain is observed.

stimulated cyclase, *AC1* in the thalamus and retina is of particular interest. This transmembrane adenylate cyclase isoform is directly activated by calcium influx through voltage sensitive calcium channels (Cooper et al., 1998). Furthermore, the activity of AC1 is negatively coupled to Gi heteromeric proteins and could thus also be modulated by the 5-HT1B receptor (Mendez et al., 1999; Sari, 2004). In the adult brain, the expression of AC1 is restricted to the hippocampus and to the cerebellum (Xia et al., 1991). However, during development AC1 expression is much broader (Matsuoka et al., 1997; Nicol et al., 2005). In particular, a strong AC1 expression is found in all the RGCs from E14 to P10 and strong transient AC1 expression is found all along the whisker to barrel pathway, including the nucleus trigeminalis principalis in the brainstem, the ventrobasal thalamic nucleus, and the barrel cortex (Nicol et al., 2005). A strain of "barrelless" mice, the *brl* of Lausanne was characterized as a spontaneous mutation of the AC1 gene, caused by the insertion of a retrotransposon in the AC1 coding sequence (Abdel-Majid et al., 1998). The *brl* mice do not develop barrels (Welker et al., 1996) and have an altered segregation of the ipsi/contralateral retinal inputs (Ravary et al., 2003). Similar phenotypes are observed after the genetic invalidation of the AC1 gene in both the barrelfield (Abdel-Majid et al., 1998) and the retinal projections (Nicol et al., in revision). A decrease in the number of "silent synapses" has been reported in the *brl* mice (Lu et al., 2003). This was interpreted as a reduced turnover of the AMPA subtype of glutamate receptor in the cortical neurons, suggesting that altered post-synaptic potentiation mechanisms could be involved in the altered barrel development. Observations in the retinotectal system indicate however that altered remodeling of the retinal axons in the colliculus is due to the AC1 deficiency in the RGCs and not in the post-synaptic neurons (Ravary et al., 2003; Nicol et al., 2006).

4. Presynaptic Mechanisms Involved in the Modulation of Axon Remodeling

Clearly the results discussed above indicate that neurotransmission and its regulation by modulatory transmitters such as serotonin play a role in "sculpting" the terminal fields of thalamic and retinal axons (Erzurumlu and Kind, 2001). They also indicate that pre-synaptic mechanisms operating in the thalamocortical and retinogeniculate synapses are probably as important as the post-synaptic mechanisms for the refinement of connections during development. How can these molecular and genetic observations be integrated in a mechanistic framework of axon terminal modelling?

Based on the previously discussed genetic evidence we can conclude that 3 molecular actors, the 5-HT1B receptor, the calcium stimulated adenylate cyclase AC1, and the TrkB receptor, are present in the thalamic and retinal neurons and control the distribution of the axon arbors. Based on the localisation of the 5-HT1B receptor and TrkB in axon terminals, one can hypothesize that this action is local, at the level of axon terminals.

4.1. Trophic Mechanisms

A possible working model that integrates these molecular actors is that the final downstream effects is a direct control of axon growth and branching. This assumption is based on the study of knockout mice showing alterations in the axon branching of the thalamic and retinal axons. The AC1 defective *brl* mice, the TrkB-KO mice and the gain of function of the 5-HT1B receptor (in the MAOA KO mice) all have defective TCA arbors. In the MAOA-KO mice, TCAs are abnormally wide, and the number of branches is significantly reduced (Rebsam et al., 2002). In the *brl* mice, axon branches are abnormally widespread (Welker et al., 1996; Hage, 2003). In the TrkB-KO the TCAs extend abnormally in the upper cortical layers (Vitalis et al., 2002). However, these genetic observations *in vivo* are difficult to interpret, since similar alterations in the axon branching phenotypes have also been observed in the NR1 cortex-specific KO, indicating that signals arising from the post-synaptic neurons modulate pre-synaptic axon branching (Lee et al., 2005). More direct evidence for an effect of presynaptic signaling pathways on axon growth have been observed *in vitro*, on cultures of retinal and thalamic neurons. 5-HT and 5-HT1B receptor agonists stimulate the growth and branching of rodent thalamic axons (Lotto et al., 1999; Lieske et al., 1999). Furthermore, activation of the cAMP pathways and of BDNF, via the TrkB receptors, has major growth promoting effects on RGCs (Meyer-Franke et al., 1998), and axons from brl mice show altered branching. In this scheme, the 5-HT1B receptor could be coupled to the heterotrimeric protein Gi, which in turn inhibits AC1. Thus, overactivation of the 5-HT1B receptors, when there is an excess of 5-HT (such as observed in the MAOA and 5-HTT KO mice), would be expected to reduce AC1 function, and to decrease the production of cAMP in the retinal and thalamic axon terminals. This hypothesis seems to be corroborated by the similar phenotype of the MAOA and AC1-KO mice. Since cAMP and PKA control the phosphorylation of a very wide number of proteins (Shabb, 2001), there are many different possibilities beyond that point to modulate axon growth and its response to guidance and trophic molecules. One of the consequences could be a reduced cycling of the TrkB receptor at the plasma membrane resulting in a reduced trophic growth support for the ingrowing axons (Figure 6). A more direct consequence of deregulated cAMP in the growth cone could be a modified response to repulsive and attractive molecules expressed in the target fields (Song and Poo, 1999,). Evidence in our laboratory argue strongly to such mechanism (Nicol et al., 2006). This effect could be via PKA-dependent phosphorylation of cytoskeletal proteins and regulatory proteins which are involved in the dynamic control of the actin and microtubule cytoskeleton in the growth cone (Dent and Gertler, 2003) (Figure 6) and thus has the potential of directly affecting growth cone motility. Finally, cAMP could have a transcriptional control on target genes containing a cAMP- responsive element (CRE) which is important for these remodeling events (West et al., 2001) (Figure 6).

4.2. Presynaptic Potentiation Mechanisms

An alternative model is that serotonin and cAMP modulate activity dependent mechanisms via a pre-synaptic control of glutamate release (Figure 6). 5-HT1B receptors are localised pre-synaptically in the thalamocortical, retinogeniculate and retinotectal synapses

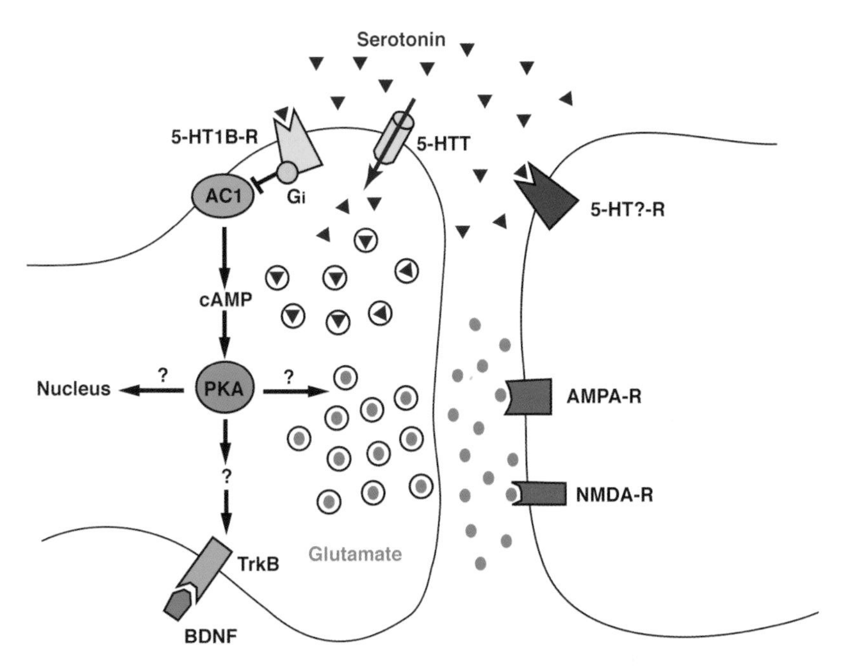

Figure 6 Models of the presynaptic mechanisms involving serotonin and adenylate cyclase1 in the thalamocortical and retinogeniculate axon terminals. Retinal and thalamic axons express 5-HTT, VMAT2, 5-HT1B receptors and AC1. Extracellular serotonin is taken up by 5-HTT and stored into synaptic vesicles by the VMAT2. On the postsynaptic site, glutamate receptors such as AMPA or NMDA receptor and other 5-HT receptor (5-HT?-R) are expressed. The 5-HT1B presynaptic receptor is coupled to a Gi protein which inhibits adenylate cyclase 1 (AC1). AC1 synthesizes cAMP which activates Protein Kinase A (PKA). Activation of PKA controls a number of cellular processes that are important for axon terminal remodeling. Non exclusive hypothesis include: 1) a control of neurotransmitter release mechanisms, 2) a control of the membrane trafficking of neurotrophin receptors such as TrkB, thereby modifying the axonal response to BDNF, 3) a control of axon dynamic behaviour, via the phosphorylation of cytoskeletal proteins, 4) a transcriptional effect via CREB.

(Bennett-Clarke et al., 1993; Boschert et al., 1994; Sari et al., 1999). In these systems, electrophysiological studies showed that 5-HT1B receptor activation results in the inhibition of glutamate release (Rhoades et al., 1994; Laurent et al., 2002). Moreover, using the thalamocortical slice preparation it was demonstrated that the effects of 5-HT1B receptors could be that of a low pass filter. 5-HT1B receptor stimulation reduces EPSCs evoked by low frequency stimulation whereas it relieves the short depression evoked by high frequency stimulation (Laurent et al., 2002). Thus, the 5-HT1B receptors could control afferent neural activity generated in the whisker pad or in the thalamus and allow activity-dependent stabilization of synapses and the consolidation of the corresponding branches. In this hypothesis the uptake of serotonin by the thalamic and retinal axons, as well as its possible co-release, would act to finely adjust the levels of 5-HT at the developing synapses (Figure 6).

Similarly, AC1 could play a role in the activity-dependent strengthening of active synapses. Indeed, AC1 is involved in long term potentiation in a number of systems including the somatosensory cortex (Lu et al. 2003). In the hippocampus, AC1 has been involved in pre-synaptic potentiation mechanisms that are observed at the mossy fiber synapse (Villacres et al., 1998). The existence of such presynaptic plasticity mechanisms has not yet been shown in the developing retinotectal and thalamocortical synapses, but could constitute another mechanism for synaptic consolidation and refinement of connections during development.

5. Conclusion

The molecular composition of the developing retinal and thalamic axons display striking similarities during the phases of axon elaboration and refinement of sensory maps. This concerns not only general growth associated proteins, but a more specific repertoire of serotonin receptors and their associated signaling pathways. Mutant mouse analyses demonstrate the requirement of such molecular pathways for the refinement of the visual and somatosensory maps. However, it remains uncertain whether this involves activity-driven mechanisms or a modulation of neural growth. These are two likely hypothesis that are not necessarily exclusive since activity-dependent and activity-independent pathways could cooperate, either in synchrony, or at different stages to refine the retino-geniculate and thalamo-cortical projections. For instance, it is conceivable that the 5-HT1B receptor and downstream AC1-mediated mechanisms could act on the growth of axons during early stages of development whereas they may act on synaptic consolidation during later stages of development. It will be the role of future studies to tease out the contribution of these mechanisms.

Acknowledgements

We are indebted to all the present and former members of the laboratory and to our collaborators who contributed to generate the data reviewed in this article in particular, Oliviez Cases, Cécile Lebrand, Xavier Nicol, Anne Ravary, Nathalic Salichon, Isabelle Seif, Tamia vitalis. We thank Luc Maroteaux for critical reading of the manuscript. The INSERM, Fondation de France, Retina France and Fondation France Telecom are acknowledged for their financial support.

References

Abdel-Majid, R.M., Leong, W.L., Schalkwyk, L.C., Smallman, D.S., Wong, S.T., Storm, D.R., Fine, A., Dobson, M.J., Guernsey, D.L., and Neumann, P.E. (1998). Loss of adenylyl cyclase I activity disrupts patterning of mouse somatosensory cortex. *Nat Genet, 19:* 289–291.

Agmon, A. and Connors, B.W. (1992). Correlation between intrinsic firing patterns and thalamocortical synaptic responses of neurons in mouse barrel cortex. *J Neurosci, 12,* 319–329.

Agmon, A., Yang, L.T., Jones, E.G., and O'Dowd, D.K. (1995). Topological precision in the thalamic projection to neonatal mouse barrel cortex. *J Neurosci, 15,* 549–561.

Agmon, A., Yang, L.T., O'Dowd, D.K., and Jones, E.G. (1993). Organized growth of thalamocortical axons from the deep tier of terminations into layer IV of developing mouse barrel cortex. *J Neurosci, 13,* 5365–5382.

Aigner, L. and Caroni, P. (1993). Depletion of 43-kD growth-associated protein in primary sensory neurons leads to diminished formation and spreading of growth cones. *J Cell Biol, 123,* 417–429.

Alvarez, C., Vitalis, T., Fon, T., Hanoun, N., Hamon, M., Seif, I., Gaspar, P., and Cases, O. (2002). Altered somatosensory thalamocortical development inmice lacking the vesicular monoamine transporter type 2, *Neuroscience, 115,* 753–764.

Armstrong-James, M. and Fox, K. (1987). Spatiotemporal convergence and divergence in the rat S1 "barrel" cortex. *J Comp Neurol, 263,* 265–281.

Auso, E., Cases, O., Fouquet, C., Camacho, M., Garcia-Velasco, J.V., Gaspar, P., and Berbel, P. (2001). Protracted expression of serotonin transporter and altered thalamocortical projections in the barrelfield of hypothyroid rats. *Eur J Neurosci, 14:* 1968.

Belford, G.R. and Killackey, H.P. (1979). Vibrissae representation in subcortical trigeminal centers of the neonatal rat. *J Comp Neurol, 183,* 305–321.

Bennett-Clarke, C.A., Leslie, M.J., Chiaia, N.L., and Rhoades, R.W. (1993). Serotonin 1B receptors in the developing somatosensory and visual cortices are located on thalamocortical axons. *Proc Natl Acad Sci U S A, 90:*153–157.

Bennett-Clarke, C.A., Leslie, M.J., Lane, R.D., and Rhoades, R.W. (1994). Effect of serotonin depletion on vibrissa-related patterns of thalamic afferents in the rat's somatosensory cortex. *J Neurosci, 14:*7594–7607.

Berardi, N., Pizzorusso, T., and Maffei, L. (2000). Critical periods during sensory development. *Curr Opin Neurobiol, 10:*138–145.

Binshtok, A.M., Fleidervish, I.A., and Gutnick, M.J. (2004). Endogenous 5-HT released from thalamic axons induces local disinhibition in developing layer 4 of mouse barrel cortex. Proceedings of the 4th Forum of European Neuroscience, Lisbon. July 2004. 225.2.

Boschert, U., Amara, D.A., Segu, L., and Hen, R. (1994). The mouse 5-hydroxytryptamine 1B receptor is localized predominantly on axon terminals. *Neuroscience, 58:*167–182.

Boulland, J.L., Qureshi, T., Seal, R.P., Rafiki, A., Gundersen, V., Bergersen, L.H., Fremeau, R.T., Jr., Edwards, R.H., Storm-Mathisen, J., and Chaudhry, F.A. (2004). Expression of the vesicular glutamate transporters during development indicates the widespread corelease of multiple neurotransmitters. *J Comp Neurol, 480:* 264–280.

Boylan, C.B., Bennett-Clarke, C.A., Chiaia, N.L., and Rhoades, R.W. (2000). Time course of expression and function of the serotonin transporter in the neonatal rat's primary somatosensory cortex. *Somatosens Mot Res, 17:*52–60.

Boylan, C.B., Kesterson, K.L., Bennett-Clarke, C.A., Chiaia, N.L., and Rhoades, R.W. (2001). Neither peripheral nerve input nor cortical NMDA receptor activity are necessary for recovery of a disrupted barrel pattern in rat somatosensory cortex. *Brain Res Dev Brain Res, 129:*95–106.

Cases, O., Seif, I., Grimsby, J., Gaspar, P., Chen, K., Pournin, S., Muller, U., Aguet, M., Babinet, C., and Shih, J.C. (1995). Aggressive behavior and altered amounts of brain serotonin and norepinephrine in mice lacking MAOA [see comments]. *Science, 268:*1763–1766.

Cases, O., Vitalis, T., Seif, I., De Maeyer, E., Sotelo, C., and Gaspar, P. (1996). Lack of barrels in the somatosensory cortex of monoamine oxidase A-deficient mice: role of a serotonin excess during the critical period. *Neuron, 16:*297–307.

Cases, O., Lebrand, C., Giros, B., Vitalis, T., De Maeyer, E., Caron, M.G., Price, D.J., Gaspar, P., and Seif, I. (1998). Plasma membrane transporters of serotonin, dopamine, and norepinephrine mediate serotonin accumulation in atypical locations in the developing brain of monoamine oxidase A knock-outs. *J Neurosci, 18:*6914–6927.

Catalano, S.M., Robertson, R.T., and Killackey, H.P. (1991). Early ingrowth of thalamocortical afferents to the neocortex of the prenatal rat. *Proc Natl Acad Sci U S A, 88:*2999–3003.

Catalano, S.M., Robertson, R.T. & Killackey, H.P. (1996) Individual axon morphology and thalamocortical topography in developing rat somatosensory cortex. *J. Comp Neurol.*, **367,** 36–53.

Chen, C. and Regehr, W.G. (2000). Developmental remodeling of the retinogeniculate synapse. *Neuron, 28:*955–966.

Chiaia, N.L., Fish, S.E., Bauer, W.R., Bennett-Clarke, C.A., and Rhoades, R.W. (1992). Postnatal blockade of cortical activity by tetrodotoxin does not disrupt the formation of vibrissa-related patterns in the rat's somatosensory cortex. *Brain Res Dev Brain Res, 66:*244–250.

Cohen-Cory, S. (1999). BDNF modulates, but does not mediate, activity-dependent branching and remodeling of optic axon arbors in vivo. *J Neurosci, 19:*9996–10003.

Cooper, D.M., Schell, M.J., Thorn, P., and Irvine, R.F. (1998). Regulation of adenylyl cyclase by membrane potential. *J Biol Chem, 273:*27703–27707.

D'Amato, R.J., Blue, M.E., Largent, B.L., Lynch, D.R., Ledbetter, D.J., Molliver, M.E., and Snyder, S.H. (1987). Ontogeny of the serotonergic projection to rat neocortex: transient expression of a dense innervation to primary sensory areas. *Proc Natl Acad Sci U S A, 84:*4322–4326.

Dawson, D.R. and Killackey, H.P. (1985). Distinguishing topography and somatotopy in the thalamocortical projections of the developing rat. *Brain Res, 349:*309–313.

Demyanenko, G.P. and Maness, P.F. (2003). The L1 cell adhesion molecule is essential for topographic mapping of retinal axons. *J Neurosci, 23:*530–538.

Dent, E.W. and Gertler, F.B. (2003). Cytoskeletal dynamics and transport in growth cone motility and axon guidance. *Neuron, 40:*209–227.

Erzurumlu, R.S. and Jhaveri, S. (1990). Thalamic axons confer a blueprint of the sensory periphery onto the developing rat somatosensory cortex. *Brain Res Dev Brain Res, 56:*229–234.

Erzurumlu, R.S., Jhaveri, S., and Benowitz, L.I. (1990). Transient patterns of GAP-43 expression during the formation of barrels in the rat somatosensory cortex. *J Comp Neurol, 292,* 443–456.

Erzurumlu, R.S. and Killackey, H.P. (1983). Development of order in the rat trigeminal system. *J Comp Neurol, 213:*365–380.

Erzurumlu, R.S. and Kind, P.C. (2001). Neural activity: sculptor of 'barrels' in the neocortex. *Trends Neurosci, 24:*589–595.

Feldman, D.E., Nicoll, R.A., and Malenka, R.C. (1999). Synaptic plasticity at thalamocortical synapses in developing rat somatosensory cortex: LTP, LTD, and silent synapses. *J Neurobiol, 41*, 92–101.

Fukuda, T., Kawano, H., Ohyama, K., Li, H.P., Takeda, Y., Oohira, A., and Kawamura, K. (1997). Immunohistochemical localization of neurocan and L1 in the formation of thalamocortical pathway of developing rats. *J Comp Neurol, 382*, 141–152.

Gaspar, P., Cases, O. and Maroteaux, L. (2003). The developmental role of serotonin: news from mouse molecular genetics. *Nat Rev Neurosci., 4*:1002–1012

Goodman, C.S. and Shatz, C.J. (1993). Developmental mechanisms that generate precise patterns of neuronal connectivity. *Cell, 72 Suppl*, 77–98.

Gras, C., Herzog, E., Bellenchi, G.C., Bernard, V., Ravassard, P., Pohl, M., Gasnier, B., Giros, B., and El Mestikawy, S. (2002). A third vesicular glutamate transporter expressed by cholinergic and serotoninergic neurons. *J Neurosci, 22*, 5442–5451.

Hage, I. (2003). Early postnatal development of thalamocortical axons to barrel cortex in NOR and barrelless mice a morphological tracing study. Ph D. Thesis Université de Lausanne, p. 1–97.

Hashimoto, K. and Kano, M. (2003). Functional differentiation of multiple climbing fiber inputs during synapse elimination in the developing cerebellum. *Neuron, 38*:785–796.

Henderson, T.A., Woolsey, T.A., and Jacquin, M.F. (1992). Infraorbital nerve blockade from birth does not disrupt central trigeminal pattern formation in the rat. *Brain Res Dev Brain Res, 66*:146–152.

Herzog, E., Gilchrist, J., Gras, C., Muzerelle, A., Ravassard, P., Giros, B., Gaspar, P., and El Mestikawy, S. (2004). Localization of VGLUT3, the vesicular glutamate transporter type 3, in the rat brain. Localization of VGLUT3, the vesicular glutamate transporter type 3, in the rat brain. *Neuroscience, 123*:983–1002.

Jeanmonod, D., Rice, F.L., and Van der, L.H. (1981). Mouse somatosensory cortex: alterations in the barrelfield following receptor injury at different early postnatal ages. *Neuroscience, 6*:1503–1535.

Jensen, K.F. and Killackey, H.P. (1987). Terminal arbors of axons projecting to the somatosensory cortex of the adult rat. II. The altered morphology of thalamocortical afferents following neonatal infraorbital nerve cut. *J Neurosci, 7*:3544–3553.

Jhaveri, S., Erzurumlu, R.S., and Crossin, K. (1991). Barrel construction in rodent neocortex: role of thalamic afferents versus extracellular matrix molecules. *Proc Natl Acad Sci U S A, 88*:4489–4493.

Jung, M., Petrausch, B., and Stuermer, C.A. (1997). Axon-regenerating retinal ganglion cells in adult rats synthesize the cell adhesion molecule L1 but not TAG-1 or SC-1. *Mol Cell Neurosci, 9*:116–131.

Killackey, H.P., Rhoades, R.W., and Bennett-Clarke, C.A. (1995). The formation of a cortical somatotopic map. *Trends Neurosci, 18*:402–407.

Krug, K., Smith, A.L., and Thompson, I.D. (1998). The development of topography in the hamster geniculo-cortical projection. *J Neurosci, 18*:5766–5776.

Laurent, A., Goaillard, J.M., Cases, O., Lebrand, C., Gaspar, P., and Ropert, N. (2002). Activity dependent presynaptic role of the serotonin 1B receptors on the thalamocortical transmission in the somatosensory pathway of the neonatal mice. *J Neuroscience, 22*:886–900.

Lebrand, C., Cases, O., Adelbrecht, C., Doye, A., Alvarez, C., El Mestikawy, S., Seif, I., and Gaspar, P. (1996). Transient uptake and storage of serotonin in developing thalamic neurons. *Neuron, 17:*823–835.

Lebrand, C., Cases, O., Wehrle, R., Blakely, R.D., Edwards, R.H., and Gaspar, P. (1998). Transient developmental expression of monoamine transporters in the rodent forebrain. *J Comp Neurol, 401:*506–524.

Lebrand, C., Gaspar, P, Nicolas, D, and Hornung, JP. Tansitory uptake of serotonin in the developing sensory pathways of the common marmoset *(callithrix jacchus).* J. Comp. Neurol. (in press).

Lee, L.J., Iwasato, T., Itohara, S., and Erzurumlu, R.S. (2005). Exuberant thalamocortical axon arborization in cortex-specific NMDAR1 knockout mice. *J Comp Neurol, 485:*280–292.

Lidov, H.G. and Molliver, M.E. (1982). An immunohistochemical study of serotonin neuron development in the rat: ascending pathways and terminal fields. *Brain Res Bull, 8:*389–430.

Lieske, V., Bennett-Clarke, C.A., and Rhoades, R.W. (1999). Effects of serotonin on neurite outgrowth from thalamic neurons in vitro. *Neuroscience, 90*, 967–974.

Lohof, A.M., Quillan, M., Dan, Y., and Poo, M.M. (1992). Asymmetric modulation of cytosolic cAMP activity induces growth cone turning. *J Neurosci, 12*, 1253–1261.

Lopéz-Bendito, G. and Molnár, Z. (2003). Thalamocortical development: how are we going to get there? *Nat Rev Neurosci, 4*, 276–289.

Lotto, B., Upton, L., Price, D.J., and Gaspar, P. (1999). Serotonin receptor activation enhances neurite outgrowth of thalamic neurones in rodents. *Neurosci Lett, 269:*87–90.

Lu, H.C., She, W.C., Plas, D.T., Neumann, P.E., Janz, R., Crair, M.C. (2003) Adenylyl cyclase I regulates AMPA receptor trafficking during mouse cortical'barrel' map development.*Nat Neurosci.* 6:939–47.

Ma, P.M. (1993). Barrelettes–architectonic vibrissal representations in the brainstem trigeminal complex of the mouse. II. Normal post-natal development. *J Comp Neurol, 327:*376–397.

Maier, D.L., Mani, S., Donovan, S.L., Soppet, D., Tessarollo, L., McCasland, J.S., and Meiri, K.F. (1999). Disrupted cortical map and absence of cortical barrels in growth- associated protein (GAP)-43 knockout mice. *Proc Natl Acad Sci U S A, 96:*9397–9402.

Marotte, L.R., Leamey, C.A., and Waite, P.M. (1997). Timecourse of development of the wallaby trigeminal pathway: III. Thalamocortical and corticothalamic projections. *J Comp Neurol, 387:*194–214.

Masana, Y., Wanaka, A., Kato, H., Asai, T., and Tohyama, M. (1993). Localization of trkB mRNA in postnatal brain development. *J Neurosci Res, 35:*468–479.

Matsuoka, I., Suzuki, Y., Defer, N., Nakanishi, H., and Hanoune, J. (1997). Differential expression of type I, II, and V adenylyl cyclase gene in the postnatal developing rat brain. *J Neurochem, 68:*498–506.

Mendez, J., Kadia, T.M., Somayazula, R.K., El Badawi, K.I., and Cowen, D.S. (1999). Differential coupling of serotonin 5-HT1A and 5-HT1B receptors to activation of ERK2 and inhibition of adenylyl cyclase in transfected CHO cells. *J Neurochem, 73:*162–168.

Menna, E., Cenni, M.C., Naska, S., and Maffei, L. (2003). The anterogradely transported BDNF promotes retinal axon remodeling during eye specific segregation within the LGN. *Mol Cell Neurosci, 24*, 972–983.

Meyer-Franke, A., Wilkinson, G.A., Kruttgen, A., Hu, M., Munro, E., Hanson, M.G., Jr., Reichardt, L.F., and Barres, B.A. (1998). Depolarization and cAMP elevation rapidly recruit TrkB to the plasma membrane of CNS neurons. *Neuron, 21*:681–693.

Molnár, Z., Adams, R., and Blakemore, C. (1998). Mechanisms underlying the early establishment of thalamocortical connections in the rat. *J Neurosci, 18*:5723–5745.

Mooney, R., Penn, A.A., Gallego, R., and Shatz, C.J. (1996). Thalamic relay of spontaneous retinal activity prior to vision. *Neuron, 17*:863–874.

Mooney, R.D., Shi, M.Y., and Rhoades, R.W. (1994). Modulation of retinotectal transmission by presynaptic 5-HT1B receptors in the superior colliculus of the adult hamster. *J Neurophysiol, 72*:3–13.

Naegele, J.R., Jhaveri, S., and Schneider, G.E. (1988). Sharpening of topographical projections and maturation of geniculocortical axon arbors in the hamster. *J Comp Neurol, 277*:593–607.

Nicol, X., Muzerelle, A., Bachy, I., Ravary, A., and Gaspar, P. (2005). Spatiotemporal localisation of the calcium stimulated cyclase, AC1 and AC8: during mouse brain development. *J Comp Neurol*, 486:281–94.

Nicol, X., Muzerelle, A., Rio, J.P., Metin, C. and Gaspar, P. (2006) Requirement of adenylate cyclase 1 for the ephrin-A5-dependent retraction of exubenant retinal axons. *J. Neurosci., 26*:862–872.

Nicol, X., Bennis, M., Ishikawa, Y., Chan, G., Repérant, J., Storm, D., Gaspar, P. Role of the calcium modulated cyclases in the development of the retinal projections. *European Journal of Neuroscience* (in revision).

Niell, C.M., Meyer, M.P., and Smith, S.J. (2004). In vivo imaging of synapse formation on a growing dendritic arbor. *Nat Neurosci, 7*:254–260.

O'Leary, D.D., Ruff, N.L., and Dyck, R.H. (1994). Development, critical period plasticity, and adult reorganizations of mammalian somatosensory systems. *Curr Opin Neurobiol, 4*, 535–544.

Osterheld-Haas, M.C., Van der, L.H., and Hornung, J.P. (1994). Monoaminergic afferents to cortex modulate structural plasticity in the barrelfield of the mouse. *Brain Res Dev Brain Res, 77*:189–202.

Ozdinler, P.H. and Erzurumlu, R.S. (2002). Slit2, a branching-arborization factor for sensory axons in the Mammalian CNS. *J Neurosci, 22*:4540–4549.

Persico, A.M., Mengual, E., Moessner, R., Hall, S.F., Revay, R.S., Sora, I., Arellano, J., DeFelipe, J., Gimenez-Amaya, J.M., Conciatori, M., Marino, R., Baldi, A., Cabib, S., Pascucci, T., Uhl, G.R., Murphy, D.L., Lesch, K.P., and Keller, F. (2001). Barrel pattern formation requires serotonin uptake by thalamocortical afferents, and not vesicular monoamine release. *J Neurosci, 21*:6862–6873.

Pollock, G.S., Robichon, R., Boyd, K.A., Kerkel, K.A., Kramer, M., Lyles, J., Ambalavanar, R., Khan, A., Kaplan, D.R., Williams, R.W., and Frost, D.O. (2003). TrkB receptor signaling regulates developmental death dynamics, but not final number, of retinal ganglion cells. *J Neurosci, 23*:10137–10145.

Purves, D. and Lichtman, J.W. (1980). Elimination of synapses in the developing nervous system. *Science, 210*, 153–157.

Ravary, A., Muzerelle, A., Darmon, M., Murphy, D.L., Moessner, R., Lesch, K.P., and Gaspar, P. (2001). Abnormal trafficking and subcellular localization of an N-terminally truncated serotonin transporter protein. *Eur J Neurosci, 13*:1349–1362.

Ravary, A., Muzerelle, A., Herve, D., Pascoli, V., Ba-Charvet, K.N., Girault, J.A., Welker, E., and Gaspar, P. (2003). Adenylate cyclase 1 as a key actor in the refinement of retinal projection maps. *J Neurosci, 23*:2228–2238.

Rebsam, A., S.I., and Gaspar, P. (2005). Dissociating barrel development and plasticity in the mouse somatosensory cortex. *J Neurosci, 25:*706–710.

Rebsam, A., Seif, I., and Gaspar, P. (2002). Refinement of thalamocortical arbors and emergence of barrel domains in the primary somatosensory cortex: role of 5-HT. *J Neurosci, 22:*8541–8552.

Rhoades, R.W., Bennett-Clarke, C.A., Chiaia, N.L., White, F.A., McDonald, G.J., Haring, J.H., and Jacquin, M.F. (1990). Development and lesion induced reorganization of the cortical representation of the rat's body surface as revealed by immunocytochemistry for serotonin. *J Comp Neurol, 293:*190–207.

Rhoades, R.W., Bennett-Clarke, C.A., Lane, R.D., Leslie, M.J., and Mooney, R.D. (1993). Increased serotoninergic innervation of the hamster's superior colliculus alters retinotectal projections. *J Comp Neurol, 334,* 397–409.

Rhoades, R.W., Bennett-Clarke, C.A., Shi, M.Y., and Mooney, R.D. (1994) Effects of 5-HT on thalamocortical synaptic transmission in the developing rat. *J. Neurophysiol.*, 72, 2438–2450.

Rice, F.L. and Van der Loos, H. (1977). Development of the barrels and barrel field in the somatosensory cortex of the mouse. *J Comp Neurol, 171:*545–560.

Salichon, N., Gaspar, P., Upton, A.L., Picaud, S., Hanoun, N., Hamon, M., De Maeyer, E., Murphy, D.L., Mossner, R., Lesch, K.P., Hen R. and Seif, I. (2001). Excessive activation of serotonin (5-HT) 1B receptors disrupts the formation of sensory maps in monoamine oxidase a and 5-ht transporter knock-out mice. *J Neurosci., 21:*884–896

Sanes, J.R. and Lichtman, J.W. (1999). Development of the vertebrate neuromuscular junction. *Annu Rev Neurosci, 22:*389–442.

Sari, Y. (2004). Serotonin1B receptors: from protein to physiological function and behavior. *Neurosci Biobehav Rev, 28:*565–582.

Schlaggar, B.L., De Carlos, J.A., and O'Leary, D.D. (1993a). Acetylcholinesterase as an early marker of the differentiation of dorsal thalamus in embryonic rats. *Brain Res Dev Brain Res, 75:*19–30.

Schlaggar, B.L., Fox, K., and O'Leary, D.D. (1993b). Postsynaptic control of plasticity in developing somatosensory cortex. *Nature, 364,* 623–626.

Senft, S.L. and Woolsey, T.A. (1991). Growth of thalamic afferents into mouse barrel cortex. *Cerebral Cortex, 1:*308–335.

Shabb, J.B. (2001). Physiological substrates of cAMP-dependent protein kinase. *Chem Rev, 101:*2381–2411.

Shatz, C.J. (1996). Emergence of order in visual system development. *Proc Natl Acad Sci U S A, 93:*602–608.

Shatz, C.J. and Stryker, M.P. (1988). Prenatal tetrodotoxin infusion blocks segregation of retinogeniculate afferents. *Science, 242:*87–89.

Snider, C.J., Dehay, C., Berland, M., Kennedy, H., and Chalupa, L.M. (1999). Prenatal development of retinogeniculate axons in the macaque monkey during segregation of binocular inputs. *J Neurosci, 19:*220–228.

Song, H.J. and Poo, M.M. (1999). Signal transduction underlying growth cone guidance by diffusible factors. *Curr Opin Neurobiol, 9:*355–363.

Toborg, C.L. and Feller, M.B. (2005). Spontaneous patterned activity and the refinement of retinal projections. *Prog. Neurobiol. 76:*213–235.

Ugolini, G., Cremisi, F., and Maffei, L. (1995). TrkA, TrkB and p75 mRNA expression is developmentally regulated in the rat retina. *Brain Res, 704,* 121–124.

Ulupinar, E., Datwani, A., Behar, O., Fujisawa, H., and Erzurumlu, R. (1999). Role of semaphorin III in the developing rodent trigeminal system. *Mol Cell Neurosci, 13:*281–292.

Upton, A.L., Salichon, N., Lebrand, C., Ravary, A., Blakely, R., Seif, I., and Gaspar, P. (1999). Excess of serotonin (5-HT) alters the segregation of ispilateral and contralateral retinal projections in monoamine oxidase A knock-out mice: possible role of 5-HT uptake in retinal ganglion cells during development. *J Neurosci, 19:*7007–7024.

Upton, A.L., Ravary, A., Salichon, N., Moessner, R., Lesch, K.P., Hen, R., Seif, I., and Gaspar, P. (2002). Lack of 5-HT(1B) receptor and of serotonin transporter have different effects on the segregation of retinal axons in the lateral geniculate nucleus compared to the superior colliculus. *Neuroscience, 111:*597–610.

Van der Loos, H. and Woolsey, T.A. (1973). Somatosensory cortex: structural alterations following early injury to sense organs. *Science, 179:*395–398.

Verney, C., Lebrand, C., and Gaspar, P. (2002). Changing distribution of monoaminergic markers in the developing human cerebral cortex with special emphasis on the serotonin transporter. *Anat Rec, 267:*87–93.

Villacres, E.C., Wong, S.T., Chavkin, C., and Storm, D.R. (1998). Type I adenylyl cyclase mutant mice have impaired mossy fiber long-term potentiation. *J Neurosci, 18:*3186–3194.

Vitalis, T., Cases, O., Gillies, K., Hanoun, N., Hamon, M., Seif, I., Gaspar, P., Kind, P., and Price, D.J. (2002). Interactions between TrkB signaling and serotonin excess in the developing murine somatosensory cortex: a role in tangential and radial organization of thalamocortical axons. *J Neurosci, 22:*4987–5000.

Vitalis, T., Cases, O., Callebert, J., Launay, J.M., Price, D.J., Seif, I. and Gaspar, P. (1998). Effects of monoamine oxidase A inhibition on barrel formation in the mouse somatosensory cortex: determination of a sensitive developmental period. *J Comp Neurol. 393:*169–184.

Welker, E., Armstrong-James, M., Bronchti, G., Ourednik, W., Gheorghita-Baechler, F., Dubois, R., Guernsey, D.L., Van der, L.H., and Neumann, P.E. (1996). Altered sensory processing in the somatosensory cortex of the mouse mutant barrelless. *Science, 271:*1864–1867.

West, A.E., Chen W.G., Dalva M.B., Dolmetsch R.E., Kornhauser J.M., Shaywitz A.J., Takasu M.A., Tao X., Greenberg M.E. (2001). Calcium regulation of neuronal gene expression. *Proc Natl Acad Sci U S A, 98:*11024–11031.

White, E.L., Weinfeld, L., and Lev, D.L. (1997). A survey of morphogenesis during the early postnatal period in PMBSF barrels of mouse SmI cortex with emphasis on barrel D4. *Somatosens Mot Res, 14,* 34–55.

Wiencken-Barger, A.E., Mavity-Hudson, J., Bartsch, U., Schachner, M., and Casagrande, V.A. (2004). The role of L1 in axon pathfinding and fasciculation. *Cerebral Cortex, 14,* 121–131.

Witte, S., Stier, H., and Cline, H.T. (1996). In vivo observations of timecourse and distribution of morphological dynamics in Xenopus retinotectal axon arbors. *J Neurobiol, 31:*219–234.

Xia, Z.G., Refsdal, C.D., Merchant, K.M., Dorsa, D.M., and Storm, D.R. (1991). Distribution of mRNA for the calmodulin-sensitive adenylate cyclase in rat brain: expression in areas associated with learning and memory. *Neuron, 6:*431–443.

Zhu, Q. and Julien, J.P. (1999). A key role for GAP-43 in the retinotectal topographic organization. *Exp Neurol, 155:*228–242.

12

Cellular Mechanisms Underlying the Remodeling of Retinogeniculate Connections

William Guido

A fundamental issue in developmental neurobiology is to elucidate the cellular mechanisms underlying the establishment and refinement of synaptic connectivity between developing sense organs and their central targets. For the past two decades, the mammalian retinogeniculate pathway has served as an important model for demonstrating how patterned spontaneous activity shapes synaptic connections (see Shatz, 1990; Cramer and Sur, 1995; Shatz, 1996). In more recent years, the rodent visual system has been the focus of intense inquiry largely because transgenic mouse models are used with increasing regularity to unravel the molecular mechanisms underlying the remodeling process (see Grubb and Thompson, 2004, see chapters by Rebsam and Gaspar; Hooks and Chen). While some information about the rodent retinogeniculate pathway exists (Godement et al., 1984; Mooney et al., 1993; Mooney et al., 1996; MacLeod et al., 1997; Chen and Regehr 2000) a detailed examination of ontogeny is lacking. In this review we present our working model of the developing rodent retinogeniculate pathway (Jaubert-Miazza et al., 2005; Ziburkus and Guido, 2005) and propose a mechanism we think contributes to the refinement of retinogeniculate connections.

1. Structural Organization of the Developing Retinogeniculate Pathway

The topographic representation of visual fields in the retina and the brain is cardinal feature of vision. In mammals the most distinguishing feature of central visual maps is the segregation of inputs from the two eyes. For example, retinal projections to the lateral geniculate nucleus (LGN) of the dorsal thalamus are segregated and terminate into discrete non-overlapping territories. In carnivores and primates, retinal projections from the two eyes are partitioned by cytoarchitectural boundaries or laminae. In rodents the LGN lacks an obvious lamination pattern (Fig.1A) but retinal projections are still organized into

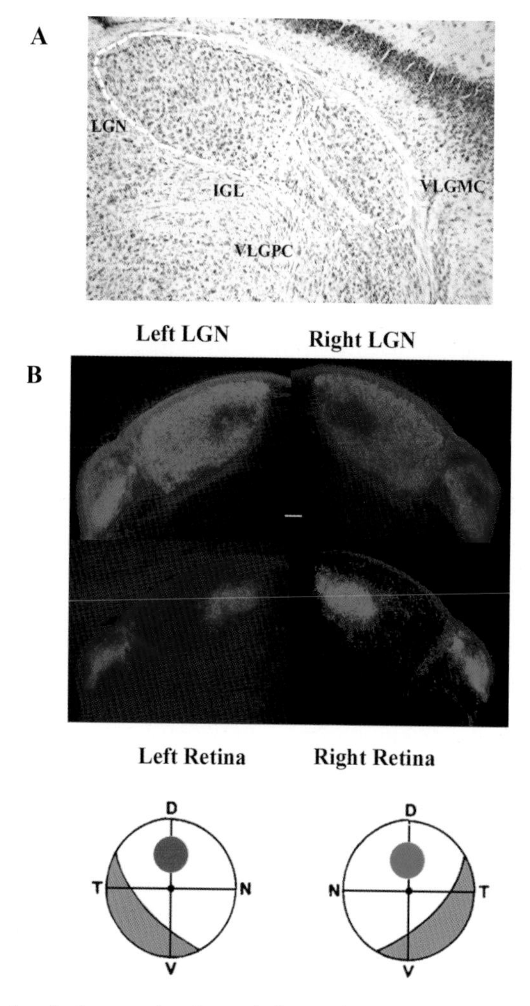

Figure 1 Anatomical organization of the rodent lateral geniculate nucleus (LGN). **A**. Coronal section through the LGN using a nissl stain. The LGN can be distinguished from the intrageniculate leaflet (IGL) and the ventral geniculate nuclei (VLGP and VLGMC). Boundaries are outlined in white. Note the cytoarchitecture of the LGN lacks an eye-specific laminar pattern. **B**. Anterograde labeling of retinal projections with fluorescent conjugates of cholera toxin B (CTB) reveals eye-specific organization. Shown are coronal sections through the LGN of the left and right hemisphere. Alexa Fluor 594 (red), injected into the left eye labels the terminal fields of uncrossed (ipsilateral) retinal axons in the left hemisphere and crossed (contralateral) axons in the right hemisphere. Alexa Fluor 488 (green) injected into the right eye labels contralateral axons of the left hemisphere and ipsilateral axons in the right hemisphere. Scale bar = 100 μm. Below the fluorescent images is a diagram depicting the quadrants of the retina and corresponding Alexa Fluor. The gray region is the "temporal crescent" and represents the location and origin of the uncrossed pathway. Abbreviations: D = dorsal, N = nasal, T = temporal, V = ventral quadrants of the retina.

eye-specific domains (see Reese, 1988; Sefton and Dreher, 1994). Such eye-specific patterning can be visualized by the anterograde labeling of retinal ganglion cells (Godement et al., 1984; Jeffery, 1984). In recent years, the use of the cholera toxin β subunit (CTB) has proven to be an effective and reliable tracer, labeling very thin axons even at early postnatal ages (Angelucci et al., 1996; Muir- Robinson et al., 2002; Torborg and Feller, 2004; Jaubert-Miazza et al., 2005; Ziburkus and Guido, 2005) By making eye injections of CTB conjugated to different fluorescent dyes (Alexa Fluor 488 and Alex Flour 594) it is possible to visualize retinal projections from both eyes simultaneously in single sections of the LGN (Fig. 1B). In the adult, axons from nasal retina and most of temporal retina cross at the optic chiasm and project contralaterally to the lateral and ventral regions of LGN (Reese and Jeffery, 1983; Reese and Cowey, 1987). Crossed or contralateral projections occupy as much as 85–90% of the total area in LGN (Fig 1B). A much smaller group of retinal axons arising from ventro-temporal regions of retina (i.e., the "temporal crescent"), do not cross at the optic chiasm, but remain uncrossed and project ipsilaterally into the antero-medial region of LGN (Reese and Cowey, 1983; Reese and Jeffery, 1983). Uncrossed or ipsilateral projections form a cylinder that runs through LGN and occupies about 10–15% of the nucleus (Fig 1B). This form of eye specific patterning is not present during development but emerges sometime during early postnatal life (Fig. 2; Jaubert-Miazza et al., 2005). Between P3-5, the inputs from the two eyes share a substantial amount of terminal space in LGN. By P7, retinal projections from the two eyes begin to show clear signs of segregation, and certainly by the time of natural eye opening (P 12-14), they are well segregated and resemble the pattern found at older ages. An analysis of the spatial extent of the terminal fields in LGN reveals that the bulk of anatomical rearrangements occur among ipsilateral projections (Fig. 2B). At P3 they occupy about 60% of the LGN and overlap with contralateral ones by as much as 57%. By P7, the ipsilateral projections begin to recede but are still fairly robust (26%), sharing about 18% of LGN with contralateral projections. By P 12-14, the time of natural eye opening, ipsilateral inputs resemble the adult profile, occupying about 12% of LGN and sharing little (<2%) if any territory with contralateral projections.

→

Figure 2 Pattern of retinogeniculate projections in the developing mouse. **A.** Retinogeniculate axon segregation in the developing mouse revealed by the anterograde transport of CTB conjugated two different fluorescent probes. Alexa Fluor 594 (red) labels crossed (contralateral eye) projections and Alexa Fluor 488 (green) labels uncrossed (ipsilateral eye) projections. Panels from left to right depict red and green fluorescence labeling of the same section of LGN, the superimposed fluorescence pattern, and corresponding pseudo-colored image. For the latter, pixel intensity is "normalized" so that every pixel ≥ to a defined threshold level is assigned a value of 255. Pixels that contain both red and green fluorescence are considered as areas of overlap and represented as yellow. Scale

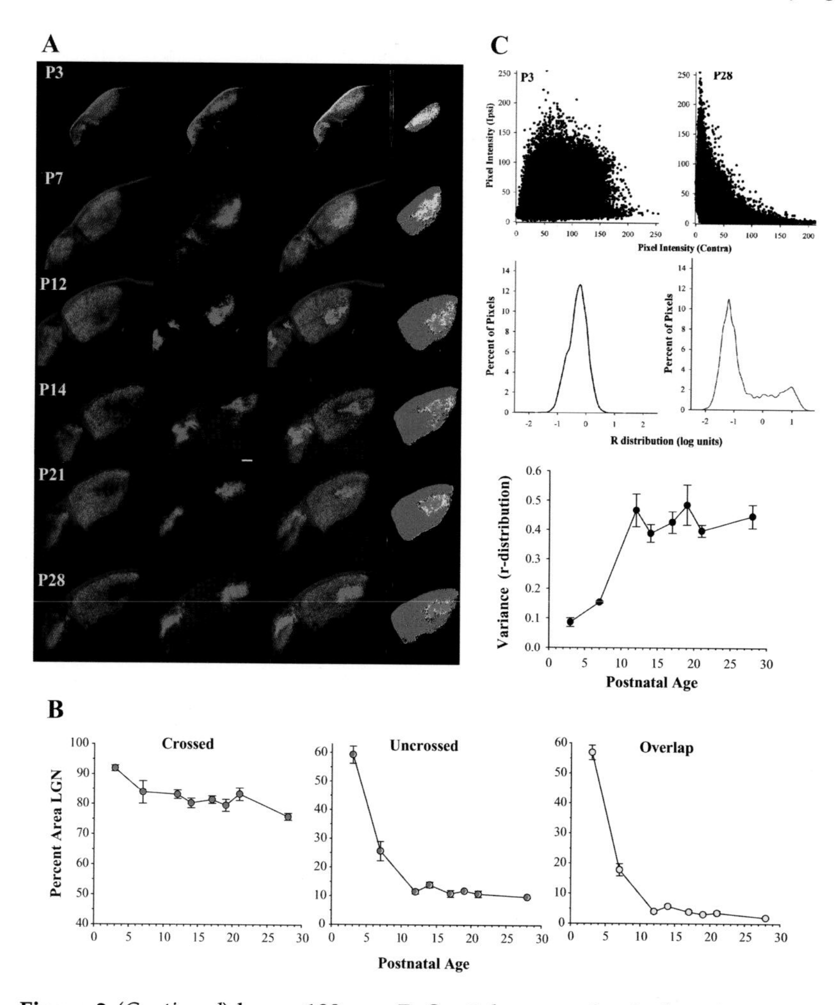

Figure 2 (*Continued*) bar = 100 μm. **B**. Spatial extent of retinal projections in the developing LGN. Graphs plot the percent area in LGN occupied by the crossed (contralateral eye), uncrossed (ipsilateral eye), and overlapping terminal fields at different ages. Each point represents the mean and SEM at P3 (n = 3), P7(n = 3), P12 (n = 5), P14 (n = 9), P17 (n = 4), P19 (n = 5), P21 (n = 4), P28 (n = 8). Crossed projections show modest changes with age. Uncrossed ones undergo substantial retraction between P3-12 and account for the high degree of overlap at P3 and P7. Analysis of pixel intensity. *Top*: Scatterplots of pixel intensity for a single section of LGN at P3 and P28. Each point represents a pixel in which the fluorescence intensity of the contralateral projection is plotted against the intensity of the ipsilateral projection. At P3, the projections from the two eyes overlap and pixel intensities are positively correlated. At P28, the inputs are segregated and the points are inversely related. *Middle*: R-distributions of pixel intensity. For each pixel (see **A**), the logarithm of the intensity ratio, R ($\log_{10} F_I/F_C$), is plotted as a frequency histogram (bin size = 0.1 log units). Narrow r-distributions (P3) reflect unsegregated patterns and wide ones (P28) show segregated ones. *Bottom*: Graph showing the variance values obtained from R-distributions at different ages. Each point represents the mean and SEM for same animals in **A**. Variance increases with age and indicates a progressive increase in the degree of eye specific segregation between P3-12. Adopted from (Jaubert-Miazza et al., 2005).

The difference between unsegregated and segregated retinal projections can be further quantified by analyzing the fluorescence intensity of individual pixels (Fig. 2C). These measures have advantages over those that rely on measures of spatial extent because they provide an unbiased or threshold-independent index of segregation (Muir-Robinson et al., 2002; Torborg and Feller, 2004; Jaubert-Miazza et al., 2005). Scatterplots of pixel intensity reveal important differences between unsegregated and segregated patterning (Fig. 2C). For these, each point on the scatterplot represents a pixel in which the fluorescence intensity of the contralateral projection is plotted against the intensity of the ipsilateral one. When contralateral and ipsilateral inputs share terminal space in LGN (at P3), pixel intensities show a positive correlation (i.e., high contralateral and ipsilateral intensities in the same pixels). In contrast, when the projections from the two eyes no longer overlap (at P28), intensities become inversely related (i.e., high contralateral and low ipsilateral intensities in the same pixels, or visa versa). Pixel intensity can also be expressed as the logarithm of the ratio fluorescence intensities representing the ipsilateral and contralateral projections ($R = \log_{10} F_I/F_C$). R-values representing each pixel of LGN can then be plotted as a frequency distribution (Fig. 2C). Narrow, unimodal R-distributions, which are found at young postnatal ages, indicate that the majority of pixels have intensity values that are not dominated by one projection or the other. In contrast, wide distributions, which are seen at older ages are bimodal. There is a large peak (between -2 and -1) that represents the great majority of pixels dominated by a high contralateral intensity and a smaller one (between 0.5 and 1) that corresponds to pixels dominated by a high ipsilateral value. The variance of R-distributions can be used statistically to compare the relative widths of distributions either across mice of different ages (Fig 2C) or to compare wild-type mice against transgenic strains that lack an element suspected to play a role in remodeling of retinogeniculate connections (Torborg and Feller, 2004; Torborg et al., 2005). In the case of retinogeniculate development, a significant increase in variance occurs between P3 and P12 (Fig. 2C). This pattern coincides with estimates of spatial extent (Fig. 2A-B) and indicates that axon segregation stabilizes at about the time of natural eye opening (P12-14).

A closer inspection of our developmental results suggest that segregation occurs more gradually than previously recognized (Jeffery, 1984; Godement et al., 1984), and in two stages (Muir Robinson et al., 2002; Jaubert-Miazza et al., 2005). During the first postnatal week a more macroscopic level of organization is achieved where the final positioning of contralateral and ipsilateral projections are established and the initial pruning of arbors begin. Between the first and second week, a more focal form of retraction occurs, as ipsilateral projections undergo extensive attrition and discrete non-overlapping fields are formed. These events appear to be regulated by two types of spontaneous retinal activity; an early phase (P0-8) of cholinergic transmission that contributes to a large-scale establishment of eye specific territories, and a late one

(P10-P14) involving glutamate signaling that drives local patterns of segregation (Feller, 2002; Muir-Robinson et al., 2002).

2. Functional Organization of the Developing Retinogeniculate Pathway

To explore the functional implications of these anatomical rearrangements, we examine the pharmacology and underlying circuitry of the synaptic responses of developing LGN cells by utilizing an in vitro isolated brainstem recording preparation (Hu, 1993; Lo et al., 2002). This explant is especially suited for the study of retinogeniculate transmission because unlike a conventional slice preparation, in the isolated brainstem, retinal axons innervating LGN as well as the intrinsic circuitry of LGN remain intact. Additionally, by sparing large segments of each optic nerve and applying separate stimulation to them, we can determine the degree to which inputs from the two eyes converge onto single LGN cells (Ziburkus et al., 2003; Jaubert-Miazza et al., 2005; Ziburkus and Guido, 2005).

In a mature thalamic relay cell, retinal stimulation evokes an excitatory postsynaptic potential (EPSP) that is followed by inhibitory postsynaptic (IPSP) activity (Fig. 3B). These EPSP/IPSP pairs reveal that retinal axons make excitatory connections with relay cells (Fig. 5). Additionally, retinal axons possess collaterals that form excitatory connections with neighboring interneurons (Fig. 5), which in turn form feed- forward inhibitory connections with relay cells (Lindstrom, 1982; Ziburkus et al., 2003; Blitz and Regehr, 2005). IPSP activity serves many functions in LGN, from shaping the receptive field structure of relay cells to establishing the overall gain of signal transmission (Sherman and Guillery, 1996).

These inhibitory aspects of synaptic circuitry are not present during early postnatal development (Ziburkus et al., 2003). Instead, the majority of synaptic responses (70%) at early ages are purely excitatory (Fig. 3B-D). Pharmacology experiments indicate excitatory responses are mediated by glutamate receptor activation and involve the coincident activation of two receptor subtypes, conventionally classified as N-methyl-D-aspartate (NMDA) and non-NMDA (Scharfman et al., 1990). At early ages, the excitatory response is comprised largely of NMDA activity (Ramoa and McCormick, 1994; Ramoa and Prusky, 1997; Chen and Regehr, 2000). The unique voltage dependency of NMDA receptors figures prominently in development because it allows for an influx of Ca^{2+} ions (along with a Na^{+2} influx and K^{+} efflux) during periods of heightened neural activity. It is the activity dependent sequestration of Ca^{2+} that triggers a cascade of intracellular signaling events responsible for the eventual consolidation of adult patterns of connectivity (Constantine-Paton et al., 1990; Goodman and Shatz, 1993). After the first few postnatal days of pure excitatory activity, inhibitory responses begin to emerge, but the full complement of IPSP activity is not evident

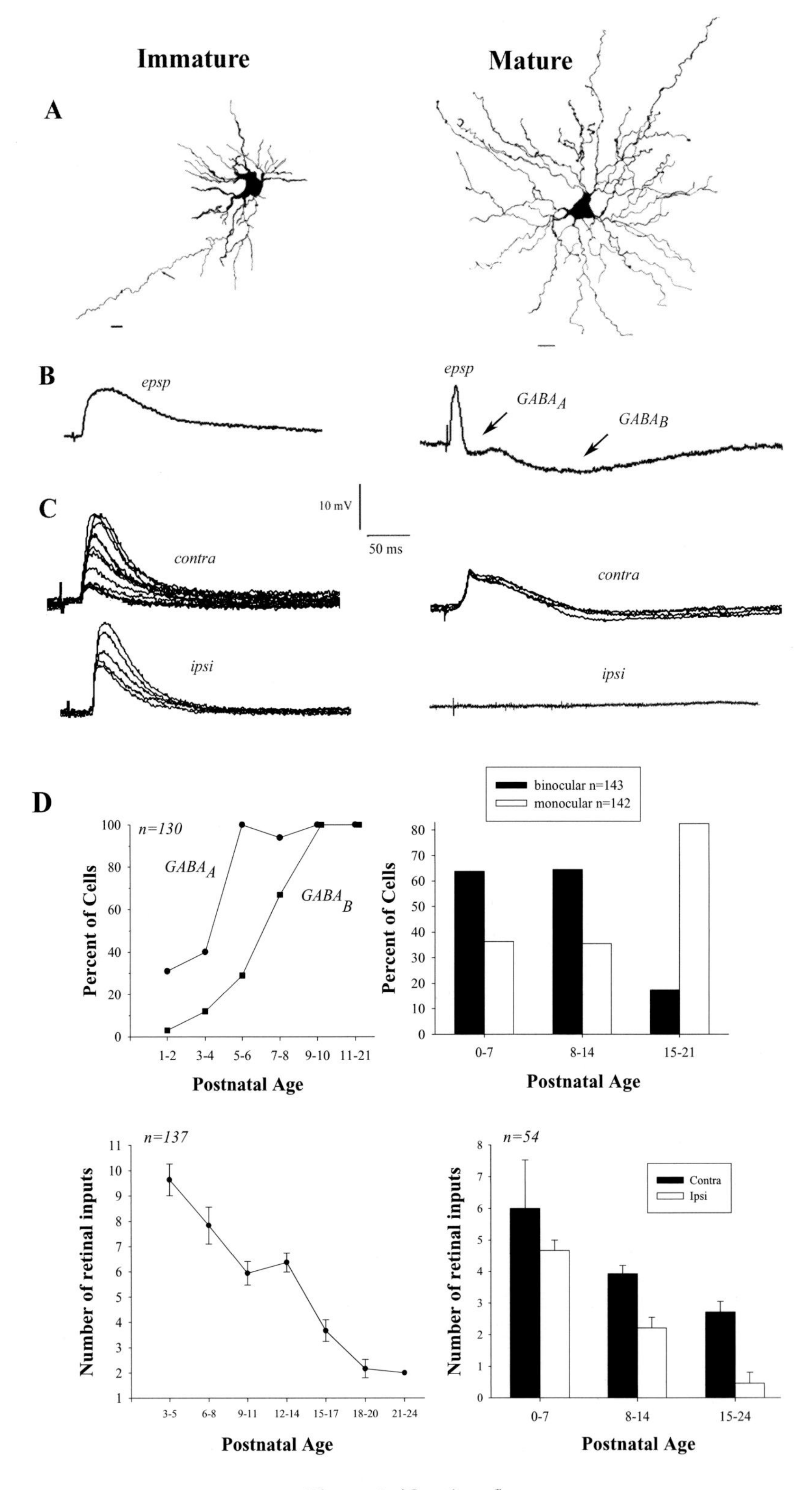

Figure 3 (*Continued*)

until P10 (Fig. 3D). Inhibitory responses are mediated by two types of GABA receptors (Crunelli et al., 1988; Ziburkus et al., 2003). The first to appear is an early, fast hyperpolarizing response which involves a Cl^- conductance acting through the $GABA_A$ receptor subtype. These early fast $GABA_A$ mediated IPSPs also affect EPSP activity, often times curtailing the late NMDA component of the excitatory response (Fig. 3B). A second type of IPSP emerges near the end of the first week and involves a G-protein activated K^+ conductance through a $GABA_B$ receptor subtype. It follows the $GABA_A$ IPSP and is slower and long-lasting. Thus, excitatory and inhibitory synapses in LGN develop at different rates, with inhibitory ones maturing more slowly than excitatory ones (Fig. 3B-D). The functional significance of this sequence is not clear but the delayed onset of inhibitory activity may promote an increased level of excitatory postsynaptic events implicated in synaptic remodeling (e.g., NMDA and high threshold voltage-gated Ca^{2+} channel activity).

Accompanying these age related changes in postsynaptic receptor function, are changes in the pattern of synaptic connectivity (Fig 3C). As one would expect from the eye-specific patterning in the adult LGN, mature relay cells are monocular, receiving input from one or the other eye

<———

Figure 3 (*Continued*) Synaptic responses of developing LGN cells. **A**. Camera lucida drawings of an immature and mature relay cells labeled with biocytin during intracellular recording. At early and late ages, cells have relatively large somata and multipolar dendritic arbors consistent with those of class A thalamocortical cells (Grossman et al., 1973). Note the dendritic tree of immature cells is sparse with arbors having fewer and shorter branches (Parnavelas et al., 1977; Ziburkus et al., 2003). Beneath each cell are representative examples of synaptic responses evoked by electrical stimulation of the optic tract (**B**) and optic nerves (**C**). Immature activity (*left*) is purely excitatory and has a large NMDA component. ON stimulation at different stimulus intensities evokes binocular responses. The graded responses evoked at different stimulus intensities reflect multiple retinal inputs. Mature activity (*right*) consists of a short duration EPSP followed by two IPSPs, one mediated by $GABA_A$ and the other by $GABA_B$ receptor activation. ON stimulation at different stimulus intensities evokes a response of constant amplitude and reflects a single monocular input. Recordings in **C** were done in the presence of GABA antagonists. **D**. Graph showing the incidence of EPSPs followed by $GABA_A$ and $GABA_B$ IPSPs at different ages. "Pure EPSPs" which lack inhibitory activity, prevail at young ages (P1-4). IPSP activity emerges during the first week so that by P9-10 postsynaptic activity is comprised of EPSP/IPSP pairs. **B**. Graph showing the incidence of binocular and monocular excitatory responses at P0-7, P8-14, and P15-21. Binocular responses are frequently encountered between P0-14. After P14 binocular responses are rare and the majority of responses are monocular. Summary graphs depicting the age related changes in retinal convergence. *Left*: Plot showing means and SEMs for the total number of inputs cells receive at different ages. *Right*: Histogram showing the average number of inputs a cell receives from the contralateral and ipsilateral eye between P0-7, P8-14, P15-24. There is a decrease in retinal convergence with age, due largely to the loss of inputs arising from the ipsilateral eye. All recordings were conducted in regions of the LGN that in the adult receive input exclusively from the contralateral eye.

(Reese and Jeffery 1983; Reese, 1988; Sefton and Dreher, 1994; Ziburkus and Guido 2005; but see Grieve, 2005). However, given the diffuse nature of early retinal projections, we expect and do indeed find a high incidence of binocular responses (70%) during the early phases of axon segregation (Ziburkus et al., 2003; Jaubert-Miazza et al., 2005 Ziburkus and Guido, 2005). When recording in regions of the LGN that in the mature state receive input exclusively from the contralateral eye, separate and distinct EPSPs are readily evoked by stimulation of either optic nerve (figure 3C-D). After P14, there is a rapid decrease in the incidence of excitatory binocular responses. By P18 recorded responses are monocular (Jaubert-Miazza et al., 2005; Ziburkus and Guido, 2005). Interestingly, what remains in a subset of mature relay cells is a binocularly mediated inhibitory response (Fig. 5; Ziburkus et al., 2003). Binocular inhibitory responses seem to arise from interneurons that receive input from one eye and then inhibit relay cell activity from the other eye (Alhsen et al., 1985). Our observations in the rodent are consistent with those made in the cat and monkey (Alhsen et al., 1985; Guido et al., 1989; Schroeder et al., 1990) and suggest that binocular inhibitory interactions are a fundamental (albeit ignored) feature of mammalian geniculate circuitry.

Another transient feature in synaptic connectivity is the prevalence of responses that reflect the convergence of multiple retinal ganglion cell inputs onto a single LGN cell (Chen and Regehr, 2000; Ziburkus and Guido, 2003; Jaubert-Miazza et al., 2005; Ziburkus and Guido 2005, see chapter by Hooks and Chen). To estimate the number of retinal inputs converging onto a signal relay cell, the optic nerves are electrically shocked at various levels of stimulus intensity and the amplitude of evoked EPSPs are measured. In developing LGN cells that receive multiple inputs, a progressive increase in stimulus intensity gives rise to a step-wise increase in EPSP amplitude (Fig. 3C). These graded changes reflect the successive recruitment of active inputs innervating a single cell. In adult LGN cells, the amplitude of EPSPs remain relatively constant when increasing levels of stimulation are used (Fig. 3C). Our estimates indicate mature cells receive monocular input from 1–3 retinal ganglion cells (Lo et al., 2002; Jaubert-Miazza et al., 2005; Ziburkus and Guido 2005). In contrast, cells recorded between P0-7 receive at least 3–6 inputs from each eye. In fact our estimates seem conservative; some LGN cells are reported to receive in excess of 20 retinal inputs even during the second postnatal week (Chen and Regehr, 2000). In our lab we find that between P15-21, the degree of retinal convergence rapidly declines to resemble the mature state, with the most significant attrition occurring among ipsilateral eye inputs (Fig. 3D).

Our results indicate the anatomical rearrangements occurring in the developing LGN translate directly into functional changes in connectivity (Fig. 5). Initially, overlapping projections from the two eyes lead to a high degree of retinal convergence. As retinal projections from the two eyes recede and overlapping territories dissipate, synapses are eliminated and cells receive far fewer inputs from just one eye. This form of synaptic refinement also correlates well with the maturation of receptive field properties reported in a number of mammalian species (Shatz and Kirkwood 1984; Tootle and Friedlander, 1986; Sefton and Dreher, 1994; Tavazoie and Reid, 2000; Grubb and Thompson, 2003; Grieve, 2005). Immature receptive fields are typically binocular, quite large and irregularly shaped, and lack distinct on- and off- subregions. In contrast, mature fields are monocular, much smaller, and have well defined concentric center-surround organization.

Finally, it is worth noting the discovery of a somewhat novel synaptic event that occurs in LGN during early postnatal life. Strong activation of optic tract fibers with either a single or repetitive (25–100 Hz) shock often evokes EPSPs that gave rise to a high amplitude (25–40 mV), long-lasting (300–1300 msec), slow decaying depolarization (Fig. 4). These "plateau potentials" have a voltage dependency and pharmacology consistent with the activation of high-threshold L-type Ca^{2+} channels activation (Kammermeier and Jones, 1998; Lo et al., 2002). The L-type (long lasting) channel is a voltage gated, high threshold Ca^{2+} channel that is found in many different neuronal structures and cell types. These channels have been implicated in a variety of cellular function including activity dependent gene expression, cellular excitability, synaptic plasticity, and cell survival (Lipscombe et al., 2004). In thalamic relay nuclei such as the LGN, L-type channels are prevalent, although their role in synaptic integration has been largely ignored. They are localized primarily on somata and proximal dendrites (Budde et al., 1998), thus placing them in an ideal location to modulate retinally evoked EPSPs (Wilson et al., 1987).

Plateau potentials recorded in LGN are encountered far more frequently between P0-7, then decline gradually with age so that by P18-21 they are rarely recorded (Fig. 4 D). There are at least two factors that contribute to the developmental regulation of plateau potentials. First, the high degree of retinal convergence and heightened NMDA activity seen at early ages favors the spatial and temporal summation of EPSPs (Fig. 4C). Such sustained levels of synaptically induced depolarizations greatly increase the likelihood that high-threshold L-type channels are activated (Fig. 4C). Secondly, the density of L-type Ca^{2+} channels found among LGN cells varies with age. Using an antibody that recognizes and labels the pore forming α_{1C} subunit of the L-type channel, we found expression to peak between P0-7, but then declines gradually so by P28 there is a four-fold reduction in the density of labeled cells (Jaubert-Miazza et al., 2005).

3. Early Retinal Activity Shapes the Developing Retinogeniculate Pathway

As discussed in detail by other chapter contributors, the refinement of retinogeniculate connections depends on the coordinated firing patterns of developing retinal ganglion cells. Even before photoreceptors

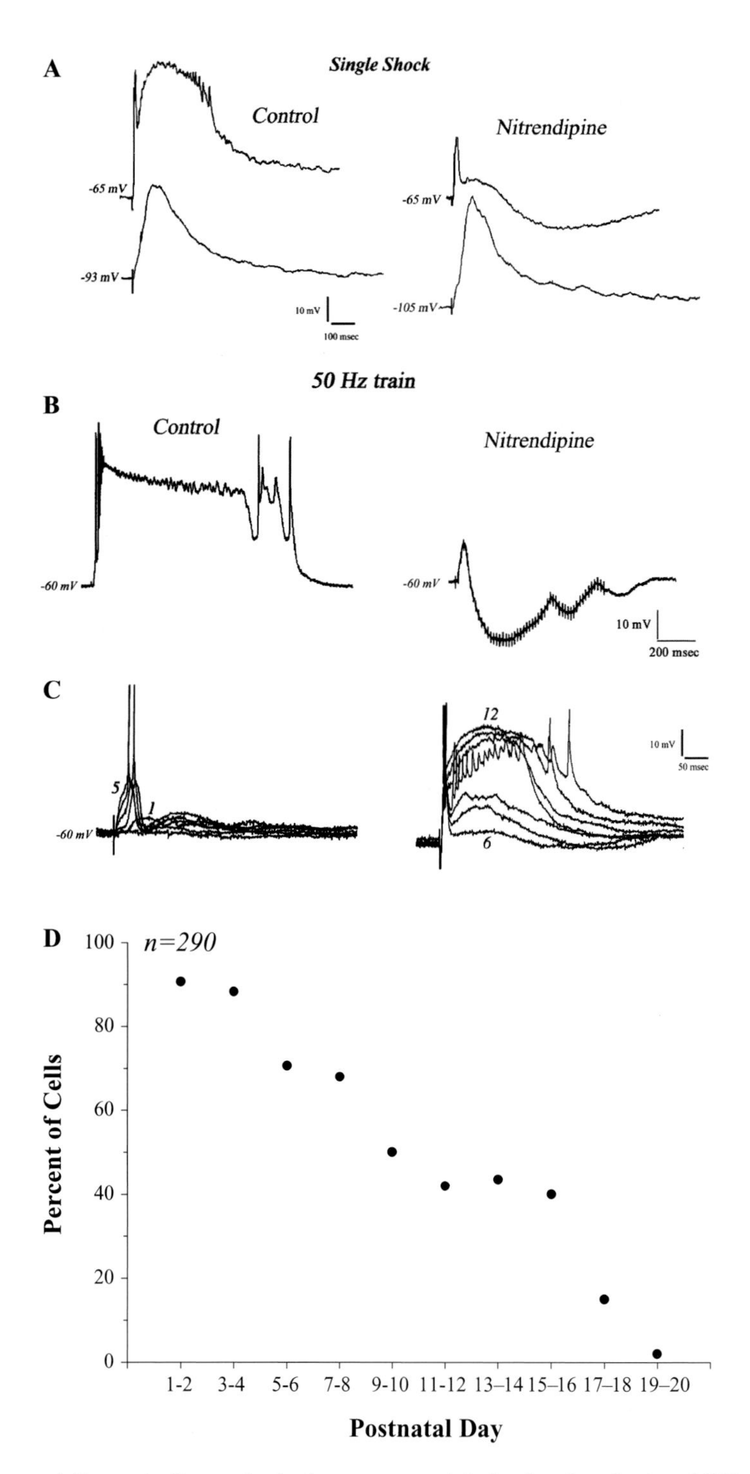

Figure 4 Synaptically evoked plateau potentials in the developing LGN. **A**. Pharmacology and voltage dependency of the plateau potential. Shown are the synaptic responses evoked by single or repetitive stimulation of optic tract. At −65 mV, a single shock at high stimulus intensity evokes a large plateau

form functional connections, groups of neighboring retinal ganglion cells fire spontaneously in rhythmic bursts of activity that travel across the retina in wave-like fashion (see Wong 1999; Demas et al., 2003). These spontaneous discharges are of sufficient strength to generate prolonged bursts of action potentials in LGN (Mooney et al., 1996; Weliky and Katz, 1999). In fact, they also seem ideally suited to activate L-type mediated plateau potentials (Lo et al., 2002). Repetitive stimulation of retinal afferents in a manner that approximates the high frequency discharge of spontaneously active retinal ganglion cells leads to a massive summation of EPSP activity and triggers robust plateau-like activity (Fig. 3B).

When early spontaneous retinal activity is blocked or the wave like patterns severely altered, retinal axon arbors in LGN fail to segregate and maintain a diffuse projection pattern (see Shatz 1990, Shatz 1996; Goodman and Shatz, 1993; Chapman 2004, see chapter by Huberman and Chapman). Perhaps the most celebrated model for explaining how the activity of immature neurons can form orderly connections is the Hebb (1949) synapse. In this model (Fig. 6A), high levels of coincident activity between pre- and postsynaptic elements leads to a strengthening and consolidation of synapses. A corollary of this principle is that low levels of activity result in synapse weakening and elimination (Stent, 1973; Bear et al., 1987; Constantine-Paton et al., 1990; Cramer and Sur, 1995). A proposed substrate for activity dependent remodeling is based on forms of synaptic plasticity first demonstrated in the hippocampus, in which the degree of frequency pairing between pre- and postsynaptic elements leads to a long-term potentiation (LTP) or depression (LTD) in synaptic strength (see Bear and Malenka, 1994; Malenka and Bear, 2004). In this model, NMDA receptor activation is needed for the induction of changes in synaptic efficacy (Collingridge, 1992). The voltage dependency of NMDA receptors enables them to act as "coincident detectors". That is, Ca $^{2+}$ entry through NMDA receptors only occurs when there is sufficient depolarization. A large increase in the intracellular concentration of Ca^{2+} (high levels of NMDA receptor activation)

Figure 4 (*Continued*)depolarization (*control*). At a more hyperpolarized level (−93 mV), the same stimulation fails to evoke one, but results in a large postsynaptic response. In the presence of the L-type Ca^{2+} channel antagonist nitrendipine, the plateau potential is abolished, but what remains is an underlying EPSP and IPSP. At −105 mV, a large postsynaptic potential is present, indicating nitrendipine does not impede synaptic transmission. **B**. A large long-lasting plateau potential is evoked by repetitive stimulus train (50Hz). Nitrendipine application blocks the plateau potential but has no effect on the underlying EPSP/IPSP. **C**. Summation of convergent retinal input evokes a plateau potential. Progressive increase in stimulus intensity leads to a step-wise increase in EPSP amplitude. At high levels of stimulation the graded responses give rise to plateau potentials. Numbered traces depict different retinal inputs. **D**. Summary graph plotting the incidence of plateau potentials at different ages. Each point depicts the percentage of cells exhibiting a plateau response. The age related decrease coincides with the period of retinal geniculate axon segregation. Adopetd from (Lo et al., 2002).

triggers a distinct signaling cascade that leads to the strengthening of co-active elements. Modest or low levels of intracellular Ca^{2+} (low levels of NMDA receptor activation) trigger a different signaling cascade that leads to the weakening and eventual loss of less active, asynchronous ones. In the developing neocortex, there is evidence indicating that LTP and LTD exists and an influx of Ca^{2+} through NMDA receptors contributes to the formation of orderly connections (Kirkwood and Bear, 1994a; 1994b). Such long-term modifications in synaptic strength may therefore embody the synaptic rearrangements occurring during the time of retinogeniculate axon segregation, when afferents from the two eyes are competing for synaptic space with the dendrites of relay cells. To test for this possibility we examined the synaptic responses of LGN cells before and after high frequency stimulation (HFS) of a single optic nerve. The tetanus protocol, which consists of six 1-sec trains of 50 Hz stimulation delivered every 30 sec for 3 min, is designed to mimic (at least in the temporal domain) the intrinsic firing patterns of developing retinal ganglion cells (Wong et al., 1993; Wong and Oakley, 1996). This form of stimulation produces robust changes in synaptic strength (figure 6B). In cells that receive monocular input from the contralateral eye, HFS of the contralateral optic nerve produces a long-term, "homosynaptic" form of potentiation (>150%). In cells that receive input from the two eyes, HFS of the contralateral optic nerve produces both homo- and heterosynaptic changes in synaptic strength. That is, EPSPs evoked by a single shock delivered to the tetanized, contralateral nerve are increased (homosynaptic potentiation) while those responses evoked by stimulation of the untetanized, ipsilateral optic nerve are reduced (>50%) (heterosynaptic depression). Thus, heightened activity along one pathway leads to a increase in synaptic strength as well as a concomitant decrease in strength along a less active pathway.

Another important aspect of these results is the underlying pharmacology. Many examples of synaptic plasticity in the hippocampus and the neocortex seem to rely on NMDA receptor activation (Bear and Malenka, 1994; Constantine-Paton et al., 1990; Cramer and Sur, 1995; Malenka and Bear 2004)). However, the plasticity we observe in LGN seems to rely on the activation of a high threshold L-type Ca^{2+} channel. HFS of the optic nerve triggers a sustained L-type mediated plateau potential, which is likely activated by massive spatial and temporal summation of EPSPs evoked by HFS (Fig. 4B). When this activity is blocked pharmacologically, HFS fails to induce any changes in synaptic strength (Fig. 6B).

The L-type activity recorded in the rodent LGN is identical to the synaptically-evoked plateau potentials recorded in neurons of the developing rodent superior colliculus (Lo and Mize, 2000) and brainstem trigeminal nuclei (Lo and Erzurumlu, 2002) and shares some similarities to plateau related activity reported in the rodent brainstem (Rekling and Feldman, 1997), spinal cord (Kien and Eken 1998), and invertebrate motor neurons (Dicaprio, 1997). Thus, this event may reflect a highly conserved mechanism by which cells can acquire large amounts of Ca^{2+} in an activity dependent manner.

A role for L-type Ca^{2+} channels in synaptic plasticity has been well documented. Activity through these channels can induce long term depression and potentiation in the hippocampus (Magee and Johnston, 1997), superior colliculus (Lo and Mize, 2000) and the principal nucleus of brainstem (Guido et al., 2001). Retinal axons also fail to segregate properly in the developing LGN and superior colliculus of transgenic mice that show reduced levels of L-type channel activity (Guido, unpublished results, Cork et al., 2001). Finally, the Ca^{2+} influx associated with the synaptically evoked plateau potential could also contribute to signaling events and gene expression involved in the stabilization of developing connections (Ghosh and Greenberg, 1995; Greenberg and Ziff, 2001). For example, Ca^{2+} entry via L-type channels favors signaling cascades brought on by heightened periods of neural activity (Mermelstein et al., 2000). One in particular involves the cAMP response element (CRE/CREB) transcription pathway. The CRE binding protein (CREB) is a calcium and cAMP regulated transcriptional activating protein shown to be important for thalamic circuit development and retinogeniculate axon segregation (Pham et al., 2001).

4. Conclusions

The retinogeniculate synapse undergoes a significant period of remodeling during early postnatal life (Fig. 5). At birth, retinal axons from the two eyes share common terminal space in LGN but prior to the time of natural eye opening (P12) they segregate to form distinct and separate eye specific domains. These structural rearrangements are accompanied by changes in postsynaptic receptor function and patterns in synaptic connectivity. At young ages, synaptic responses are largely excitatory, dominated by NMDA receptor activity and plateau-like depolarizations mediated by L-type Ca $^{2+}$ channel activation. Additionally, LGN cells are binocularly responsive, receiving input from several different retinal ganglion cells. As retinal inputs from the two eyes segregate into non-overlapping territories, NMDA and L-type activity subsides and inhibitory activity emerges. There is also a loss of binocular responsiveness and a decrease in retinal convergence. During the period of synaptic remodeling, changes in synaptic strength can be induced by the high frequency stimulation of retinal fibers in a manner that approximates their spontaneous activity. These alterations last several minutes, rely on the activation of the L-type Ca^{2+} channels, and are consistent with a Hebbian model of activity-dependent synaptic plasticity.

Acknowledgments

I would like to thank past and present lab members including Jokubas Ziburkus, Fu-Sun Lo, Erick Green, Lisa Jaubert- Miazza, Jeremy Mills, and Kim Bui for their contributions. This work was supported

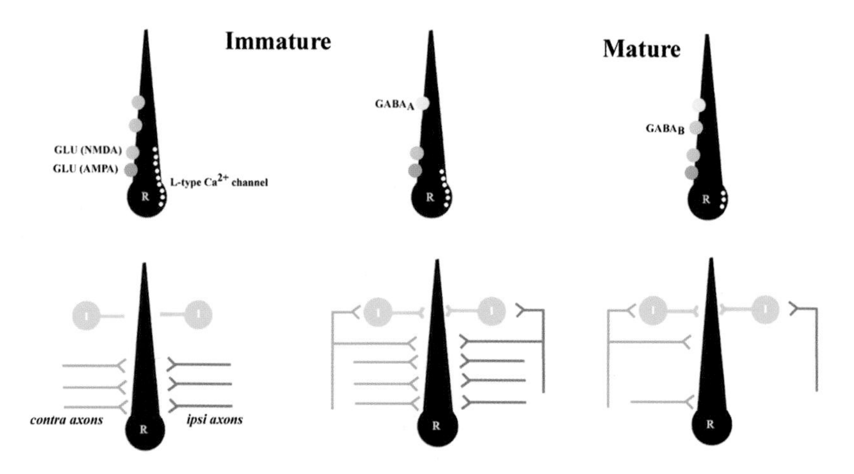

Figure 5 Developmental remodeling at the retinogeniculate synapse. Drawings of immature and mature relay cells (R) summarizing the major developmental changes that occur in receptor function, synaptic connectivity, and L-type channel expression. Immature relay neurons display robust excitatory activity via NMDA receptor activation, receive multiple inputs from the two eyes, and have a high density of L-type Ca^{2+} channels that give rise to plateau-like depolarizations. During maturation, inhibitory responses emerge through a feed-forward circuit involving intrinsic interneurons (I). There is a loss of binocular responsiveness, a reduction in retinal convergence, and a decline in the expression of L-type Ca^{2+} channels which all contribute to the eventual elimination of plateau potential activity. Note that a subset of mature relay neurons continue to display binocular inhibitory responses.

Figure 6 LTP and LTD at the retinogeniculate synapse. **A**. The Hebbian model of synaptic plasticity. Schematic showing how retinal activity leads to long-term changes in synaptic strength and the stabilization of retinogeniculate connections. Shown are two retinal axons competing for terminal space on an LGN cell. Retinal activity is illustrated above each input as spike trains. Heightened retinal activity evokes robust postsynaptic activity in LGN. The coincident pairing of heightened pre- and post-synaptic activity leads to a long-term potentiation (LTP) in subsequent synaptic activity. Accompanying this is a large Ca^{2+} influx through the NMDA iontophore and/or voltage gated Ca^{2+} channels. Increased levels of intracellular Ca^{2+} triggers a series of signaling events that leads to the strengthening and eventual consolidation of the active synapse. In contrast, low levels of retinal activity evokes smaller postsynaptic excitatory responses, weaker and less synchronous pairing of pre- and post-synaptic activity, a long-term depression (LTD) of subsequent synaptic activity, and less Ca^{2+} influx. Low levels of intracellular Ca $^{2+}$ activates a separate set of signaling events that results in a weakening and eventual elimination of the less active synapse. **B**. Activity dependent modifications in the synaptic strength of LGN cells. Examples of synaptic responses in three different LGN cells (**A-C**) recorded before (pre-tetanus, *left*) and after (post-tetanus, *right*) high frequency stimulation of the contralateral optic nerve (ON). Representative responses are obtained 5 min. before and 10 min. after tetanus. Corresponding plots show changes in EPSP amplitude before and after tetanus. Values are expressed as a percentage of the average baseline response and reflect an average obtained from responses

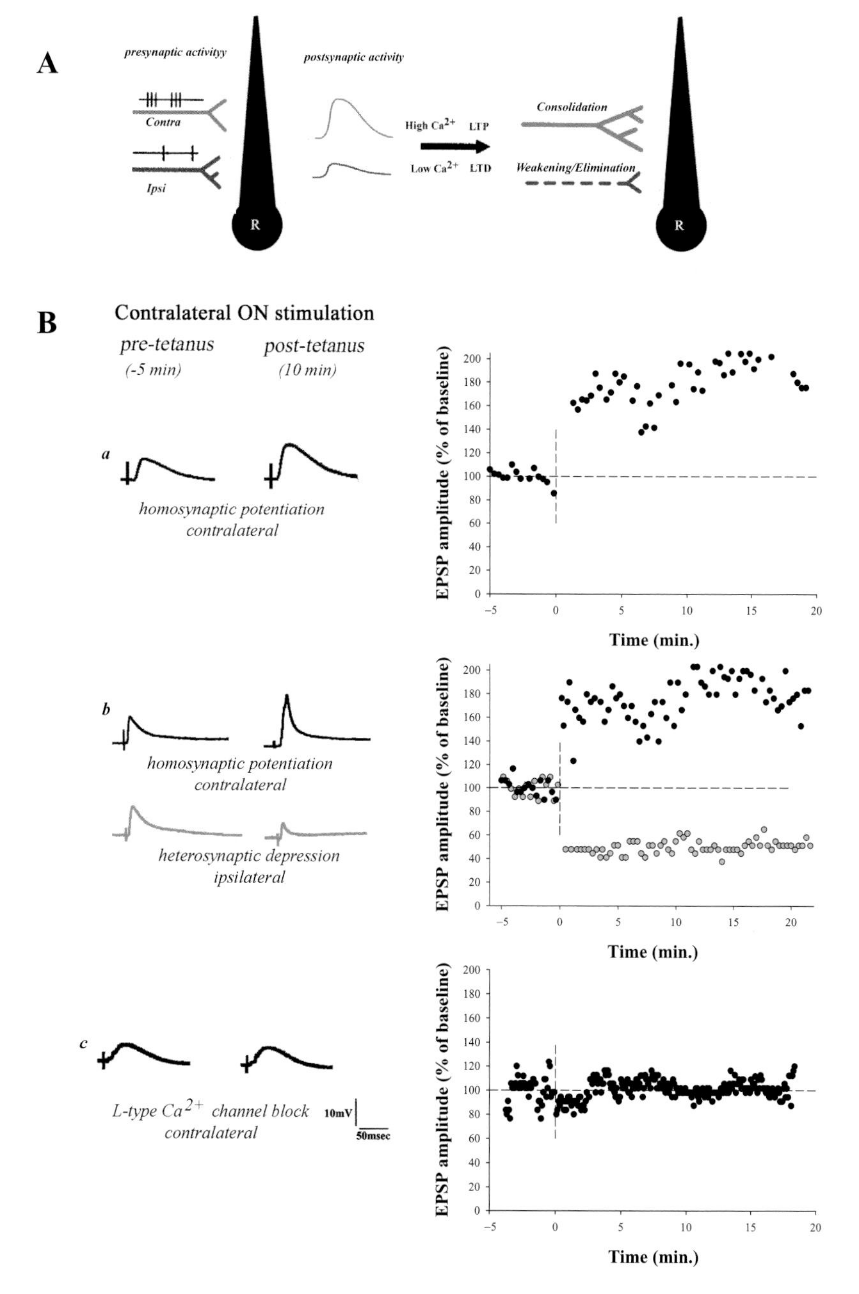

Figure 6 (*Continued*) evoked once every 5–15 sec. **A**. Synaptic responses along the contralateral, tetanized pathway leads to a homosynaptic potentiation. **B**. In a binocular cell the synaptic responses along along the contralateral tetanized pathway leads to homosynaptic potentiation (*black symbols*) and a heterosynaptic depression of responses along the ipsilateral, untetanized pathway (*grey symbols*). **C**. The blockade of L-type Ca^{2+} activity during tetanus results in no change in synaptic strength.

by a grants from the Whitehall Foundation and National Eye Institute (EY12716).

5. References

Ahlsén G., Lindström S., and Lo F-S (1985). Interaction between inhibitory pathways to principal cells in the lateral geniculate nucleus of the cat. *Exp. Brain Res.* 58: 134–143.

Angelucci A., Clasca F., and Sur M. (1996) Anterograde axonal tracing with the subunit B of cholera toxin: a highly sensitive immunohistochemical protocol for revealing fine axonal morphology in adult and neonatal brains. *J. Neurosci. Meth.* 65: 101–112.

Bear M.F., Cooper L.N., and Ebner F.F. (1987). A physiological basis for a theory of synapse modification. *Science* 237: 42–48.

Bear M.F. and Malenka R.C. (1994) Synaptic plasticity, LTP and LTD. *Curr Opin Neurobiol.* 4: 389–399.

Blitz, D.M. and Regehr W.G. (2005) Timing and specificity of feedforward inhibition within the LGN. *Neuron* 45: 917–928.

Budde T., Munsch T., and Pape H-C. (1998). Distribution of L-type calcium channels in rat thalamic neurons. *Eur. J. Neurosci.* 10: 586–597.

Chapman B. (2004). The role of eye specific segregation iin the retino-geniculostriate pathway, In: Chalupa, L.M. and Werner, J.S. (eds.) *The Visual Neurosciences*, Volume 1. MIT Press, Cambridge MA, pp 108–116.

Chen C. and Regehr W.G. (2000) Developmental remodeling of the retinogeniculate synapse. *Neuron* 28: 955–966.

Collingridge G. (1992). The mechanism of induction of receptor-dependent long-term potentiation in the hippocampus. *Exp. Physiol.* 77: 771–797.

Constantine-Paton M, Cline HT, and Debski E. Patterned activity, synaptic convergence, and the NMDA receptor in developing visual pathways. Annual Review of Neuroscience 13: 129–154, 1990.

Cork R.J., Namkung Y., Shin H.S., and Mize R.R. (2001). Development of the visual pathway is disrupted in mice with a targeted disruption of the calcium channel beta(3)-subunit gene. *J. Comp. Neurol.* 440: 177–91.

Cramer, K.S. and Sur, M. (1995). Activity dependent remodeling of connections in the mammalian visual system. *Curr. Opin. Neurobiol.* 5: 106–111.

Crunelli V., Haby M., Jassik-Gersfeld, Leresche N., and Pirchio M. (1988) Cl^- and K^+ dependent inhibitory postsynaptic potentials evoked by interneurones of the rat lateral geniculate nucleus. *J. Physiol.* 399: 153–176.

Demas J., Eglen, S.J., and Wong, R.O. (2003). Developmental loss of synchronous spontaneous activity in the mouse retina is independent of visual experience. *J. Neurosci.* 23, 2851–2860.

Dicaprio, R. (1997). Plateau potentials in motor neurons in the ventilatory system of the crab. *J. Exp. Biol.* 200: 1725–1736.

Feller, M.B. (2002). The role of nAChR-mediated spontaneous retinal activity in visual system development. *J. Neurobiol.* 53: 556–567.

Ghosh A. and Greenberg M.E. (1993) Calcium signaling in neurons: molecular mechanisms and cellular consequences. *Science* 268: 239–247, 1995.

Godement, P., Salaun, J., and Imbert, M. (1984). Prenatal and postnatal development of retinogeniculate and retinocollicular projections in the mouse. *J. Comp. Neurol.* 230: 552–575.

Goodman C.S. and Shatz C.J. (1993) Developmental mechanisms that generate precise patterns of neuronal connectivity. *Cell* (Suppl) 72: 77–98, 1993.
Grieve, K.L. (2005). Binocular visual responses in cells of the rat dLGN *J. Physiol.* 566: 119–124.
Grossman A, Lieberman AR, Webster KE. (1973) A Golgi study of the rat lateral geniculate nucleus. *J. Comp. Neurol* . 150: 441–446.
Grubb, M.S. and Thompson, I.D. (2003). Quantitative characterization of visual response properties in the mouse dorsal lateral geniculate nucleus. *J. Neurophysiol.* 90: 3594–3607.
Grubb, M.S., and Thompson, I.D. (2004). The influence of early experience on the development of sensory systems. *Cur. Opin. Neurobiol.* 14: 503–512.
Guido W, Lo F-S, and Erzurumlu R.S. (2001) Synaptic plasticity in the trigeminal principal nucleus during the period of barrelette formation and consolidation. *Dev. Brain Res.* 132: 97–102.
Guido W, Tumosa N, and Spear P. D. (1989) Binocular interactions in the cat's dorsal lateral geniculate nucleus. I. Spatial-frequency analysis of responses of X, Y and W cells to nondominant-eye stimulation. *J. Neurophysiol.* 62: 526–543.
Hebb, D.O. (1949) The organization of behavior. Wiley, New York.
Hu, B. (1993). Membrane potential oscillations and corticothalamic connectivity in rat associational thalamic neurons in vitro. *Acta Physiol. Scand.* 148: 109–113.
Jaubert-Miazza, L, Green, E., Lo, F-S, Bui, K., Mills, J., and Guido, W. (2005) Structural and functional composition of the developing retinogeniculate pathway in the mouse. Visual Neuroscience, 22:661–676.
Jeffery, G. (1984). Retinal ganglion cell death and terminal field retraction in the developing rodent visual system. *Dev. Brain Res.* 13: 81–96.
Kammermeier P.J. and Jones S.W. (1998) Facilitation of L-type calcium current in thalamic neurons. *J. Neurophysiol.* 79: 410–417.
Kiehn, O. and Eken, T. (1998). Functional role of plateau potentials in vertebrate motor neurons. *Curr. Opin. Neurosci.* 8: 746–752
Kirkwood A. and Bear M.F. (1994a) Hebbian synapses in visual cortex. *J. Neurosci.* 14: 1634–1645.
Kirkwood A. and Bear M.F. (1994b) Homosynaptic long-term depression in the visual cortex. *J. Neurosci.* 14: 3404–3412.
Kirkwood A., Dudek S.M., Aizenman C.D., and Bear M.F. (1993). Common forms of synaptic plasticity in the hippocampus and neocortex in vitro. *Science* 260: 1518–1521.
Lindström S. Synaptic organization of inhibitory pathways to principal cells in the lateral geniculate nucleus of the cat. Brain Research 234: 447–453, 1982.
Lipscombe D., Helton, T.D., Xu, W. (2004). L-type calcium channels: the low down. *J. Neurophysiol.* 92: 2633–2641.
Lo, F-S., and Erzurumlu, R.S. (2002). L-type calcium channel-mediated plateau potentials in barrelette cells during structural plasticity. *J. Neurophysiol.* 88: 794–801.
Lo F-S. and Mize R.R. (2000) Synaptic regulation of L-type Ca^{2+} channel activity and long-term depression during refinement of the retinocollicular pathway in developing rodent superior colliculus. *J. Neurosci.* 20: 1–6.
Lo F-S., Ziburkus J., and Guido W. (2002) Synaptic mechanisms regulating the activation of a Ca^{2+} -mediated plateau potential in developing relay cells of the lateral geniculate nucleus. *J. Neurophysiol.* 87: 1175–1185.
MacLeod, N., Turner, C., and Edgar, J (1997) Properties of developing lateral geniculate neurones in the mouse. *Int. J. Dev. Neurosci.* 15: 205–224.

Magee, J.C. and Johnston, D. (1997). A synaptically controlled, associative signal for Hebbian plasticity in hippocampal neurons. *Science* 275(5297), 209–213.

Malenka, R.C. and Bear, M.F. (2004). LTP and LTD: an embarrassment of riches. *Neuron* 44: 5–21.

Mermelstein, P.G., Bito, H., Deisseroth, K., and Tsien, R.W. (2000). Critical dependence of cAMP response element-binding protein phosphorylation on L-type calcium channels supports a selective response to EPSPs in preference to action potentials. *J. Neurosci.* 20: 266–273.

Mooney, R., Madison, D.V., and Shatz, C.J. (1993). Enhancement of transmission at the developing retinogeniculate synapse. *Neuron* 10: 815–825.

Mooney, R., Penn, A.A., Gallego, R., and Shatz, C.J. (1996). Thalamic relay of spontaneous retinal activity prior to vision. *Neuron* 17: 863–874.

Muir Robinson, G., Hwang, B.J., and Feller, M.B. (2002). Retinogeniculate axons undergo eye specific segregation in the absence of eye specific layers. *J. Neurosci.* 22: 5259.

Parnavelas J.G., Mounty E.J., Bradford R., and Lieberman A.R. (1975) The postnatal development of neurons in the dorsal lateral geniculate nucleus of the rat: a golgi study. *J. Comp. Neurol.* 171: 481–500.

Pham, T.A., Rubenstein, J.L., Silva, A.J., Storm, D.R., and Stryker, M.P. (2001). The CRE/CREB pathway is transiently expressed in thalamic circuit development and contributes to refinement of retinogeniculate axons. *Neuron* 31(3), 409–420.

Rekling, J.C. and Feldman, J.L. (1997). Calcium-dependent plateau potentials in rostral ambiguous neurons in the newborn mouse brain stem in vitro. *J. Neurophysiol.* 78: 2483–2492.

Ramoa A.S. and McCormick D.A. (1994). Enhanced activation of NMDA receptor responses at the immature retinogeniculate synapse. *J. Neurosci.* 14: 2098–2105.

Ramoa A.S. and Prusky G. (1997) Retinal activity regulates developmental switches in functional properties and ifenprodil sensitivity of NMDA receptors in the lateral geniculate nucleus. *Dev. Brain Res.* 101: 165–176.

Reese, B.E. (1988). 'Hidden lamination' in the dorsal lateral geniculate nucleus: the functional organization of this thalamic region in the rat. *Brain Res.* 472: 119–137.

Reese B.E. and Cowey A. (1983) Projection lines and the ipsilateral retinogeniculate pathway in the hooded rat. *Neurosci* 10: 1233–1247.

Reese B.E. and Cowey A. (1987). The crossed projection from the temporal retina to the dorsal lateral geniculate nucleus in the rat. *Neurosci.* 20: 951–959.

Reese B.E. and Jeffrey G. (1983) Crossed and uncrossed visual topography in dorsal lateral geniculate nucleus of the pigmented rat. *J. Neurophysiol.* 49: 878–885.

Scharfman H.E., Lu S-M, Guido W., Adams P.R. and Sherman S.M. (1990) N-Methyl-D-aspartate (NMDA) receptors contribute to excitatory postsynaptic potentials of cat lateral geniculate neurons recorded in thalamic slices. *Proc. Nat Acad. Sci. (USA)* 87: 4548–4552.

Schroeder C.E., Tenke C.E., Arezzo J.C. and Vaughan H.G. (1990) Binocularity in the lateral geniculate nucleus of the alert macaque. *Brain Res.* 52: 303–310.

Shatz, C.J. (1996) Emergence of order in visual system development. *Proc. Nat. Acad. Sci. (USA)* 93: 602–608.

Shatz C.J. (1990) Impulse activity and the patterning of connections during CNS development. *Neuron* 5: 745–756.

Shatz, C.J. and Kirkwood, P.A. (1984). Prenatal development of functional connections in the cat's retinogeniculate pathway. *J. Neurosci.* 3: 482–489.

Sefton AJ. and Dreher B. (1995) Visual system. In: Rat Nervous System, Chap. 32, Paxinos, G. (ed.). Academic Press, London, pp. 833–898.

Sherman S.M. and Guillery R. (1996) The functional organization of thalamocortical relays. Journal of Neurophysiology 76: 1367–1395.

Stent G.S. (1973). A physiological mechanism for Hebb's postulate of learning. *Proc. Nat. Acad.Sci. (USA)* 70: 997–1001.

Tavazoie S.F. and Reid R.C. (2000). Diverse receptive fields in the lateral geniculate nucleus during thalamocortical development. *Nature Neurosci.* 3: 606–616.

Torborg, C.L. and Feller, M.B. (2004). Unbiased analysis of bulk axonal segregation patterns. *J. Neurosci. Meth.* 135: 17–26.

Torborg, C.L., Hansen, K.A., and Feller, M.B. (2005). High frequency, synchronized bursting drives eye-specific segregation of retinogeniculate projections. *Nature Neuroscience* 8(1), 72–78.

Tootle, J.S. and Friedlander, M.J. (1986) Postnatal development of receptive field surround inhibition in kitten dorsal lateral geniculate nucleus. *J. Neurosci.* 56: 523–541.

Wilson J.R., Friedlander M.J., and Sherman S.M. (1984) Fine structural morphology of identified X- and Y-cells in the cat's lateral geniculate nucleus. *Proc. Royal Soc. Lond. Biol. Sci.* 221: 441–486.

Weliky M. and Katz L.C. (1999) Correlational structure of spontaneous neuronal activity in the developing lateral geniculate nucleus in vivo. *Science* 285: 599–604.

Wong R.O.L. (1999) Retinal waves and visual system development. *Ann. Rev. Neurosci.* 22: 29–47.

Wong R.O.L., Miester M., and Shatz C.J. (1993) Transient period of correlated bursting activity during the development of the mammalian retina. *Neuron* 11: 923–938.

Wong ROL and Oakley DM. (1996) Changing patterns of spontaneous bursting activity of on- and off- retinal ganglion cells during development. *Neuron* 16: 1087–1095.

Ziburkus, J. and Guido, W. (2005). Loss of binocular responses and reduced retinal convergence during the period of retinogeniculate axon segregation. *J. Neurophysiol.*, accepted with revision.

Ziburkus, J., Lo, F-S., and Guido, W. (2003). Nature of inhibitory postsynaptic activity in developing relay cells of the lateral geniculate nucleus. *J. Neurophysiol.* 90: 1063–1070.

13

A Model for Synaptic Refinement in Visual Thalamus

Bryan M. Hooks and Chinfei Chen

Abstract

How the developing brain specifies precise neural connectivity has long interested neuroscientists. Because of the immense number of cells and synapses in the CNS, it seems unlikely that each individual cell's identity and connectivity is intrinsically or genetically specified. Both neural activity and molecular cues provide possible mechanisms by which network properties of the developing brain can emerge from relative disorder. Emphasizing the visual system as a model for synaptic development, we review the role of various factors in synaptic maturation, including sensory activity. Furthermore, we elucidate why the mouse visual system could prove advantageous for investigation of mechanisms governing circuit development.

A Unilinear View of Synaptic Development

The development of synaptic connections in the central nervous system (CNS) can be divided into several stages. Axons from the presynaptic cell must map to postsynaptic neurons, distinguishing them from inappropriate targets. Once axons have reached a set of potentially appropriate targets, they form synaptic connections. Initial connections tend to be relatively weak and redundant, with connections made to a large number of postsynaptic cells. Over development, these connections are refined, as some inputs are eliminated while others are strengthened (Figure 1). While it is unlikely that all CNS connections develop in this manner, this canonical view of synaptic development helps provide a framework upon which we can begin to understand the mechanisms that underlie synapse maturation and circuit development.

One model synapse for studying functional changes in the maturing CNS is the retinogeniculate synapse in the visual system. This synapse is the connection between the principal output layer of retina, the retinal ganglion cells, and its targets in thalamus, the dorsal lateral geniculate nucleus (dLGN) relay neurons that project to the visual cortex.

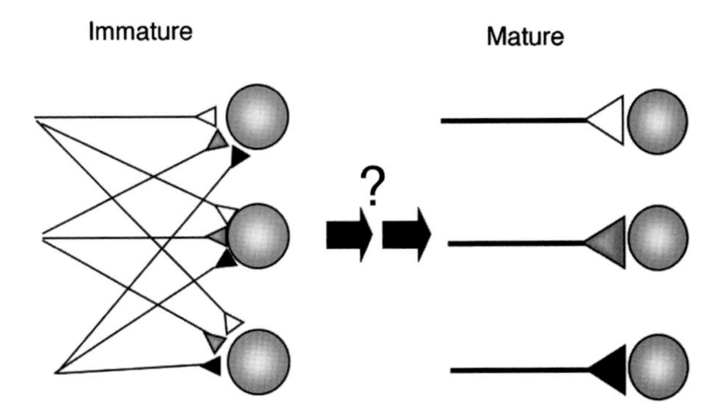

Figure 1 Synaptic refinement during development. Early in development, synaptic connections in the central nervous system are often weak, imprecise and redundant (immature). Over development, these connections are refined as some inputs are strengthened and others are eliminated (mature).

Developmental pruning and strengthening of retinogeniculate connections follows a unilinear pattern, beginning with many inputs and terminating with as few as one afferent (Mouse: (Chen and Regehr, 2000); Ferret: (Tavazoie and Reid, 2000); Rat: (Ziburkus and Guido, 2005) and see this volume, Guido, 2005). However, details of the mechanisms underlying these later steps are not clearly understood. Since the retinogeniculate system recapitulates typical developmental patterns seen in the strengthening of a strong driving input in the CNS, we review the advantages of studying the visual system for understanding sensory circuit development and synaptic maturation, with particular attention to the mouse as a model organism.

The Visual Thalamus As A Model For Studying Synaptic Development

The retinogeniculate synapse shares many general features with other CNS connections and thus provides a useful model for the study of development of CNS synaptic transmission. First, it is an excitatory glutamatergic connection, with the postsynaptic membrane characterized by both major subtypes of glutamate receptors, AMPA and NMDA receptors (AMPAR and NMDAR). The contribution of each subtype changes over development, with NMDAR contribution to early transmission being greater, sometimes to the exclusion of AMPAR (Chen and Regehr, 2000). Furthermore, the subunit composition of the receptors may change during maturation as well: consistent with this, NMDAR excitatory synaptic current timecourse accelerates during development (Carmignoto and Vicini, 1992; Monyer et al., 1994; Chen and Regehr, 2000).

Previous experiments have revealed several of the factors governing the anatomical development of this connection, making this one of the best-studied CNS connections. Retinal ganglion cells form a precise long distance connection to the dLGN. The molecular guidance cues responsible for directing optic tract formation, such as the Ephrin/Eph receptor signaling pathway, are thought to place the axon terminal in a general area of potentially appropriate targets in the LGN and superior colliculus ((Feldheim et al., 1998; Lyckman et al., 2001; Oster and Sretavan, 2003)). Once the axon reaches an appropriate target region, it forms connections to many relay neurons. For example, in developing cat LGN, HRP-labeled retinogeniculate axons are seen to initially form sparse synaptic contacts in both ipsi- and contra-lateral eye layers (Campbell and Shatz, 1992).

However, as the animal matures, more elaborate arbors are seen in the appropriate layer (ipsi- or contra-), while inappropriate connections retract and disappear (Sretavan et al., 1988). In the dLGN, this results in the segregation of projections to eye-specific LGN laminae, with afferents to each layer exclusively from the ipsi- or contra-lateral eye. The process is activity-dependent, and ipsi- and contra-lateral segregation can be blocked by intraventricular infusion of TTX (Shatz and Stryker, 1988; Sretavan et al., 1988). Indeed, the blockade of activity does not result in stasis in axon morphology, but instead TTX-treated axons continue to arborize extensively in both appropriate and inappropriate areas, resulting in an anatomy that is unlike normal axons at any stage of development (Sretavan et al., 1988). Later axon refinement into ON/OFF sublaminae is also activity-mediated, depending on NMDA receptor function (Hahm et al., 1999). Since segregation occurs before birth or eye-opening in a variety of mammals, spontaneous waves of retinal activity provide the *in vivo* source of activity required for lamination (Galli and Maffei, 1988; Meister et al., 1991; Wong et al., 1995; Penn et al., 1998; Stellwagen and Shatz, 2002), though the specific features of the pattern of spontaneous activity necessary for establishing appropriate connections is debated (Huberman et al., 2003).

Synaptic structures continue to develop even after eye-specific layers are complete. Light and electron microscopic studies of retinal afferents in cat LGN indicate that refinement of the terminal arborization continues for several weeks after eye-opening, including an increase in terminal size and involution into glomerular structures (Mason, 1982a, b). This complements functional data, which show an increase in the amplitude of postsynaptic response, a change in the AMPAR/NMDAR current ratio, and NMDAR current timecourse, as well as a change in the number of afferents connected to each dLGN target. Indeed, synaptic maturation proceeds in the canonical manner, with many weaker functional inputs onto postsynaptic cells pruning to several strong ones after eye opening (Chen and Regehr, 2000; Tavazoie and Reid, 2000).

This previous work sets the table for investigations of mechanisms of maturation in several key ways. First, the normal timecourse and trajectory of synaptic maturation are well defined, both morphologically and functionally. Second, a number of the molecular players of early stages in development are identified. Lastly, the pattern of "pre-sensory" spontaneous activity and visually-evoked activity have been extensively studied.

Mouse as a Model for Studying the Visual System

The synaptic organization of the mouse visual system is grossly similar to that in other mammals, such as cats, ferrets, and primates, used in vision research. Recent research suggests that mice and rats are evolutionarily more closely related to primates than cats (grouping Glires and Primates within Euarchontoglires) (Amrine-Madsen et al., 2003), though there is no consensus on placental mammal higher order relationships. With divergence from the felines dating to greater than 90 mya, it is possible that shared neurological features represent convergence instead of homology. One recent paper underscoring this phylogenetic distinction between cat and human shows convergence of center-surround inputs to form orientation-selective cells, typical of cat LGN to layer 4 striate cortex connections, occurs at the layer 4 to layer 2/3 projection in tree shrew—also more closely related to primates than cat (Mooser et al., 2004). Information flows from the retinal ganglion cells (RGCs) in the eye to thalamocortical projection neurons in the dLGN; these in turn project to layer 4 of visual cortex. Since mice, like ferrets, have eyes positioned on the side of their heads and consequently small arcs of binocular vision, the visual system is more monocular than binocular. Thus, instead of multiple eye-specific thalamic layers, the mouse dLGN is divided into one large region innervated by RGCs of the contralateral eye, with a small region devoted to ipsilateral projections. This contrast persists in cortex, where layer 4 pyramidal cell responses are dominated by the contralateral eye.

At a cellular level in the mouse LGN, the structure is also similar to that in primates and cats. Intrathalamic inhibitory circuits include both intrinsic interneurons and thalamic reticular nucleus (NRT) neurons, which provide fast ionotropic inhibition to thalamic relay cells. The dLGN is the only sensory nucleus of rodent LGN that contains intrinsic interneurons (Steriade et al., 1997). These neurons form triadic connections with RGC afferents and their thalamocortical targets, which provide inhibition to relay cells from the presynaptic dendrites (PSD) of intrinsic interneurons when excited by retinal afferents. Glomeruli have been studied at the EM level in cat, and are present in rodents as well (Rafols and Valverde, 1973). One cellular difference in rodents is that the presence of X and Y cells in cat (alpha and beta cells in primate) has not yet been demonstrated, although almost all relay neurons exhibit linear spatial summation responses consistent with a predominance of X cells (Grubb and Thompson, 2003).

One model system for studying synapse development is the retinogeniculate slice preparation developed in rat (Turner and Salt, 1998) and mouse (Chen and Regehr, 2000). This model sacrifices *in vivo* recording for the ability to study synaptic changes at the cellular level.

Experimentally, the retinogeniculate slice preparation has the advantage of geographic and pharmacological separation of inputs. GABAergic connections can be blocked, avoiding complications of intrinsic interneuron activity also excited by retinal afferents. Although retinogeniculate and corticothalamic synapses share similar pharmacology (both are glutamatergic), they are easily separable by stimulation of the optic tract only (retinal axons) using a bipolar stimulation configuration. Additionally, retinal afferents make contact on proximal spines, compared to more distal corticothalamic connections, allowing effective space clamp of the cell (in whole cell voltage clamp mode) even for relatively large (5nA) and fast (decay <10ms) synaptic events. Distinguishing between these afferents is important, since they may follow distinct cell type-specific rules for synaptic maturation. Because of the size of the mouse brain and the thickness of the slice (250 um), it is possible to capture most of the nucleus in a single slice. Thus, the retinogeniculate slice preparation gives anatomical access to one specific class of afferents, as well as neurophysiological access for clean recording.

Synaptic Strengthening and Pruning of the Mouse Retinogeniculate Synapse

Although axon morphological development appears macroscopically complete once eye-specific layers have formed (largely complete by P8 in mice), functional (electrophysiological) responses at the retinogeniculate synapse continue to develop after the time of eye opening (P14) (Godement et al., 1984; Chen and Regehr, 2000).

Figure 2 demonstrates the normal developmental changes that occur at the retinogeniculate synapse over a period ranging from P10 to P32. Retinal fibers in the optic tract are activated with gradually increasing intensities, and the synaptic current elicited at −70 mV and +40 mV are recorded. All recordings are obtained from relay neurons located in the ventral-lateral area of the dLGN, a region that is largely monocularly innervated by p7-8 (Muir Robinson et al., 2002; Jaubert-Miazza et al, 2005; see this volume, Guido, 2005). In Figure 2A, the peak currents evoked from a relatively young animal, before eyes are open, plotted as a function of the stimulus intensity is shown on the left, while on the right, currents elicited at all stimulus intensities from the cell are overlaid. The total amount of current increases with the increase in stimulus intensity, consistent with the wiring model shown in Figure 1(left).

In contrast, a similar experiment performed on thalamic relay neurons in older animals, two weeks after eye opening, revealed dramatic changes in the synaptic response (Figure 2C). Rather than a graded increase in synaptic current in response to increased stimulus intensity, in many cells, there appears to be only one step-like increase in synaptic current. This functional change is consistent with a refinement of the

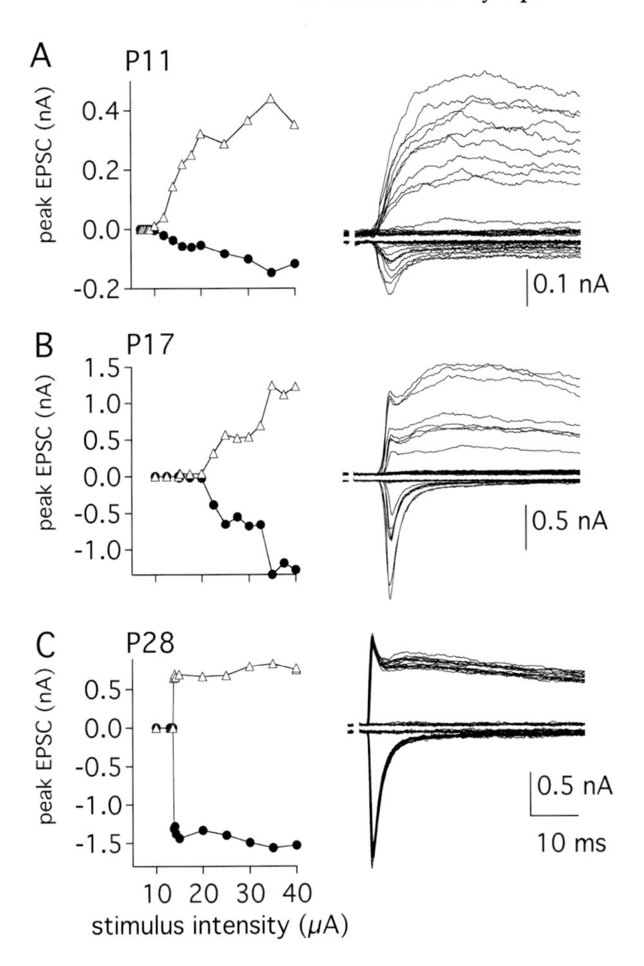

Figure 2 Representative responses to incremental increase in stimulus intensities from animals of different ages. (Left panels) Plots of the peak amplitudes of the AMPAR (black circles) and NMDAR (triangles) components of the synaptic current elicited as a function of stimulus intensity. (Right panels) Superposition of the synaptic currents elicited over the range of stimulus intensities while alternating between holding potentials of +40 mV (outward currents, NMDAR) and −70 mV (inward currents, AMPAR). Adapted, with permission, from (Chen and Regehr, 2000; Copyright 2000 by Elsevier).

synaptic inputs to the recorded cell over development as predicted with the wiring model shown in Figure 1.

These results indicate that, at the time of eye-opening, the LGN principal cells of mice are still innervated by approximately 15–20 inputs (Chen and Regehr, 2000). However, over the next two weeks, the number of afferents is reduced to about one to three. In addition to the pruning phenomenon, each individual retinogeniculate afferent activates a larger current (synaptic strengthening by more than 50-fold). Furthermore, the ratio of excitatory current carried by two subclasses of glutamate receptors changes: in younger animals, the NMDA receptor dominates; the AMPA receptor current component increases in more mature animals. Lastly, the timecourse of NMDA receptor mediated current changes, suggesting a change in NR2 subunit composition, possibly from NR2B to NR2A (Carmignoto and Vicini, 1992; Monyer et al., 1994).

The disparity in timing of the completion of morphological development (eye-specific layering) and electrophysiological maturation raises the question of whether the same mechanisms govern both processes. Specifically, is neural activity involved in the process of functional retinogeniculate synaptic maturation? The influence of activity on development in this electrophysiological assay may be different from the results with eye-specific layer formation for several reasons. First, at different developmental stages, different sources of activity (spontaneous retinal waves early, or visually-evoked spikes late) prevail, perhaps conveying distinct information. Second, in contrast to earlier work in eye-specific layer formation, these electrophysiological recordings focus on the region of LGN that is innervated almost exclusively by contralateral afferents. Although synaptic remodeling found in the monocular region appears very similar to that in binocularly innervated areas of the LGN (Ziburkus et al., 2003; Ziburkus and Guido, 2005; see this volume, Guido, 2005), some of the rules that govern afferent pruning and development in a region where retinal ganglion cells of the same eye compete may differ. This is suggested by the finding that retinal ganglion cell activity block disrupts the contralateral projections to the binocular but not monocular region of the LGN (see figure 2C,D of Penn et al., 1998). Finally, the electrophysiological assay can reveal functional changes that occur at the synaptic or molecular level of axon terminals that are not visible in light microscopy of entire axon branches.

Possible Mechanisms Underlying Synaptic Maturation in the Visual System

Maturation of Receptive Fields after Eye-Opening

Specific studies mapping LGN receptive fields indicate that completion of eye-specific layer formation does not mark the end of synaptic development. Studies in ferret demonstrate that, shortly after eyes open (P35), receptive fields of principal dLGN neurons take a variety of nonspecific shapes that are diffuse and irregular (Daniels et al., 1978; Blakemore and Vital-Durand, 1986; Tavazoie and Reid, 2000). However, over the course of two weeks following eye opening, the receptive fields sharpen into the classical circular center-surround structure. This developmental pattern is consistent with the changes in the strength and innervation of retinogeniculate connections described at the synaptic level in mice (Chen and Regehr, 2000). The receptive field changes and synaptic maturation that occur in the days following eye-opening may depend visually-evoked patterned activity, spontaneous activity, and activity-independent mechanisms, and it will be interesting to test the degree to which each contributes.

Recent work shows that visually experience does play a role in some components of visual system development, permitting, for instance, segregation of ON/OFF dendritic arbors in retinal ganglion cells (Tian and Copenhagen, 2003). Yet in cat dLGN, dark rearing does not change the orientation and direction selectivity of relay cells (Zhou et al., 1995); this finding corresponds with the failure to find morphological changes in X and Y retinogeniculate cell arbors (Garraghty et al., 1987). However, activity does play a part in distinguishing ON and OFF cells in the LGN. Intraocular injection of TTX postnatally to silence all retinal activity leads to unusual LGN cells that respond to both ON and OFF transients, as well as binocular responses from cells in layers corresponding to the injected eye (Archer et al., 1982). More recent work demonstrates that visual experience through closed eyelids (i.e., before eye opening) can also play a role in segregating ON/OFF responses as well as tuning orientation selectivity in dLGN, suggesting a role of visual experience in dLGN development (Akerman et al., 2002).

By comparison, the striate cortex is much more plastic than the LGN. Early studies in the development of the visual system by Hubel and Wiesel showed that a developmental critical period exists, during which monocular deprivation can reduce responsiveness of cortical cells to inputs from one eye or the other. This effect is due to imbalanced competition between inputs, as binocular deprivation does not reduce responses to either eye (reviewed in Hubel and Wiesel, 1998). Further studies revealed that orientation and direction selectivity could be reduced in dark-rearing (Blakemore and Van Sluyters, 1975), while only responses from direction-selective cells (sensitive to motion) were reduced in strobe rearing (Cynader and Chernenko, 1976). The development of cortical maps for ocular dominance and orientation proceeds normally in binocularly deprived cats, but maintenance of these maps requires visual experience (Crair et al., 1998). These findings suggest that experiential factors may be important in the final refinement of the synaptic circuitry moreso than in their initial establishment. Thus, patterned vision is important for the proper maturation of cortical circuits, though debate continues on the degree to which visual experience contributes to the full developmental process.

In other regions of the visual system, such as the superior colliculus, eye opening causes an acceleration in the normal reduction of neuronal inputs that synapse onto superficial collicular neurons in rat (Lu and Constantine-Paton, 2004). Because these cells receive inputs from the cortex, retina and brainstem that cannot be distinguished in the brain slice preparation, the specific class of inputs that exhibits sensitivity to visual activity during development is not yet clear. Since cortical and subcortical synapses appear to obey different rules of plasticity, it will be important to identify the specific presynaptic cell populations in order to better understand the role of activity in circuit development.

Neural Activity and Activity-Independent Cues

Several mechanisms could contribute to the retinogeniculate pruning and synaptic strengthening that is illustrated schematically in Figure 1. First, development may proceed in an activity-independent manner, with pruning and synaptic maturation governed by the interactions of cell surface proteins on the retinogeniculate axon and its target cell. These processes could also be genetically programmed. Second, spontaneous activity (such as retinal waves) may play a role in pruning and strengthening. For the purposes of the synaptic maturation time course, it is important to note that, while retinal wave activity persists after eye opening, in mouse it is found to be largely gone by P15 and completely extinct by P21 (Demas et al., 2003). Though this activity ends many days before synaptic refinement is complete, it may initiate molecular mechanisms that then proceed in an activity-independent manner, functioning even days or weeks after this activity subsides. Indeed, a recent study in the spinal cord demonstrated interactions between activity and molecular cues; reduction of spontaneous activity prevented the normal expression of molecular pathfinding cues such as ephrins and NCAM (Hanson and Landmesser 2004). Lastly, the specific activity from visually-evoked stimuli may be responsible for refining the projection and driving it to maturation. Figure 3 illustrates the timeline of key functional and anatomical changes in the mouse visual system over development, and compares these events to the time periods during which different sources of activity occur.

Arguing in favor of the activity independence of early synaptogenesis, studies in the visual cortex find early formation of ocular dominance columns, even in the absence of retinal input (Crowley and Katz, 1999, 2000). However, contributions from intrinsic thalamic activity cannot be ruled out. Alternatively, other investigators argue for an important role

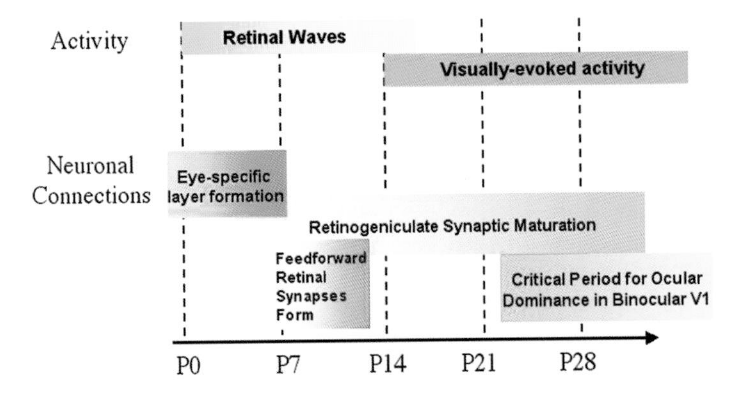

Figure 3 Timeline of major events in visual system. Developmental periods in the rodent associated with changes in the anatomy and function of synaptic circuits in different areas of the visual system are plotted with respect to the postnatal age. Time windows for different forms of activity are superimposed for comparison.

of activity in the developing visual system. The source of this activity, then, is of interest: early in development, spontaneous activity mediated by cholinergic retinal waves contributes to retinal ganglion cell firing patterns; later, spontaneous waves cease and excitation of photoreceptors begins to drive ganglion cell spiking. Some specific features of high frequency presynaptic activity required for normal eye-specific layer segregation have been identified (Torborg et al 2005), though other work challenges the notion that any pattern of activity is required (Huberman et al 2003). Thus, depending on the time in development in which they occur, changes in activity may have a large effect on either axon anatomy or postsynaptic cell physiology. For example, in cat visual cortex, blocking activity late in development produces a shift in cortical ocular dominance profiles, but no anatomic change is detected in dLGN axonal arborizations (reviewed in Katz and Shatz 1996). It is likely that future experiments will reveal a cooperative role of both activity and gene expression in the strengthening and pruning of synaptic connections over development, much like what has recently been described for the determination of the shape, location and segregation of eye-specific layers (Pfeiffenberger et al, 2005; Huberman et al, 2005; see this volume, Huberman and Chapman, 2005).

Hebbian versus Homeostatic Mechanisms

A leading theory for the role of activity in synapse formation, derived from Donald Hebb, is that synapses are strengthened when pre- and post-synaptic cells are simultaneously excited (reviewed in (Katz and Shatz, 1996)). This principle, and a time window critical for its occurrence was demonstrated in tadpole; it was even shown that weak subthreshold inputs can be strengthened by pairing with either postsynaptic depolarization or the simultaneous excitation of a suprathreshold input (Zhang et al., 1998). Long lasting changes in synaptic function, such as LTP and LTD, have also been extensively studied due to their possible role in learning and memory. These mechanisms may also be present in different synapses of the visual system, including the retinogeniculate synapse (Artola and Singer, 1987, 1993; Mooney et al., 1993; Bear and Malenka, 1994).

Opposing the synaptic strengthening, however, is the requirement for a homeostatic mechanism to keep Hebbian mechanisms from strengthening inputs ad infinitum, as well as to keep weaker inputs from becoming stronger lest they simply are activated simultaneously with a stronger, more frequently active input. Synaptic scaling, as described by Turrigiano and colleagues (Turrigiano et al., 1998; Desai et al., 2002), is a form of global non-synapse-specific plasticity that allows individual neurons to scale the strength of postsynaptic response to the overall level of synaptic inputs. How do Hebbian and homeostatic mechanisms coexist? One hypothesis is proposed by Murthy and colleagues: they show that altering the amount of activity in cultured neurons has different effects if it occurs before or after synapse formation (Burrone et al.,

2002). Thus, while activity is important both in determining synaptic strength of developing connections, the developmental timing of the activity with respect to synaptogenesis is also important. While exploring the rules by which activity governs synaptic strengthening and pruning *in vivo*, it is necessary to bear in mind mechanisms previously proposed to govern activity-dependent changes in synaptic strength.

Comparison to the Development of Other Synapses

Lessons from the Neuromuscular Junction

Our understanding of synaptic refinement comes from a number of studies at a variety of synapses. The best understood system for studying synaptic refinement is the neuromuscular junction of rat (reviewed by Sanes and Lichtman, 1999; Wyatt and Balice-Gordon, 2003). Electrophysiological studies recording from skeletal muscle and incrementally stimulating the motor nerve demonstrated a decrease in the number of inputs from 2–3 to 1 between the ages of p0 and p14 (Redfern, 1970; Brown et al., 1976; O'Brien et al., 1978). Initially, the multiple inputs are relatively equal in size, and then over time, a disparity grows between the strength of the inputs until one remains (Colman et al., 1997). Presynaptically, a decrease in release probability at weaker inputs contributes to this disparity (Kopp et al., 2000) and precedes elimination of the input. Monitoring of the postsynaptic ACh receptor with fluorescently labeled bungarotoxin revealed that in some instances, postsynaptic receptors also disappear prior to the elimination of the input innervating the region (Balice-Gordon et al., 1993).

More recently, optical and genetic tools have allowed the fluorescent labeling of a few motor neurons, and thus more detailed visualization of changes that occur as the neuromuscular junction undergoes developmental refinement (reviewed in Lichtman and Sanes, 2003). Images of the retraction of axon terminals away from the muscle during development demonstrated that the functional elimination of synaptic current correspond to an active process of atrophy and withdrawal of the presynaptic axons (Colman et al., 1997). Moreover, mice containing GFP and CFP labeled motor neurons innervating the same motor fiber provided insight into the competition between these inputs. These studies demonstrated a takeover of the postsynaptic territory of the eliminated input by the strengthening terminal in cases when the postsynaptic receptors did not disappear from the territory of the eliminated input (Walsh and Lichtman, 2003).

Activity contributes to synaptic refinement at the neuromuscular junction. Increasing or decreasing action potential activity in motor neurons leads to an enhancement or retardation of synaptic refinement, respectively (O'Brien et al., 1978; Thompson et al., 1979; Ribchester and Taxt, 1983; Thompson, 1983). The relative activity between competing motor neurons appears to be more important than the absolute levels of activity. Consistent with the hypothesis that the more active neuron wins the refinement competition is the result of a study using a genetically titratable knockout of the choline acetyltransferase (ChAT; biosynthetic enzyme of acetylcholine, the NMJ neurotransmitter) to control the amount of neurotransmission at individual motor neurons. In all cases of doubly innervated neuromuscular junctions, the more active neuron (ChAT+) occupied more than 50% of the site (Buffelli et al., 2003). However, in motor fibrils doubly innervated by ChAT- motor neurons, synaptic connections are maintained, suggesting that inactive neurons are still competent to maintain synaptic contact and it is specifically their reduction in activity relative to ChAT+ fibers that causes withdrawal.

It is still unclear the extent or degree that activity contributes to synaptic refinement at the neuromuscular junction. Synaptic refinement at the neuromuscular junction starts prenatally in rat and mice when up to 15 motor neurons are estimated to innervate a single motor fibril (personal communication, Wylie and Lichtman); no previous experiments blocking activity have tested whether all initial inputs can be maintained in the absence of competition. Whether activity drives or modulates the full extent of synaptic refinement, or only the later (postnatal) phase of this process has yet to be determined. However, these studies in the neuromuscular junction provide a framework within which the mechanisms important in synaptic refinement in CNS synapses can be compared.

Development at Other CNS Synapses

A number of synapses in the central nervous system have also been shown to refine during development. These include the chick cochlear nerve to nucleus magnocellularis synapse and the climbing fiber to Purkinje cell (CF-PC) synapse. At both synapses, the extent of input pruning documented postnatally is much less than that observed at the retinogeniculate synapse, decreasing from four to one inputs over a one-week period. However, some mechanisms underlying this process are likely to be shared with other central and peripheral synapses. Consistent with synaptic refinement at the neuromuscular junction, the strength of the CF-PC input, which correlates with the glutamate concentration transients in the synaptic cleft, is predictive of the input that will be eliminated (Hashimoto and Kano, 2003).

Molecules that contribute to synaptic refinement have also been identified at the CF-PC synapse. Insulin-like growth factor-I was found to enhance both synaptic current amplitudes and the degree of multiple fiber innervation of Purkinje cells by climbing fibers (Kakizawa et al., 2003), suggesting that loss of survival factors is associated with synapse elimination. The IGF-1 results are particularly interesting, as chronic IGF-1 application starting at P8 but not P12 was effective in increasing synaptic currents and blocking synaptic pruning, even though development was monitored until P24-P36. Thus, events during an early developmental time window may continue to exert their effects in phenomena weeks later. The same authors propose a role for activation of mGluR1 in pruning, thus implicating activity as well (Kano et al.,

1997); furthermore, they propose that IGF and mGluR1 could mediate early and late phases of synaptic pruning respectively. Evidence points to a postsynaptic site of action for this metabotropic glutamate receptor, possibly involving an intracellular signal transduction cascade that includes protein kinase C, the guanine nucleotide binding protein, Gaq, phospholipase Cb4 and P/Q-type calcium channels (Kano et al., 1995; Offermanns et al., 1997; Kano et al., 1998; Miyazaki et al., 2004).

Inhibitory CNS synapses have also been shown to undergo synaptic remodeling during development (Sanes and Friauf, 2000; Kandler, 2004). Synaptic refinement occurs at the glycinergic connection between the medial nucleus of the trapezoid body and the lateral superior olive in the rat auditory system during the first postnatal week. A four-fold decrease in the input area to the LSO is accompanied by a 12-fold increase in synaptic strength, resulting in a two-fold sharpening of the tonotopic map (Kim and Kandler, 2003). It is also worth noting that this refinement occurs at a developmental time when the ionic currents associated with inhibitory glycinergic channels are depolarizing to immature cells. In addition, some inhibitory inputs actually release glutamate as the neurotransmitter (Gillespie et al., 2005). In contrast to the neuromuscular junction, there appears to be a temporal dissociation between functional development, which occurs during the first postnatal week, when compared to morphological refinement, which takes place after the onset of hearing, in the third postnatal week (Sanes and Siverls, 1991). Therefore patterned auditory activity does not seem to drive the functional refinement of this connection.

Multiple Forms of Synaptic Refinement in the CNS

Functional refinement at different types of synapses may differ both in terms of the final synaptic configuration (i.e., many presynaptic afferents vs. a few) as well as the underlying cellular mechanisms mediating the result. The organization and architecture of the mature retinogeniculate synapse represents only one of many synaptic configurations that have been described in the CNS. An example of alternative synaptic organization can be seen in the other glutamatergic input to relay neurons, the corticothalamic projections to the LGN. Many cortical neurons connect to a given relay neuron, each with relatively weak synaptic strength (Deschenes et al., 1984; Granseth and Lindstrom, 2003; Li et al., 2003; Reichova and Sherman, 2004). The difference in the synaptic organization between the two glutamatergic inputs likely arises from their functional roles. While the retinogeniculate input is often referred to as a "driver" input, one that carries primary information, the corticothalamic input is considered a "modulator" of that information (Sherman and Guillery, 1998; Alitto and Usrey, 2003). Thus, it would not be surprising that rules governing synaptic refinement at the corticogeniculate synapse would be quite different from that of the retinogeniculate connection (Clasca et al., 1995; Reichova and Sherman, 2004).

Supporting this idea are recent studies in the cortex illustrating that many synaptic connections develop with striking specificity, without a period of exuberant innervation that is subsequently pruned (Callaway and Lieber, 1996; Bender et al., 2003; Bureau et al., 2004). What, then, is the link between the synapses that have been shown to exhibit considerable functional synaptic remodeling during development? The retinogeniculate synapse, the excitatory and inhibitory auditory synapses, the neuromuscular junction, and the climbing fiber to purkinje cell synapse in the cerebellum, could all be classified as strong projections that convey primary information or instruction. Thus the ultimate function of the synaptic connection could dictate the specific rules of synaptic development in the CNS. The degree to which these disparate cell types share molecular mechanisms governing synapse pruning and elimination will become clear from future studies.

Conclusions

The great diversity of connections in the nervous system affords the testing of a large number of potential rules governing their formation, maturation, and plasticity. The retinogeniculate synapse of the mouse visual system provides one model system where the anatomical development and patterns of presynaptic activity are well-known and synaptic refinement is robust and large. The features of this experimental model provide an opportunity to identify the molecules and rules important in strengthening and remodeling at strong, precise, driving inputs of the central nervous system.

References

Akerman, C. J., Smyth, D. and Thompson, I. D. (2002). Visual experience before eye-opening and the development of the retinogeniculate pathway. *Neuron* 36, 869–79.

Alitto, H. J. and Usrey, W. M. (2003). Corticothalamic feedback and sensory processing. *Current Opinion in Neurobiology* 13, 440–5.

Amrine-Madsen, H., Koepfli, K. P., Wayne, R. K. and Springer, M. S. (2003). A new phylogenetic marker, apolipoprotein B, provides compelling evidence for eutherian relationships. *Molecular Phylogenetics & Evolution* 28, 225–40.

Archer, S. M., Dubin, M. W. and Stark, L. A. (1982). Abnormal development of kitten retino-geniculate connectivity in the absence of action potentials. *Science* 217, 743–5.

Artola, A. and Singer, W. (1987). Long-term potentiation and NMDA receptors in rat visual cortex. *Nature* 330, 649–52.

Artola, A. and Singer, W. (1993). Long-term depression of excitatory synaptic transmission and its relationship to long-term potentiation. *Trends in Neurosciences* 16, 480–7.

Balice-Gordon, R. J., Chua, C. K., Nelson, C. C. and Lichtman, J. W. (1993). Gradual loss of synaptic cartels precedes axon withdrawal at developing neuromuscular junctions. *Neuron* 11, 801–15.

Bear, M. F. and Malenka, R. C. (1994). Synaptic plasticity: LTP and LTD. *Current Opinion in Neurobiology* 4, 389–99.
Bender, K. J., Rangel, J. and Feldman, D. E. (2003). Development of columnar topography in the excitatory layer 4 to layer 2/3 projection in rat barrel cortex. *Journal of Neuroscience* 23, 8759–70.
Blakemore, C. and Van Sluyters, R. C. (1975). Innate and environmental factors in the development of the kitten's visual cortex. *Journal of Physiology* 248, 663–716.
Blakemore, C. and Vital-Durand, F. (1986). Organization and post-natal development of the monkey's lateral geniculate nucleus. *Journal of Physiology* 380, 453–91.
Brown, M. C., Jansen, J. K. and Van Essen, D. (1976). Polyneuronal innervation of skeletal muscle in new-born rats and its elimination during maturation. *Journal of Physiology* 261, 387–422.
Buffelli, M., Burgess, R. W., Feng, G., Lobe, C. G., Lichtman, J. W. and Sanes, J. R. (2003). Genetic evidence that relative synaptic efficacy biases the outcome of synaptic competition. *Nature* 424, 430–4.
Bureau, I., Shepherd, G. M. G. and Svoboda, K. (2004). Precise development of functional and anatomical columns in the neocortex. *Neuron* 42, 789–801.
Burrone, J., O'Byrne, M. and Murthy, V. N. (2002). Multiple forms of synaptic plasticity triggered by selective suppression of activity in individual neurons. *Nature* 420, 414–8.
Busetto, G., Buffelli, M., Tognana, E., Bellico, F. and Cangiano, A. (2000). Hebbian mechanisms revealed by electrical stimulation at developing rat neuromuscular junctions. *Journal of Neuroscience* 20, 685–95.
Callaway, E. M., Soha, J. M. and Van Essen, D. C. (1987). Competition favouring inactive over active motor neurons during synapse elimination. *Nature* 328, 422–6.
Callaway, E. M. and Lieber, J. L. (1996). Development of axonal arbors of layer 6 pyramidal neurons in ferret primary visual cortex. *Journal of Comparative Neurology* 376, 295–305.
Campbell, G. and Shatz, C. J. (1992). Synapses formed by identified retinogeniculate axons during the segregation of eye input. *Journal of Neuroscience* 12, 1847–58.
Carmignoto, G. and Vicini, S. (1992). Activity-dependent decrease in NMDA receptor responses during development of the visual cortex. *Science* 258, 1007–11.
Chen, C. and Regehr, W. G. (2000). Developmental remodeling of the retinogeniculate synapse. *Neuron* 28, 955–966.
Clasca, F., Angelucci, A. and Sur, M. (1995). Layer-specific programs of development in neocortical projection neurons. *Proceedings of the National Academy of Sciences of the United States of America* 92, 11145–9.
Colman, H., Nabekura, J. and Lichtman, J. W. (1997). Alterations in synaptic strength preceding axon withdrawal. *Science* 275, 356–61.
Constantine-Paton, M., Cline, H. T. and Debski, E. (1990). Patterned activity, synaptic convergence, and the NMDA receptor in developing visual pathways. *Annual Review of Neuroscience* 13, 129–54.
Constantine-Paton, M. and Cline, H. T. (1998). LTP and activity-dependent synaptogenesis: the more alike they are, the more different they become. *Current Opinion in Neurobiology* 8, 139–48.
Crair, M. C., Gillespie, D. C. and Stryker, M. P. (1998). The role of visual experience in the development of columns in cat visual cortex. *Science* 279, 566–70.

Crowley, J. C. and Katz, L. C. (1999). Development of ocular dominance columns in the absence of retinal input. *Nature Neuroscience* 2, 1125–1130.

Crowley, J. C. and Katz, L. C. (2000). Early development of ocular dominance columns. *Science* 290, 1321–1324.

Cynader, M. and Chernenko, G. (1976). Abolition of direction selectivity in the visual cortex of the cat. *Science* 193, 504–5.

Daniels, J. D., Pettigrew, J. D. and Norman, J. L. (1978). Development of single-neuron responses in kitten's lateral geniculate nucleus. *Journal of Neurophysiology* 41, 1373–93.

Demas, J., Eglen, S. J. and Wong, R. O. (2003). Developmental loss of synchronous spontaneous activity in the mouse retina is independent of visual experience. *Journal of Neuroscience* 23, 2851–60.

Desai, N. S., Cudmore, R. H., Nelson, S. B. and Turrigiano, G. G. (2002). Critical periods for experience-dependent synaptic scaling in visual cortex. *Nature Neuroscience* 5, 783–9.

Deschenes, M., Paradis, M., Roy, J. P. and Steriade, M. (1984). Electrophysiology of neurons of lateral thalamic nuclei in cat: resting properties and burst discharges. *Journal of Neurophysiology* 51, 1196–219.

Feldheim, D. A., Vanderhaeghen, P., Hansen, M. J., Frisen, J., Lu, Q., Barbacid, M. and Flanagan, J. G. (1998). Topographic guidance labels in a sensory projection to the forebrain. *Neuron* 21, 1303–13.

Galli, L. and Maffei, L. (1988). Spontaneous impulse activity of rat retinal ganglion cells in prenatal life. *Science* 242, 90–1.

Garraghty, P. E., Frost, D. O. and Sur, M. (1987). The morphology of retinogeniculate X- and Y-cell axonal arbors in dark-reared cats. *Experimental Brain Research* 66, 115–27.

Gillespie, D. C., Kim, G. and Kandler, K. (2005). Inhibitory synapses in the developing auditory system are glutamatergic. *Nature Neurosci* 8, 332–8.

Godement, P., Salaun, J. and Imbert, M. (1984). Prenatal and postnatal development of retinogeniculate and retinocollicular projections in the mouse. *Journal of Comparative Neurology* 230, 552–75.

Granseth, B. and Lindstrom, S. (2003). Unitary EPSCs of corticogeniculate fibers in the rat dorsal lateral geniculate nucleus in vitro. *Journal of Neurophysiology* 89, 2952–60.

Grubb, M. S. and Thompson, I. D. (2003). Quantitative characterization of visual response properties in the mouse dorsal lateral geniculate nucleus. *Journal of Neurophysiology* 90, 3594–607.

Hahm, J. O., Cramer, K. S. and Sur, M. (1999). Pattern formation by retinal afferents in the ferret lateral geniculate nucleus: developmental segregation and the role of N-methyl-D-aspartate receptors. *Journal of Comparative Neurology* 411, 327–45.

Hanson M. G., and Landmesser L. T. (2004). Normal patterns of spontaneous activity are required for correct motor axon guidance and the expression of specific guidance molecules. *Neuron* 43: 687–701.

Hashimoto, K. and Kano, M. (2003). Functional differentiation of multiple climbing fiber inputs during synapse elimination in the developing cerebellum. *Neuron* 38, 785–796.

Hubel, D. H. and Wiesel, T. N. (1998). Early exploration of the visual cortex. *Neuron* 20, 401–12.

Huberman, A. D., Wang, G. Y., Liets, L. C., Collins, O. A., Chapman, B. and Chalupa, L. M. (2003). Eye-specific retinogeniculate segregation independent of normal neuronal activity. *Science* 300, 994–8.

Huberman A. D., Murray K. D., Warland D. K., Feldheim D. A. and Chapman B. (2005). Ephrin-As mediate targeting of eye-specific projections to the lateral geniculate nucleus. *Nature Neurosci.* 8:1013–1021.

Jaubert-Miazza, L., Green, E., Lo, F-S., Bui, K., Mills, J., and Guido, W. (2005). Structural and functional composition of the developing retinogeniculate pathway in the mouse. *Visual Neuroscience*, in press.

Kakizawa, S., Yamada, K., Iino, M., Watanabe, M. and Kano, M. (2003). Effects of insulin-like growth factor I on climbing fibre synapse elimination during cerebellar development. *European Journal of Neuroscience* 17, 545–54.

Kandler, K. (2004). Activity-dependent organization of inhibitory circuits: lessons from the auditory system. *Current Opinion in Neurobiology* 14, 96–104.

Kano, M., Hashimoto, K., Chen, C., Abeliovich, A., Aiba, A., Kurihara, H., Watanabe, M., Inoue, Y. and Tonegawa, S. (1995). Impaired synapse elimination during cerebellar development in PKC gamma mutant mice. *Cell* 83, 1223–31.

Kano, M., Hashimoto, K., Kurihara, H., Watanabe, M., Inoue, Y., Aiba, A. and Tonegawa, S. (1997). Persistent multiple climbing fiber innervation of cerebellar Purkinje cells in mice lacking mGluR1. *Neuron* 18, 71–9.

Kano, M., Hashimoto, K., Watanabe, M., Kurihara, H., Offermanns, S., Jiang, H., Wu, Y., Jun, K., Shin, H. S., Inoue, Y., Simon, M. I. and Wu, D. (1998). Phospholipase cbeta4 is specifically involved in climbing fiber synapse elimination in the developing cerebellum. *Proceedings of the National Academy of Sciences of the United States of America* 95, 15724–9.

Katz, L. C. and Shatz, C. J. (1996). Synaptic activity and the construction of cortical circuits. *Science* 274, 1133–8.

Kim, G. and Kandler, K. (2003). Elimination and strengthening of glycinergic/GABAergic connections during tonotopic map formation. *Nature Neuroscience* 6, 282–290.

Kopp, D. M., Perkel, D. J. and Balice-Gordon, R. J. (2000). Disparity in neurotransmitter release probability among competing inputs during neuromuscular synapse elimination. *Journal of Neuroscience* 20, 8771–9.

Li, J., Guido, W. and Bickford, M. E. (2003). Two distinct types of corticothalamic EPSPs and their contribution to short-term synaptic plasticity. *Journal of Neurophysiology* 90, 3429–40.

Lichtman, J. W. and Sanes, J. R. (2003). Watching the neuromuscular junction. *Journal of Neurocytology* 32, 767–75.

Lu, W. and Constantine-Paton, M. (2004). Eye opening rapidly induces synaptic potentiation and refinement. *Neuron* 43, 237–49.

Lyckman, A. W., Jhaveri, S., Feldheim, D. A., Vanderhaeghen, P., Flanagan, J. G. and Sur, M. (2001). Enhanced plasticity of retinothalamic projections in an ephrin-A2/A5 double mutant. *Journal of Neuroscience* 21, 7684–90.

Mason, C. A. (1982). Development of terminal arbors of retino-geniculate axons in the kitten–II. Electron microscopical observations. *Neuroscience* 7, 561–82.

Mason, C. A. (1982). Development of terminal arbors of retino-geniculate axons in the kitten–I. Light microscopical observations. *Neuroscience* 7, 541–59.

Meister, M., Wong, R. O., Baylor, D. A. and Shatz, C. J. (1991). Synchronous bursts of action potentials in ganglion cells of the developing mammalian retina. *Science* 252, 939–43.

Miyazaki, T., Hashimoto, K., Shin, H. S., Kano, M. and Watanabe, M. (2004). P/Q-type Ca2+ channel alpha1A regulates synaptic competition on developing cerebellar Purkinje cells. *Journal of Neuroscience* 24, 1734–43.

Monyer, H., Burnashev, N., Laurie, D. J., Sakmann, B. and Seeburg, P. H. (1994). Developmental and regional expression in the rat brain and functional properties of four NMDA receptors. *Neuron* 12, 529–40.

Mooney, R., Madison, D. V. and Shatz, C. J. (1993). Enhancement of transmission at the developing retinogeniculate synapse. *Neuron* 10, 815–25.

Mooser, F., Bosking, W. and Fitzpatrick, D. (2004). A morphological basis for orientation tuning in primary visual cortex. *Nature Neurosci* 2004, 872–9.

O'Brien, R. A., Ostberg, A. J. and Vrbova, G. (1978). Observations on the elimination of polyneuronal innervation in developing mammalian skeletal muscle. *Journal of Physiology* 282, 571–82.

Offermanns, S., Hashimoto, K., Watanabe, M., Sun, W., Kurihara, H., Thompson, R. F., Inoue, Y., Kano, M. and Simon, M. I. (1997). Impaired motor coordination and persistent multiple climbing fiber innervation of cerebellar Purkinje cells in mice lacking Galphaq. *Proceedings of the National Academy of Sciences of the United States of America* 94, 14089–94.

Oster, S. F. and Sretavan, D. W. (2003). Connecting the eye to the brain: the molecular basis of ganglion cell axon guidance. *British Journal of Ophthalmology* 87, 639–45.

Penn, A. A., Riquelme, P. A., Feller, M. B. and Shatz, C. J. (1998). Competition in retinogeniculate patterning driven by spontaneous activity. *Science* 279, 2108–12.

Pfeiffenberger C., Cutforth T., Woods G., Yamada J., Renteria R. C., Copenhagen D. R., Flanagan J. G. and Feldheim D. A. (2005). Ephrin-As and neural activity are required for eye-specific patterning during retinogeniculate mapping. *Nature Neurosci.* 8:1022–1027.

Rafols, J. A. and Valverde, F. (1973). The structure of the dorsal lateral geniculate nucleus in the mouse. A Golgi and electron microscopic study. *Journal of Comparative Neurology* 150, 303–32.

Redfern, P. A. (1970). Neuromuscular transmission in new-born rats. *Journal of Physiology* 209, 701–9.

Reichova, I. and Sherman, S. M. (2004). Somatosensory Corticothalamic Projections: Distinguishing Drivers from Modulators. *Journal of Neurophysiology*.

Ribchester, R. R. and Taxt, T. (1983). Motor unit size and synaptic competition in rat lumbrical muscles reinnervated by active and inactive motor axons. *Journal of Physiology* 344, 89–111.

Ridge, R. M. and Betz, W. J. (1984). The effect of selective, chronic stimulation on motor unit size in developing rat muscle. *Journal of Neuroscience* 4, 2614–20.

Sanes, D. H. and Siverls, V. (1991). Development and specificity of inhibitory terminal arborizations in the central nervous system. *Journal of Neurobiology* 22, 837–54.

Sanes, D. H. and Friauf, E. (2000). Development and influence of inhibition in the lateral superior olivary nucleus. *Hearing Research* 147, 46–58.

Sanes, J. R. and Lichtman, J. W. (1999). Development of the vertebrate neuromuscular junction. *Annual Review of Neuroscience* 22, 389–442.

Shatz, C. J. and Stryker, M. P. (1988). Prenatal tetrodotoxin infusion blocks segregation of retinogeniculate afferents. *Science* 242, 87–9.

Sherman, S. M. and Guillery, R. W. (1998). On the actions that one nerve cell can have on another: distinguishing "drivers" from "modulators". *Proceedings of the National Academy of Sciences of the United States of America* 95, 7121–6.

Sretavan, D. W., Shatz, C. J. and Stryker, M. P. (1988). Modification of retinal ganglion cell axon morphology by prenatal infusion of tetrodotoxin. *Nature* 336, 468–71.

Stellwagen, D. and Shatz, C. J. (2002). An instructive role for retinal waves in the development of retinogeniculate connectivity. *Neuron* 33, 357–67.

Steriade, M., Jones, E. G. and McCormick, D. A. (1997). *Thalamus*. 2 vols, Elsevier Science Ltd, Oxford.

Tavazoie, S. F. and Reid, R. C. (2000). Diverse receptive fields in the lateral geniculate nucleus during thlamocortical development. *Nature Neuroscience* 3, 608–16.

Thompson, W., Kuffler, D. P. and Jansen, J. K. (1979). The effect of prolonged, reversible block of nerve impulses on the elimination of polyneuronal innervation of new-born rat skeletal muscle fibers. *Neuroscience* 4, 271–81.

Thompson, W. (1983). Synapse elimination in neonatal rat muscle is sensitive to pattern of muscle use. *Nature* 302, 614–6.

Tian, N. and Copenhagen, D. R. (2003). Visual stimulation is required for refinement of ON and OFF pathways in postnatal retina. *Neuron* 39, 85–96.

Torborg C. L., Hansen K. A., Feller M. B. (2005). High frequency, synchronized bursting drives eye-specific segregation of retinogeniculate projections. *Nat Neurosci.* 8:72–8.

Turner, J. P. and Salt, T. E. (1998). Characterization of sensory and corticothalamic excitatory inputs to rat thalamocortical neurones in vitro. *Journal of Physiology* 510, 829–43.

Turrigiano, G. G., Leslie, K. R., Desai, N. S., Rutherford, L. C. and Nelson, S. B. (1998). Activity-dependent scaling of quantal amplitude in neocortical neurons. *Nature* 391, 892–6.

Walsh, M. K. and Lichtman, J. W. (2003). In vivo time-lapse imaging of synaptic takeover associated with naturally occurring synapse elimination. *Neuron* 37, 67–73.

Wong, R. O., Chernjavsky, A., Smith, S. J. and Shatz, C. J. (1995). Early functional neural networks in the developing retina. *Nature* 374, 716–8.

Wyatt, R. M. and Balice-Gordon, R. J. (2003). Activity-dependent elimination of neuromuscular synapses. *Journal of Neurocytology* 32, 777–94.

Ziburkus, J. and Guido, W. (2005). Loss of binocular responses and reduced retinal convergence during the period of retiongeniculate axon segregation. *J. Neurophysiol.*, accepted with revision.

Ziburkus, J., Lo, F-S., and Guido, W. (2003). Nature of inhibitory postsynaptic activity in developing relay cells of the lateral geniculate nucleus. *J. Neurophysiol.* 90:1063–1070.

Zhang, L. I., Tao, H. W., Holt, C. E., Harris, W. A. and Poo, M. (1998). A critical window for cooperation and competition among developing retinotectal synapses. *Nature* 395, 37–44.

Zhou, Y., Leventhal, A. G. and Thompson, K. G. (1995). Visual deprivation does not affect the orientation and direction sensitivity of relay cells in the lateral geniculate nucleus of the cat. *Journal of Neuroscience* 15, 689–98.

14

Making and Breaking Eye-specific Projections to the Lateral Geniculate Nucleus

Andrew D. Huberman and Barbara Chapman

Introduction

In mammals, axons from the two eyes are segregated within their targets. A striking example of this is found in the lateral geniculate nucleus (LGN) wherein ganglion cell axons arising from the right- and left-eyes are organized into a highly stereotyped arrangement of non-overlapping domains called eye-specific layers (Jones, 1985). Rakic (1976) was the first to examine the development of eye-specific retinogeniculate projections. By injecting tritiated proline into one eye of macaque embryos *in utero*, he found that early in prenatal development, axons from the labeled eye filled the entire LGN, whereas later in prenatal life, the labeled axons were restricted to distinct portions of the LGN and were mirror-symmetric on the two sides of the brain (Rakic, 1976). This indicated that eye-specific layers emerge from a state in which axons from the two eyes initially intermingle. Subsequent experiments examined the development of eye-specific retinogeniculate projections in various species using higher sensitivity tracers (ferret: Linden et al., 1981; cat: Shatz, 1983; rat: Jeffery, 1984; mouse: Godement et al., 1984; macaque: Huberman et al., 2005a). Although the timing of eye-specific segregation was found to vary depending on the species under investigation and the sensitivity of tracer used, the results of all these experiments confirmed that i) eye-specific projections emerge from a state in which axons from the two eyes initially overlap and ii) the segregation process occurs before vision is possible. For example, in cat, eye-specific retinogeniculate segregation also occurs prenatally, in the darkness of the uterus (Shatz, 1983). In ferrets and mice, eye-specific segregation occurs after birth, between postnatal day 1 (P1) and P10 (Linden et al., 1981; Godement et al., 1984) but still before the onset of phototransduction, which begins around P15-P20 in these species (ferret: Akerman et al., 2002; mouse: Demas et al., 2003).

Competition and Hebbian Mechanisms

Several experiments have demonstrated a role for binocular competition in eye-specific retinogeniculate segregation. If one eye is removed at the stage of development when axons from the two eyes overlap, projections from the remaining eye end up distributed throughout the LGN (rat: Lund et al., 1973; macaque: Rakic, 1981; cat: Chalupa and Williams, 1984; ferret: Guillery et al., 1985a). Toward the late 1980's and early 1990's eye-specific retinogeniculate projections emerged as a premier model system for addressing the role of competition in development of precise neural connections. A competition-based model for eye-specific development was appealing because, in several respects it appeared to obey Hebb's postulate (Hebb, 1949). Within the context of segregating binocular connections into eye-specific layers, a Hebbian based model predicted that co-active inputs arising from the same eye onto a single LGN neuron would be more efficient at depolarizing the LGN neuron than would inputs arising from different eyes onto the a single LGN neuron. In theory, inputs arising from different eyes onto the same LGN cell would tend to be uncorrelated in their firing. Thus, whichever input was stronger would tend to be maintained whereas the weaker input would tend to be eliminated (reviewed in Shatz, 1990; 1996; Katz and Shatz, 1996). A Hebbian-based model was also compelling because, at the time, studies carried out on other brain regions (mainly the hippocampus) were beginning to identify the molecular mechanisms by which synaptic coincidence detection might occur, such as through NMDA receptor activation (for an early review see: Brown et al., 1988). Also, much attention was paid to experiments by Constantine-Paton and colleagues wherein they grafted a third eye onto a tadpole, forcing binocular projections into one lobe of the visual tectum. Within the dual-eye innervated tectum, axons from the native and the grafted eye segregated from one another into a series of alternating eye-specific stripes (Constantine-Paton and Law, 1978). Pharmacological blockade of NMDA receptors desegregated eye-specific stripes (Cline et al., 1987) and pharmacologically augmenting tectal NMDA receptor function caused formation of especially distinct stripes (Cline et al., 1987). Thus, multiple lines of evidence suggested that eye-specific retinogeniculate segregation is mediated by NMDA-receptor-dependent Hebbian mechanisms (also see: Ramoa and McCormick, 1994).

Waves Roll in

It was obvious how visual stimulation would drive correlated firing of neighboring ganglion cells located within the same eye more so than correlated firing of ganglion cells located in opposite eyes. However, because eye-specific LGN layers emerged before the onset of vision, it remained unclear how such correlations could arise spontaneously. Then, a series of remarkable experiments by Maffei and colleagues reported the presence of *en utero* correlated spontaneous retinal activity (Galli and Maffei, 1988; Maffei and Galli-Resta, 1990). By recording extracellularly from retinal ganglion cells of fetal rat embryos, they showed that ganglion cells spontaneously fire periodic bursts of action potentials (Galli and Maffei, 1988) and that neighboring ganglion cells were highly correlated in their firing (Maffei and Galli-Resta, 1990). In rats, eye-specific segregation occurs postnatally, not *en utero* (Jeffery, 1984). Nonetheless, these findings intrigued those interested in eye-specific LGN segregation. Other labs began careful documentation of the patterns of spontaneous retinal activity. The ferret was selected for these studies because, as a carnivore, it has robust eye-specific layers. However, unlike in other carnivores such cats, in ferrets eye-specific segregation occurs postnatally (from P1-P10), greatly facilitating *in vivo* manipulations. Using a multi-electrode recording array to extracellularly record from retinal explants *in vitro*, Shatz and colleagues simultaneously assessed the action potential activity of dozens of retinal ganglion cells (Meister et al., 1991). They showed that, during the stage of development when eye-specific segregation occurs, regions of excitation periodically spread across the retina in "wave"-like fashion, causing closely positioned ganglion cells to fire in synchrony. Individual waves were restricted in their size and random in their site of origin. Thus, over time, waves tiled the entire retinal surface (Meister et al., 1991; Wong et al., 1993). Based in their precise spatio-temporal properties, waves were hypothesized to play an instructive role in eye-specific segregation in the LGN, by engaging Hebbian-based plasticity at retino-LGN synapses (reviewed in: Shatz, 1996; Katz and Shatz, 1996).

Activity Block Experiments and Segregation

The first test of the role of spontaneous activity on eye-specific segregation was carried out by Shatz and Stryker (1988). They chronically infused the sodium channel blocker tetrodotoxin (TTX) into fetal cat brains during the period of eye-specific segregation and then labeled retino-LGN projections. In contrast to control kittens, which had normally segregated eye-specific layers in the LGN, eye-specific segregation did not occur in the TTX treated animals (Shatz and Stryker, 1988). Whole-eye labeling showed that axons from one eye were spread throughout the LGN and labeling of single retino-LGN axons revealed that this overlap reflected a massive increase in the size of ganglion cell arbors (Sretavan et al., 1988). Chronic intracranial infusion of TTX likely blocks all spiking activity throughout the fetal brain. However, by continuing to study the mechanisms by which retinal waves are generated, Shatz and colleagues discovered that P1-P10 waves are driven by acetylcholine (ACh) acting through nicotinic receptors (Feller et al., 1996). Since starburst amacrine cells are the only retinal neurons that synthesize ACh, they concluded that cholinergic drive arising from starburst amacrine cells drives retinal waves (Feller et al., 1996). Other groups later confirmed this finding

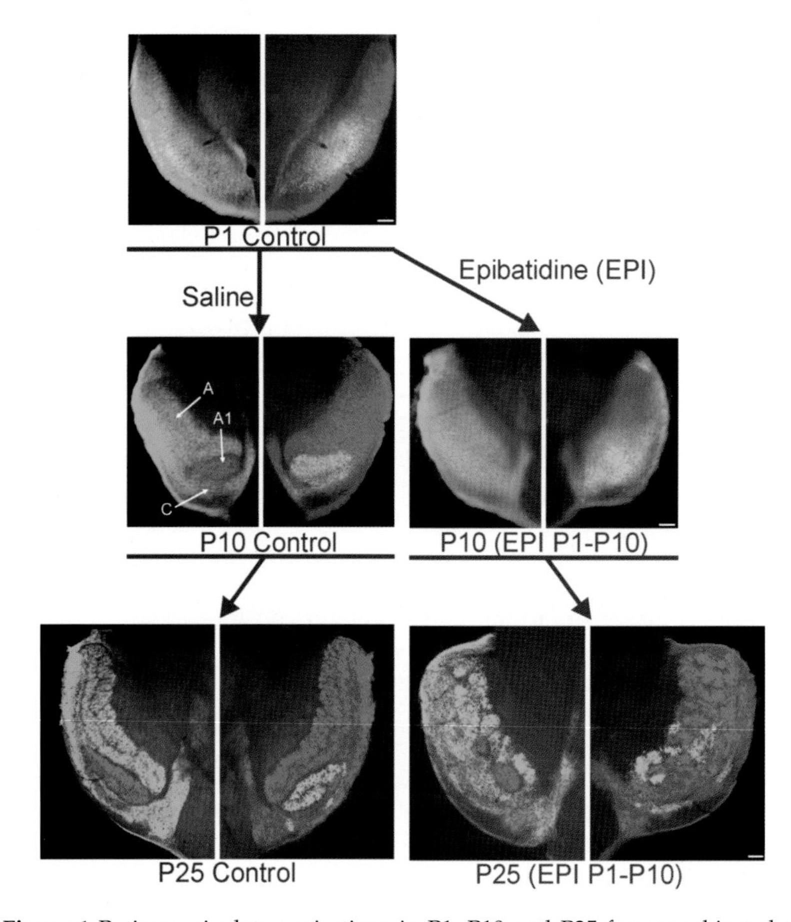

Figure 1 Retinogeniculate projections in P1, P10 and P25 ferrets subjected to control (saline) or epibatidine (EPI) injections from P1-P10. In control P1 ferrets, axons from the right eye (labeled green) and the left eye (red) are intermingled (yellow) throughout much of the LGN. In P10 control ferrets, axons from the two eyes are segregated into A (contra) and A1 (ipsi) layers. In P10 ferrets that received EPI injections from P1-P10, axons from the two eyes remain overlapping. In contrast to control P25 ferrets (in which normal segregated A and A1 layers are present) EPI treated ferrets allowed to survive until P25 or older show abnormal, patchy patterns of eye-specific projections to the LGN. Axons from the two eyes are segregated but normal A and A1 layers are not present. Scale bars (P1 = 50μm), (P10 = 75μm), (P25 = 100μm).

in rabbit (Zhou, 1998) and mouse (Bansal et al., 2000). They also found that the cholinergic agonist, Epibatidine (EPI) blocked retinal waves (through receptor desensitization) when acutely applied at high concentrations *in vitro* (Penn et al., 1998). Penn et al., (1998) found that binocular injections of EPI from P1-P10 completely prevented eye-specific segregation in the LGN (Figure 1). They also injected EPI into one eye from P1-P10 and then traced retinogeniculate projections on P10. Axons from the EPI treated eye retracted to occupy a much smaller-than-normal region of the LGN and the projection from the other eye expanded its projection. These findings represented the first demonstration that eye-specific segregation in the LGN relies on activity mediated binocular competition. Although the morphology of single axons in EPI treated animals was not assessed in these studies, comparison of the results obtained from the monocular versus binocular EPI treatment strongly suggests that their do not simply reflect drug-induced aberrant axon growth; binocular EPI treatment caused axons from both eyes to expand whereas monocular EPI treatment caused axons from the treated eye to shrink. Such opposite effects are difficult to reconcile with the idea that EPI simply causes aberrant axon growth.

The next demonstration of a role for activity-mediated binocular competition in segregation of eye-specific retinogeniculate projections came from an elegant study by Stellwagen and Shatz. They had previously discovered that the size, speed and frequency of retinal waves are mediated by adenosine acting through a cAMP-dependent pathway and that pharmacologic agents that increase intracellular levels of cAMP (such as forskolin) significantly increased the size, speed and frequency of retinal waves in vitro (Stellwagen et al., 1999). So, by injecting forskolin into one or both eyes of ferrets from P1-P10, Stellwagen and Shatz (2002) tested the role of binocular competition in eye-specific segregation by elevating, as opposed to blocking, wave activity. When wave activity was increased in both eyes, they observed no effect on patterning of eye-specific retinogeniculate projections. However, when wave activity was increased only in one eye, axons from the more active eye acquired more LGN territory than the projection from the normally active, untreated eye. Thus, the relative level of activity in the two eyes that is the key parameter for patterning of eye-specific projections.

Maintenance of Eye-specific Projections to the LGN

As mentioned above, in ferrets and mice, eye-specific segregation occurs between P1 and P10. However, retinal waves do not only occur during the eye-segregation phase of development; waves are also present prenatally and they continue until ~P25-P30 (ferret: Wong et al., 1993; mouse: Demas, 2003). From P1-P10 spontaneous retinal activity is driven by ACh, whereas waves that occur from P12-P30 are driven by glutamate released from bipolar cells (Wong et al., 2000) and possibly photoreceptors as well (Johnson et al., 2000). To test whether the spontaneous retinal activity is important for maintenance of eye-specific retinogeniculate projections, Chapman (2000) used intraocular injections of aminophosphobutyric acid (APB) to silence spontaneous all retinal activity from P10-P25. In mature ferrets, APB selectively blocks the activity of ON-center ganglion cells (Chapman and Godecke, 2000). In ferret retinae younger than P30, however, APB completely blocks all calcium waves and ganglion cell spiking (Chapman, 2000). Remarkably, even though ganglion cell projections to the LGN are completely segregated in the P10 ferret LGN (the age when the APB treatment began), blocking retinal activity from P10-P25 caused axons from the two eyes to completely desegregate within then LGN. The pattern of overlap observed in these animals was notably different from that observed in previous experiments (Stryker and Shatz, 1988; Penn et al., 1998). Rather than causing axons from the two eyes to spread out across the LGN, blocking activity from P10-P25 caused axons from contralateral eye to remain in their normal location within the inner segment of the LGN (layer "A") whereas axons from the ipsilateral eye abandoned their normal territory in the outer segment of the LGN ("A1") and translocated into layer A, wherein they intermingled with axons from the contralateral eye. These results showed that spontaneous retinal activity is necessary for the maintenance of eye-specific segregation and it also suggested that activity-independent cues favor lamina A as the target for arborization of afferents from both eyes (Chapman, 2000). In these experiments APB was injected into both eyes, so whether maintenance of eye-specific segregation relies on binocular competition was not addressed. However, taken together with the results of Penn et al., (1998) and Stellwagen and Shatz (2002), the results of Chapman (2000) indicated that spontaneous neural activity of retinal origin is a key parameter for both the establishment and maintenance of eye-specific segregation in the LGN.

Challenges to the Hebbian Model

Several features of eye-specific retinogeniculate segregation cannot be easily explained by the Hebbian model. First, the Hebbian model cannot, by itself, explain the highly stereotyped patterning of eye-specific layers in the LGN. Eye-specific layers consist of two main features: 1) non-overlapping regions of afferents from the two eyes and 2) eye-specific cellular laminae formed by LGN neurons (Jones, 1985). Within a given species, both of these features are remarkably stereotyped in terms of their shape, size and position within the LGN. For example in carnivores such as ferrets and cats the axons from the contralateral eye always occupy the innermost LGN (layer A) and axons from the ipsilateral eye always project to the more outer LGN (layer A1). Although, theoretically, Hebbian models can explain how inputs from the two eyes segregate from one another on the basis of activity mediated binocular competition, Hebbian models cannot explain how the ipsilateral eye axons always segregate into layer A1 and contralateral axons always segregate into layer A. Simply put, there must be a bias for one or the other eye to win a given piece of LGN real estate. Also, Hebbian models cannot explain the differentiation of cellular eye-specific lamination in the LGN. The development of cellular layers occurs after afferents from the two eyes segregate and normal development of cellular eye-specific layers relies, at least in part, on retinal ganglion cell axons (Linden et al., 1981; Cucchiaro and Guillery, 1984; Hutchins and Casagrande, 1990). If the pattern of retinogeniculate afferents is abnormal, the cytoarchitecture of the LGN directly reflects these abnormal inputs. For example, in monocularly or binocularly enucleated animals eye-specific cellular laminae do not develop (Brunso-Bechtold and Casagrande, 1981; Rakic, 1981; Guillery et al., 1985a; 1985b; Sretavan and Shatz, 1986b; Garraghty et al., 1988a; Morgan and Thompson, 1993). Moreover, in coat color mutants, where the density of the ipsilateral-eye projection to the LGN is reduced, the cellular laminae mirror the reduced ipsilateral input and the associated abnormal topography of the retinal projections (Guillery, 1969; 1971; Guillery and Kaas, 1971). However, non-retinal influences on eye-specific cellular lamination in the LGN have been demonstrated (Casagrande and Condo, 1988) and the factors that cause both afferent and cellular layers to develop in the same shape, size and position within the LGN were completely unknown.

Second, despite growing evidence that Hebbian mechanisms might be involved in segregation of eye-specific retino-LGN projections (Penn et al., 1998; Stellwagen and Shatz, 2002), experiments to directly test this hypothesis had not yet been carried out. To directly test the Hebbian model, one would have to find means to alter the correlational structure of spontaneous ganglion cell activity without altering activity levels and techniques. In addition, there was reason to suspect that the Hebbian model for eye segregation might not be correct. Williams et al., (1994) showed that in achiasmatic carnivores (where all axons from the eye project to the ipsilateral LGN), axons from the ganglion cells in the nasal retina formed a layer A (normally innervated by the nasally situated ganglion cells axons from the contralateral eye) and axons from the ganglion cells in the temporal portion *of the same eye* formed a layer A1. This revealed that binocular competition is not required for segregation of nasal versus temporal ganglion cell inputs to the LGN and it suggested the involvement of activity independent cues (i.e., axon guidance molecules) for patterning of left and right eye projections into their stereotyped layered arrangement. Also, TTX application to one or both retinae of postnatal ferrets caused only mild transient effects of this retinal activity block on eye-specific segregation (Cook et al., 1999). TTX completely blocks ganglion cell spiking in explants of P1-P10 ferret retina, but TTX does not prevent periodic calcium waves (Stellwagen et al., 1999). Since EPI and APB block both ganglion cell spiking and calcium waves (Penn et al., 1998; Chapman, 2000), it thus became apparent that the effects of these drugs on retinogeniculate anatomy could be due to their effects on calcium and not on spiking/synaptic events at retino-LGN synapses- the latter of which is of course central to the Hebbian model.

Segregation versus Lamination

Several studies have addressed the question of what mediates the stereotyped size, shape and positioning of eye-specific LGN layers and what is the relationship between afferents from the two eyes in patterning of cellular layers in the LGN (Brunso-Bechtold and Casagrande, 1981;

Casagrande and Condo, 1988). We carried out experiments in which we prevented the segregation of retinogeniculate inputs (by silencing spontaneous retinal activity in both eyes with EPI) and then allowed these animals an extended period of recovery, during which spontaneous retinal activity returned to normal (Huberman et al., 2002). We then assessed the effects of this manipulation on the pattern retinogeniculate inputs, the cytoarchitecture of the LGN and the physiology of LGN neurons, including their receptive field properties and retinotopic organization. As shown previously, in normal postnatal day 1 (P1) ferrets, ganglion cell axons from the two retinae overlap extensively and by P10, axons from the two eyes are segregated (Linden et al., 1981). As seen by Penn et al., in ferrets that received binocular EPI injections from P1-P10 inputs from the two eyes remain overlapped (Figure 1). In ferrets that receive binocular intravitreal injections of EPI from P1-P10, but were then allowed to survive until P25 or older (called EPI-recovery ferrets) retinogeniculate afferents end up completely segregated. However, unlike control ferrets, the spatial pattern of eye-specific retinogeniculate projections is highly aberrant. There are multiple ipsilateral projections of various shapes, positions and sizes, distributed over a significantly greater-than-normal extent of the LGN. The ipsilateral projections even extended into the region of the LGN normally occupied by only axons from the contralateral eye (Figure 1). Control experiments confirmed that the observed effects were not due to damage to the retina or residual blockade of retinal activity past P10. In fact, in ferrets that received binocular injections of EPI from P12-P25, retinogeniculate projections appeared normal, confirming that the effects were due to cholinergic block of early (P1-P10) retinal activity.

In the mature ferret, LGN cells form cytoarchitectural layers that lie in direct registration with the layers formed by the terminals of retinal afferents; they are concentrated into distinct A, and A1 laminae, as well as ON and OFF sublaminae, each separated by a cell-sparse interlaminar space (Linden et al., 1981; Stryker and Zahs, 1983; Zahs and Stryker, 1985; Hutchins and Casagrande, 1990; Hahm et al., 1999). In contrast, the LGN of the EPI-recovery animals completely lack normal patterns of cellular lamination. Clusters of cells, surrounded by cell-sparse regions are occasionally visible, but comparison of these clusters with the pattern of retinogeniculate afferents in the same tissue sections reveals that they do not correspond to eye-specific terminations zones of ganglion cell axons. (Figure 2)

To determine whether disrupting the pattern of retinal afferent lamination alters the physiology of LGN neurons, we performed multi-unit extracellular recordings in the LGN of EPI-recovery animals. All cells encountered were monocular, indicating that functional as well as anatomical segregation of eye-specific inputs to the LGN occurred following the termination of the EPI treatment. Cells in the LGN of the treated animals exhibited ON- or OFF-center responses typical of normal ferrets (Stryker and Zahs, 1983; Zahs and Stryker, 1985; Godecke and Chapman, 2000) and normal center-surround receptive field organization (Tavazoie and

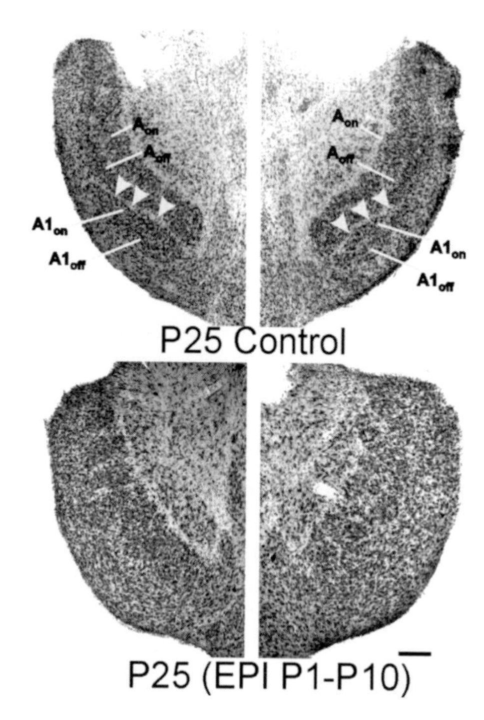

Figure 2 Cytoarchitecture of the LGN from control P25 ferrets and P25 ferrets that received intraocular injections of EPI from P1-P10 and then were allowed to recover until P25. Normal patterns of cellular lamination such as A and A1 layers and interlaminar spaces (arrowheads) are present in control P25 ferret LGN but absent from the EPI-recovery P25 LGN. Scale = 100μm.

Reid, 2000) and were present. Surprisingly, in the EPI-recovery animals, the topographic representation of the binocular visual field was mapped normally, even across the boundaries of eye-specific borders. Thus, dramatically disrupting the organization of eye-specific lamination does not affect the gross topographic representation of visual space in the LGN. This finding is unexpected given the widely varying pattern of eye-specific layers both between and within the LGN's of the EPI-recovery animals.

These experiments showed that the development of lamination in the LGN reflects three processes: the segregation of retinogeniculate afferents, the patterning of those afferents into layers, and the emergence of cellular layers that correspond to the pattern of afferent layers. Preventing spontaneous retinal activity from P1-10 prevents the formation of normal afferent and cellular layers. This indicates that there is something noteworthy about the developmental time window in which retinogeniculate segregation normally occurs for proper patterning of afferent and cellular laminae in the LGN. It is possible that the pattern of spontaneous retinal activity present from P1-P10 provides an instructive cue for eye-specific segregation. Indeed, the pattern of spontaneous retinal activity is different from P1-P10, when eye-specific segregation normally results in eye-specific lamination, than it is from P10-P25 (Wong et al., 1993), when we show that eye-specific segregation results in eye-specific patches (Huberman et al., 2002). However, during the same developmental stage when retinogeniculate afferents segregate into eye-specific laminae, retinal activity also induces ganglion cell afferents to segregate into eye-specific clusters (not layers) in the rostral superior colliculus (Thompson and Holt, 1989). Also, in studies where retinal inputs destined for the LGN were rewired into the medial geniculate nucleus (MGN) axons from the two eyes segregate into eye-specific patches, not layers (Angelucci et al., 1997). Spontaneous retinal activity is normal throughout development in these cases and yet this did not produce normal lamination. These results collectively indicate that layer formation is controlled by cues which are intrinsic and unique to the LGN, rather than by patterns of retinal activity. Moreover, it is hard to imagine how activity could give rise to highly stereotyped eye-specific layers since activity is likely to differ across animals and yet eye specific layer in the LGN always form in the same positions as layers of essentially invariant size, shape, and orientation. Patterning of layers thus almost certainly relies on the presence of signals that bias the location and boundaries of the regions into which afferents from one or the other eye segregate.

Is the Pattern of Retinal Activity Relevant for Eye-Specific Segregation?

The segregation of initially intermingled inputs to the LGN has long been hypothesized to be in response to precise spatial and temporal patterns of spontaneous ganglion cell activity. Manipulations that eliminate all spontaneous retinal activity prevent the segregation of eye-specific inputs to the LGN (Penn et al., 1998; Huberman et al., 2002), and altering the balance of retinal activity between the two eyes leads to an increase in the size of the terminal field arising from the more active eye, at the expense of the less active eye (Penn et al., 1998; Stellwagen and Shatz, 2002). However, in every experiment where spontaneous retinal activity has been blocked, all retinal activity was abolished (Penn et al., 1998; Huberman et al., 2002; Stellwagen and Shatz, 2002), and in the one experiment where retinal activity was elevated rather than eliminated (Stellwagen and Shatz, 2002), correlated ganglion cell activity was maintained. Thus, while the relative level of activity in the two eyes is important for normal retinogeniculate development, whether the normal spatio-temporal pattern (waves) of retinal activity are necessary for eye-specific segregation remained unknown.

We directly tested the idea that correlated firing of retinal ganglion cells drives eye-specific segregation. (Huberman et al., 2003). As mentioned above, during the period of eye-specific segregation, spontaneous retinal activity is driven by acetylcholine released from starburst amacrine cells (Feller et al., 1996). Therefore we injected an immunotoxin that rapidly depletes starburst amacrine cells (Gunhan et al., 2002) into the eye of postnatal day 0 (P0) ferrets. Single cell patch-clamp recordings indicated that 93% of recorded ganglion cells in control P2-P9 ferret retinae confined their spontaneous activity to periodic bursts whereas only 23% of toxin-treated cells showed periodic bursting activity, and the frequency of these bursting events was significantly reduced relative to controls. Importantly, despite the marked perturbation in ganglion cell activity patterns caused by starburst amacrine cell depletion, the mean firing rate of ganglion cells in the two treatment groups was not significantly different. To assess the correlational structure of spontaneous ganglion cell activity, we carried out dual patch-clamp recordings from neighboring ganglion cells in control and toxin-treated P2-P9 retinae. In control retinae, the spontaneous spiking activity and membrane potential changes of neighboring ganglion cells were significantly correlated, whereas the spiking and membrane potential activity of ganglion cell pairs from toxin-treated retinae were not. Cross correlation analysis confirmed that, for all the toxin-treated pairs in which both cells spiked, their spiking activity was not significantly correlated, but instead showed a distribution similar to a random spike shuffle. For the toxin treated pairs in which only one ganglion cell spiked, the membrane fluctuations of the non-spiking cell were not visibly correlated with the activity of the neighboring cell. (Figure 3)

Does disrupting the correlated firing of ganglion cells alter eye-specific segregation in the LGN? To address this we examined the pattern of retinogeniculate connections in P10 ferrets that had received toxin injections on P0. In every case, retinal projections in these animals were indistinguishable from those observed in control P10 animals: there was a clear gap in the contralateral projection that was filled by the more circumscribed projection from the ipsilateral eye. Thus, while numerous experiments have shown that blocking spontaneous activity can prevent the formation of eye-specific retinogeniculate connections (Sretavan et al., 1988; Shatz and Stryker, 1988; Penn et al., 1998; Huberman et al., 2002; Stellwagen and Shatz, 2002; Rossi et al., 2001), we showed that if the normal patterns of spontaneous activity in individual and neighboring ganglion cells are disrupted, axons from the two eyes still segregate into non-overlapping layers in the LGN. The results indicate that the presence, but not the pattern of spontaneous ganglion cell discharges, is important for eye-specific retinogeniculate segregation.

Axon Guidance Cues and Eye-specific Segregation

The above described results (Huberman et al., 2002; 2003) corroborated previous hypothesis (Williams et al., 1994; Crowley and Katz, 1999; Chapman, 2000) that molecular cues direct sorting of binocular inputs into their stereotyped pattern of eye-specific layers in the LGN. What sort of axon guidance cues might contribute to eye-specific development? The results of Williams et al., (1994) clearly showed that the nasal

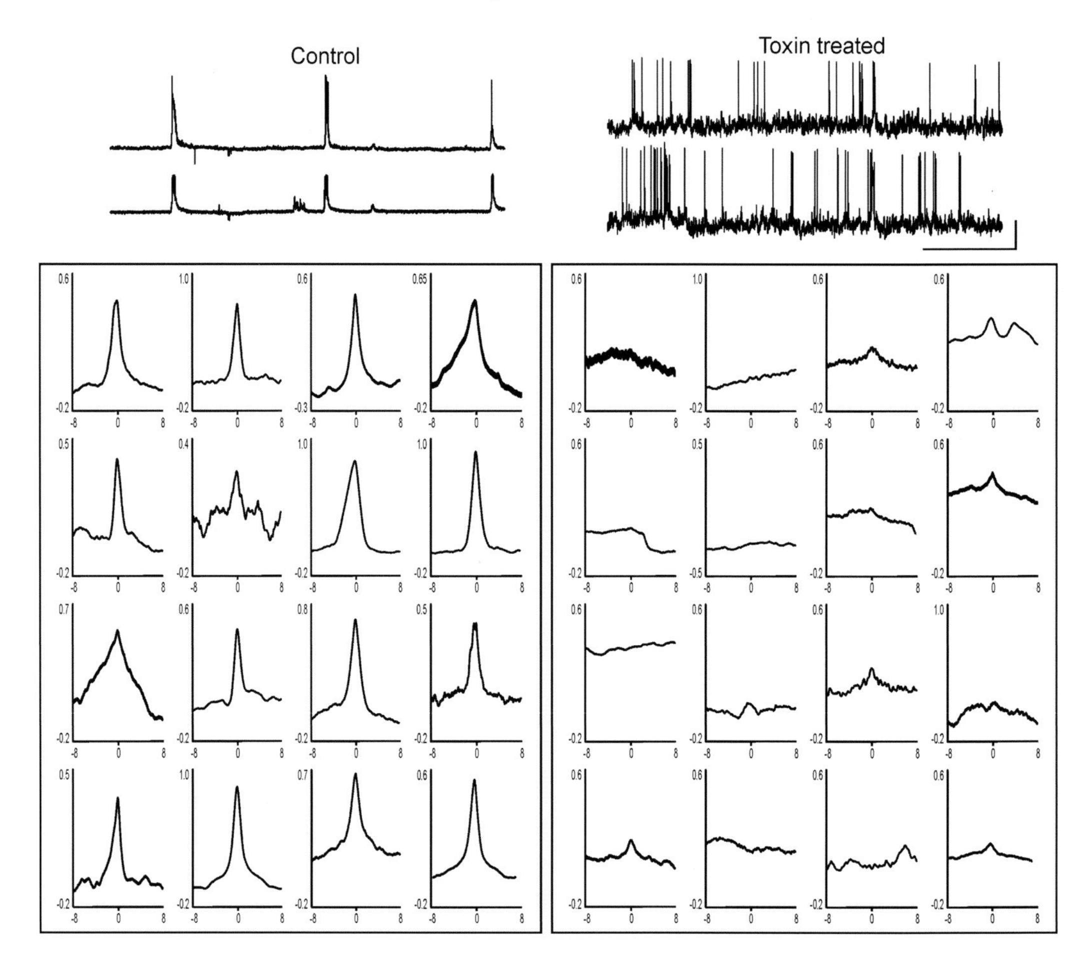

Figure 3 Paired patch clamp recordings from two neighboring retinal ganglion cells in a control ferret retina and ferret retina treated with immunotoxin to deplete cholinergic amacrine cells. In the control ganglion cell pair, firing of the two cells is highly correlated whereas in the toxin treated pair, activity does not appear correlated. Cross correlation plots of membrane potential indicate that for all 15 pairs recorded in control retinae, activity of neighboring cells is highly correlated (bottom right plot represents the mean correlation of all 15 pairs).In toxin treated pairs, none of the pairs showed significant correlations in activity (bottom right plot represents mean of all toxin pairs). Horizontal scale = 1 min. Vertical scale = 20mV.

versus temporal distinction is the essential parameter for segregating retinal inputs into A and A1 layers in the LGN. A good candidate axon guidance cue that could mediate this process were the ephrin-As because ephrin-As and their receptors (EphAs) are known to regulate topographic mapping of the nasal-temporal retina in the SC and LGN of lower vertebrates and mice (Cheng et al., 1995; Drescher et al., 1995; Nakamoto et al., 1996; Frisen et al, 1998; Feldheim et al., 1998; 2000; Feldheim, 2004). In those species, EphAs are distributed in a gradient that peaks at the far temporal retina and reaches lowest density at the far nasal retina, with matching, complementary gradients of ephrin-As distributed along the anterior-posterior axis of the target SC (Cheng et al., 1995; Drescher et al., 1995; Nakamoto et al., 1996; Frisen et al, 1998; Feldheim et al., 2000; Feldheim et al., 2004) and LGN (Feldheim et al., 1998). There are obvious differences between retinotopic maps (which are smooth and continuous) and eye-specific LGN layers (which have abrupt borders). However, the known role of ephrin-As on nasal-temporal mapping as well as on inter- and intra-areal pathfinding in other projection systems (Dufour et al., 2003) lead us to hypothesize that in species where eye-specific layers obey the nasal- versus temporal-retina distinction, ephrin-As would mediate patterning of eye-specific layers.

Using affinity-probe binding of EphA5-Alkaline Phosphatase (EphA5-AP) to detect ephrin-A proteins (Flanagan et al., 2000) and *in situ* hybridization for ephrin-A5 mRNA we examined the distribution of ephrin-As in the ferret LGN and observed the presence of an outer > inner gradient of ephrin-A5 mRNA. We also examined the pattern of EphAs and ephrinAs in the developing ferret retina and observed a central greater than peripheral (central > peripheral) gradient of EphAs proteins and mRNA and a nasal > temporal gradient of ephrin-A5 mRNA within the retinal ganglion cell layer. This nasal > temporal ephrin-A5 expression indicates that the central > peripheral retina expression of EphA5 receptor is not simply a consequence of relatively lower ganglion cell densities in the peripheral versus central retina.

Since ephrin-As have consistently been shown to be repellant toward ganglion cell axons expressing relatively higher levels of EphAs (Cheng et al., 1995; Drescher et al., 1995; Nakamoto et al., 1996; Frisen et al, 1998; Feldheim et al., 1998; 2000; 2004; Brown et al., 2000), the outer > inner gradient of ephrin-As in the LGN, combined with the central > peripheral gradient of EphAs in the retina therefore leads to a scenario whereby there are relatively higher levels of EphAs expressed in the crossed versus uncrossed ganglion cell axons that converge on a single line of projection. This could explain why the contralateral-eye layer (layer A) always maps to the inner LGN whereas the ipsilateral-eye layer (layer A1) maps to more outer LGN. For instance, consider a ferret viewing the head-neck junction of a white ferret in its left visual field; ganglion cells in the right temporal retina (orange arrow #1) view this location in the visual field and express relatively lower levels of EphAs compared to ganglion cells in the left retina (orange arrow #2) that view this same location in the visual field. These ganglion cell populations both project to the right LGN, and to the same line of projection, but the ipsilateral-eye ganglion cells maps to the more outer LGN (which contains higher concentrations of ephrin-As) than the contralateral-eye axons, which map to the more inner LGN (which contains lower concentrations of ephrin-As). (Figure 4)

To test directly whether ephrin-A:EphAs regulate eye-specific layer formation, we developed an *in vivo* retinal electroporation strategy to overexpress cDNA plasmids in ganglion cells of postnatal ferrets (Huberman et al., 2005b). Preliminary control experiments showed that

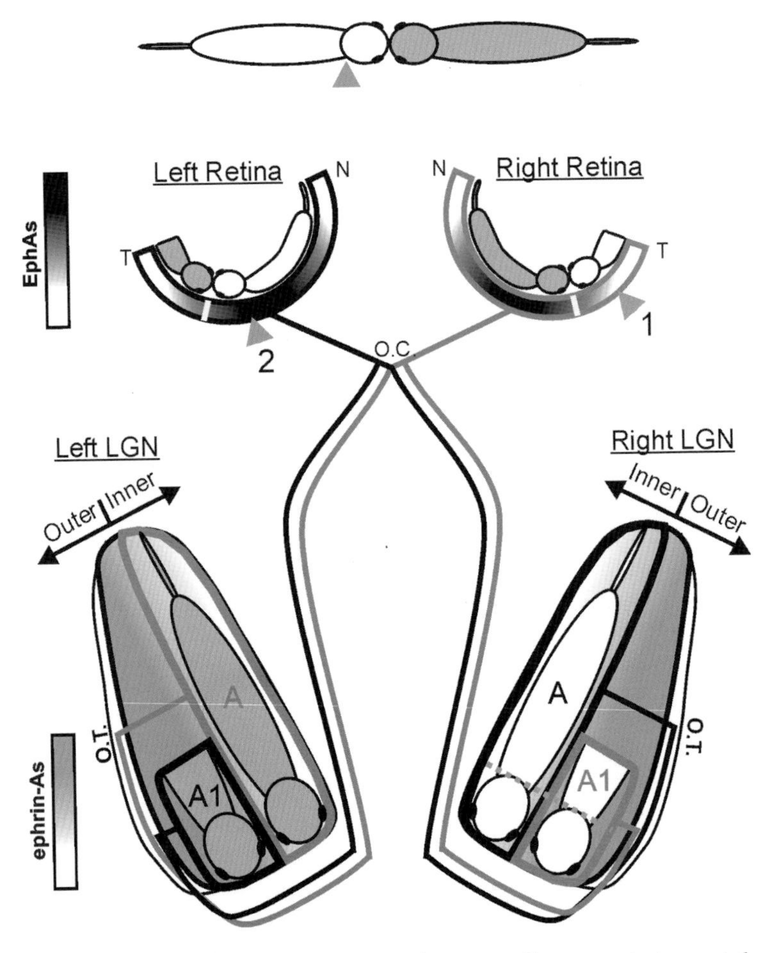

Figure 4 Schematic of eye-specific pathfinding according to central > peripheral expression of EphA receptor in the retina and outer > inner gradient of ephrin-A ligand in the LGN of the ferret.

our electroporation protocol resulted in widespread gene expression in the ganglion cell layer of the retina within 24–48 hours, that a very large percentage of ganglion cells expressed GFP and importantly, that the axons of ganglion cells transfected with GFP or other control plasmids were targeted normally in the P10 LGN. By contrast, ferrets that were electroporated with EphA3 or EphA5 cDNAs on P1, and then had their retinogeniculate projections traced on P10 showed markedly perturbed retinogeniculate projections. Axons from the ipsilateral eye were displaced to the inner LGN and into territory dominated by the contralateral-eye. This resulted in ipsilateral-eye input to the LGN that was significantly expanded along the axis perpendicular to eye specific layers (i.e., along lines of projection). In addition, axons from the contralateral-eye were found in the region of the P10 LGN normally only occupied by axons from the ipsilateral-eye (Figure 5). Thus,

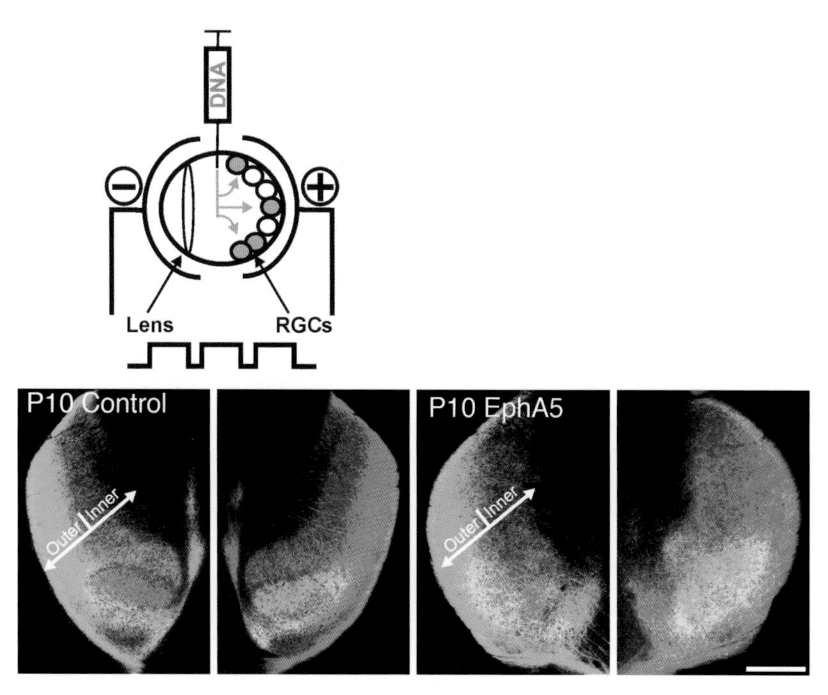

Figure 5 In vivo retinal electroporation strategy: A DNA expression plasmid is injected into the vitreal chamber of the eye and the whole eye is subjected to a series of square wave pulses using tweezer electrodes. Transient pores are created in the ganglion cell membranes, allowing DNA to access the nucleus where the construct is transcribed. Retinogeniculate projections to the LGN of a control electroporated P10 ferret and a P10 ferret electroporated with EphA5. In the EphA5 overexpressing ferret, axons from the ipsilateral eye misproject to the inner LGN, which normally only receives axons from the contralateral eye. Scale = μm. For additional details on electroporation technique see Huberman et al., (2005b).

whereas in normal and control electroporated P10 ferrets, retinogeniculate inputs are segregated into highly stereotyped eye-specific layers by P10 (Linden et al., 1981; Penn et al., 1998; Chapman 2000; Huberman et al., 2002; 2003) there was significant intermingling of contralateral- and ipsilateral-eye axons observed in the LGN of EphA3/5 electroporated P10 ferrets (Huberman et al., 2005b). The phenotype induced by EphA3/5 overexpression was almost as severe as the most extreme cases observed following complete retinal activity blockade from P1–P10 (see: Penn et al., 1998; Huberman et al., 2002). However, whereas retinal activity blockade from P0-P10 results in the maintenance of the immature pattern of binocular inputs to the LGN wherein axons from both eyes extend both anteriorly and across the outer-inner axis of the LGN (Penn et al., 1998; Huberman et al., 2002; 2003) disruption EphAs through electroporation-induced overexpression misdirected ganglion cell axons primarily along the outer-inner axis of the LGN, perpendicular to eye-specific layers. We also traced single retinogeniculate axons.

In every control ferret examined, ganglion cell axons from temporal portion of the ipsilateral retina were restricted to layer A1 within the ipsilateral LGN, as expected for animals of this age (Hahm et al., 1999). By contrast, ganglion cell axons labeled from the temporal retina of EphA5 electroporated ferrets at both ages extended much further along across the outer-inner axis of LGN than was observed in controls.

By examining the time-course of ephrin-A and EphA expression in normal ferrets we found that, whereas ephrin-A ligands are robustly expressed in the P0-P3 LGN, by P5 their levels are reduced conspicuously. Our previous work in EPI-recovery ferrets showed that normal development of stereotyped layers in the LGN is restricted to the early postnatal period, when eye-specific segregation normally occurs (Huberman et al., 2002), suggesting there may be a critical period for eye-specific layer formation. To test if the presence of ephrin-As in the LGN contributes to this critical period, we electroporated ferrets with EphA3/5 at P5. Remarkably, despite the robust overexpression induced by retinal electroporation at P5, this had no detectable effect on patterning of eye-specific retinogeniculate inputs.

These results (Huberman et al., 2005b) and those of an accompanying paper in mouse (Pfiffenberger et al., 2005) represent the first evidence for axon guidance cue-based targeting of eye-specific projections. Of course, ephrin-As likely represent one of several (and perhaps many) cues that nasal and temporal retinal axons rely on to find to their stereotyped locations in the LGN. Our ephrin-A model does not address eye-specific segregation in the C layers because eye-specificity within the C layers is not well established anatomically until ~P20. Mapping of ganglion cell inputs in the C-layers may rely on ephrins and/or non-ephrin-A cues. Microarray-based genetic screens have begun to reveal the presence of cues that distinguish between the C versus A/A1 layers of the LGN (Kawasaki et al., 2004), but these same screens did not identify eye-specific markers in the LGN. Kawasaki et al., thus suggested that the molecules that regulate eye-specific segregation may be expressed in gradients in the LGN.Our findings (Huberman et al., 2005b) directly support this hypothesis.

The issue of whether specific patterns of retinal activity (i.e., "waves") are required for eye-specific segregation in the LGN remains debated (reviewed recently in: Chalupa and Huberman, 2004; Grubb and Thompson, 2004). Torborg and Feller recently reported that eye-specific LGN segregation is absent in the β2 nAChR KO mouse, which lacks retinal waves (Torborg et al., 2005). The absence of eye-specific segregation in the LGN of P8 β2 nAChR mice is perhaps not surprising given the topographic errors present in this transgenic (McLaughlin et al., 2003; Grubb et al., 2002). In any event, it does directly contrast with our results in immunotoxin treated ferrets (Huberman et al., 2003). One likely explanation for this discepancy is a genuine species difference. Notably, in mice, ganglion cell projections to the LGN are segregated into eye-specific regions, but they are somewhat variable in shape and the LGN of rodents lacks eye-specific cellular lamination. Also, eye-specific layers in the mouse LGN are not nasal- or temporal-retina specific; each LGN receives ganglion cell axons from the entire contralateral retina (Hererra et al., 2003). Thus neural activity may play more important roles than axon guidance cues in patterning eye-specific inputs to the rodent LGN. By contrast, the pattern of eye-specific inputs to the carnivore and primate LGN is highly stereotyped and robust eye-specific cellular layers are always present. Therefore, in carnivores and primates axon guidance cues such as ephrin-As may be critical for eye-specific segregation and layer formation- whereas retinal activity may play less important roles. Indeed, the use of modern, high sensitivity tracers revealed that in the macaque visual system (which is highly similar to that of humans), eye-specific segregation occurs prior to the onset of synapse elimination in the LGN, suggesting that Hebbian plasticity does not drive segregation in this species (Huberman et al., 2005a). Regardless of species differences, it is worth considering that the β2 nAChR KO mouse is a brainwide null mutation and thus, it is unclear whether the LGN phenotype seen in the β2 nAChR KO is due directly to disruptions in retinal activity patterns or lack of β2 nAChR in the target LGN. Indeed, β2 nAChR receptors are robustly expressed in the developing and mature LGN and SC (Hill et al., 1993; Zoli et al., 1995) and cholinergic receptors on LGN neurons are critical for their firing in response to retinal input (Uhlrich et al., 1995). The studies of McLaughlin et al. (2003), Grubb et al. (2003) and Torborg et al., (2005) all cite data reported in Penn et al., (1998) that intracranial application of EPI did not alter eye-specific segregation in the ferret LGN as evidence that the phenotypes seen in the β2 AChR KO mouse are likely due to defects in patterned retinal activity and not lack of postsynaptic responsiveness. However, we have repeatedly observed that intracranial EPI injections from P1-P10 completely mimic the effects of intraocular P1-P10 EPI treatment. The source of the discrepancy between our results and those of Penn et al., (1998) are unknown. Regardless, caution should be exercised when interpreting the phenotypes observed in the β2 nAChR KO mouse (and for that matter, any phenotype observed in a brainwide null mutation transgenic). Only through the use of retinal- or LGN-specific knockouts can causal statements about the role of retinal activity patterns in eye-specific segregation be drawn.

Blocking Activity or Forcing it to Change?

There is another inherent problem with experiments involving activity perturbations. Namely, because of the difficulties associated with recording spontaneous activity *in vivo* (anesthetic shut down all spontaneous retinal activity), the field as a whole has had to assume that *in vivo* chronic application of EPI or APB or transgenic KO of receptors in fact blocks/alters activity for the full duration of the treatment. This may be a flawed assumption. In fact, the few groups that have succeeded in recording spontaneous activity *in vivo* report that removing afferent input can dramatically *increase* activity in the de-afferented target (Weliky and Katz, 1999) and recently, Demas et al., (FASEB Abstracts, 2004) examined *in vitro* spontaneous activity in retinae from a transgenic mouse in which of choline acetyltransferase (ChAT)- the rate-limiting enzyme for ACh synthesis was knocked out selectively in the retina. The logical prediction is that, because P1-P10 represents the cholinergic phase of spontaneous retinal activity, these KO mice would exhibit no ganglion cell spiking at this stage. However, Demas et al., 2004 observed a complete rescue of correlated ganglion cell firing by P5 and this activity was driven by non-cholinergic sources. Others too recently reported that in the rabbit retina, blocking ACh-mediated waves silences activity for short time but then soon after, causes a novel form of gap-junction mediated waves to emerge (Syed et al., 2004). So, while there is no evidence that genetic or pharmacologic alteration of spontaneous activity causes a compensatory rescue of activity levels or patterns *in vivo*, one should exercise caution in interpreting the results of studies that employed "chronic activity blockades" (Shatz and Stryker, 1988; Penn et al., 1998; Cook et al., 1999; Chapman et al., 2000; Rossi et al., 2001; Stellwagen and Shatz, 2002; Huberman et al., 2002; 2003). At the very least, one would be wise to assess the effects of chronic drug application *in vitro*. Regardless, to better understand the role of activity in eye-specific circuit development, there is motivation to develop strategies for chronically recording spontaneous neural activity *in vivo*.

Activity Versus Molecules?! A False Dichotomy:

There are serious challenges in assigning a mutually exclusive role for axon guidance cues versus neural activity in eye-specific segregation and layer formation. Here, we tried to emphasize that both these factors are likely to be important, albeit for different aspects of retinogeniculate development. In addition, we tried to emphasize that the mechanisms that drive eye-specific segregation and patterning may differ according to species. Indeed in mouse, both axon guidance cues and neural activity have been shown to act in parallel to induce eye-specific LGN layers (Pfieffenberger et al., 2005) whereas in carnivores and primates, axon guidance cues appear to be more important than patterned activity (Huberman et al., 2005b). In all species, however, the bulk of evidence points to a model whereby the segregation of axons from the two eyes relies, at least in part, on spontaneous retinal activity whereas the positioning of eye-specific projections into their stereotyped layered pattern relies on activity-independent (axon guidance) cues such as ephrin-A:EphAs (Huberman et al., 2005b; Pfieffenberger et al., 2005). Do neural activity and axon guidance cues interact directly to influence eye-specific pathfinding? Hanson and Landmesser (2004) recently showed that altering spontaneous activity patterns perturbs expression of axon guidance cues and, as a result, axon pathfinding in the limb bud. However, this is not seen in the retingeniculate system. EPI injections do not alter ephrin-A or EphA mRNA expression (Pfieffenberger et al., 2005). However, the field of eye-specific circuit development is in desperate need of studies that combine electrophysiology and pharmacology with molecular biology and genetics, to identify the relative contributions of neural activity and axon guidance cues.

References

Akerman CJ, Smyth D, and Thompson ID (2002) Visual experience before eye-opening and the development of the retinogeniculate pathway. *Neuron* 36: 869–79.

Akerman CJ, Tolhurst DJ, Morgan JE, Baker GE, and Thompson ID (2003) Relay of visual information to the lateral geniculate nucleus and the visual cortex in albino ferrets. J Comp Neurol. 461:217–35.

Angelucci A, Clasca F, Bricolo E, Cramer KS, and Sur M (1997) Experimentally induced retinal projections to the ferret auditory thalamus: development of clustered eye-specific projections in a novel target. *J Neurosci.* 17:2040–2055.

Bansal A, Singer JH, Hwang BJ, Xu W, Beudet A, and Feller MB (2000) Mice lacking specific nicotinic acetylcholine receptor subunits exhibit dramatically altered spontaneous activity patterns and reveal a limited role for retinal waves in forming ON and OFF circuits in the inner retina. *J Neurosci.* 20:7672–81.

Brown TH, Chapman PF, Kairiss EW, and Keenan CL (1988) Long-term synaptic potentiation. *Science* 242:724–8.

Brown A, Yates PA, Burrola, P, Ortuno D, Vaidya A, Jessell TM, Pfaff SL, O'Leary DD, and Lemke G (2000) Topographic mapping from the retina to the midbrain is controlled by relative but not absolute levels of EphA receptor signaling. *Cell* 102:77–88.

Brunso-Bechtold JK, and Casagrande VA (1981) Effect of bilateral enucleation on the development of layers in the dorsal lateral geniculate nucleus. *Neuroscience* 2:589–597.

Casagrande VA, and Condo GJ (1988) The effect of altered neuronal activity on the development of layers in the lateral geniculate nucleus. *J Neurosci.* 8: 395–416.

Chalupa LM, and Huberman (2004) New Perspectives on the Role of Neural Activity in Development of the Visual System. In: The Newest Cognitive Neurosciences, 3rd edition. MIT Press.

Chalupa LM, and Williams RW (1984) Organization of the cat's lateral geniculate nucleus following interruption of prenatal binocular competition. *Hum Neurobiol.* 3:103–7.

Chapman B (2000) Necessity for afferent activity to maintain eye-specific segregation in ferret lateral geniculate nucleus. *Science* 287:2479–2482.

Chapman B, and Godecke I (2000) Cortical cell orientation selectivity fails to develop in the absence of ON-center ganglion cell activity. *J Neurosci.* 20:1922–1930.

Cheng HJ, Nakamoto M, Bergemann AD, and Flanagan JG (1995) Complementary gradients in expression and binding of ELF-1 and Mek4 in development of the topographic retinotectal projection map. *Cell* 82:371–381.

Cline HT (1987) N-methyl-D-aspartate receptor antagonist desegregates eye-specific stripes. *Proc Natl Acad Sci USA* 84:4342–5.

Cook PM, Prusky G, and Ramoa AS (1999) The role of spontaneous retinal activity before eye opening in the maturation of form and function in the retinogeniculate pathway of the ferret. *Vis Neurosci.* 16:491–501.
Constantine-Paton M, and Law MI (1978) Eye-specific termination bands in tecta of three-eyed frogs. *Science* 202:639–41.
Crowley JC, and Katz LC (1999) Development of ocular dominance columns in the absence of retinal input. *Nat Neurosci.* 2:1125–30.
Crowley JC, and Katz LC (2000) Early development of ocular dominance columns. *Science* 290:1321–4.
Cucchiaro J, and Guillery, RW (1984) The development of the retinogeniculate pathways in normal and albino ferrets. *Proc R Soc Lond* 223:141–164.
Demas J, Eglen SJ, and Wong RO (2003) Developmental loss of synchronous spontaneous activity in the mouse retina is independent of visual experience. *J Neurosci.* 23:2851–60.
Demas J, Stacey R, Sanes JR, and Wong RO (2004) FASEB Summer Research Conferences Retinal Neurobiology and Visual Processing. Saxtons River, VT.
Drescher U, Kremoser C, Handwerker C, Loschinger J, Noda M, and Bonhoeffer F (1995) *In vitro* guidance of retinal ganglion cell axons by RAGS, a 25 kDa tectal protein related to ligands for Eph receptor tyrosine kinases. *Cell* 82:359–370.
Dufour A, Seibt J, Passante L, Depaepe V, Ciossek T, Frisen J, Kullander K, Flanagan JG, Polleux F, and Vanderhaeghen P (2003) Area specificity and topography of thalamocortical projections are controlled by ephrin/Eph genes. *Neuron* 39:453–465.
Feldheim DA, Vanderhaeghen P, Hansen MJ, Frisen J, Lu Q, Barbacid M, and Flanagan JG (1998) Topographic guidance labels in a sensory projection to the forebrain. *Neuron* 21: 1303–1313.
Feldheim DA, Kim YI, Bergemann AD, Frisen J, Barbacid M, and Flanagan JG (2000) Genetic analysis of ephrin-A2 and ephrin-A5 shows their requirement in multiple aspects of retinocollicular mapping. *Neuron* 25:563–574.
Feldheim DA, Nakamoto M, Osterfiel M, Gale NW, DeChiara TM, Rohatgi R, Yancopoulos GD, and Flanagan JG (2004) Loss-of-function analysis of EphA receptors in retinotectal mapping. *J Neurosci.* 24:2542–50.
Feller MB, Wellis DP, Stellwagen D, Werblin F, and Shatz CJ (1996) Requirement for cholinergic synaptic transmission in the propogation of spontaneous retinal waves. *Science* 272:1182–1187.
Feller MB (1999) Spontaneous correlated activity in developing neural circuits. *Neuron* 22:653–6.
Flanagan JG, and Vanderhaeghen P (1998) The ephrins and Eph receptors in neural development. *Annu Rev Neurosci.* 21:309–345.
Flanagan JG, Cheng HJ, Feldheim DA, Hattori M, Lu Q, and Vanderhaeghen P (2000) Alkaline phosphatase fusions of ligands or receptors as in situ probes for staining of cells, tissues, and embryos. *Methods Enzymol.* 327:19–35.
Frisen J, Yates PA, McLaughlin T, Friedman GC, O'Leary DD, and Barbacid M (1998) Ephrin-A5 (AL-1/RAGS) is essential for proper retinal axon guidance and topographic mapping in the mammalian visual system. *Neuron* 20:235–243.
Galli L, and Maffei L (1988) Spontaneous impulse activity of rat retinal ganglion cells in prenatal life. *Science* 242:90–1.
Garraghty PE, Shatz CJ, and Sur M (1988a) Prenatal disruption of binocular interactions creates novel lamination in the cat's lateral geniculate nucleus. *Vis Neurosci.* 1:93–102.

Godement P, Salaun J, and Imbert M (1984) Prenatal and postnatal development of retinogeniculate and retinocollicular projections in the mouse. *J Comp Neurol.* 230:552–75.
Grubb MS, Rossi FM, Changuex JP, and Thompson ID (2003) Abnormal functional organization in the dorsal lateral geniculate nucleus of mice lacking the beta 2 subunit of the nicotinic acetylcholine receptor. *Neuron* 40:1161–72.
Grubb MS, and Thompson ID (2004) The influence of early experience on the development of sensory systems. *Curr Opin Neurobiol.* 14:503–12.
Guillery RW (1969) An abnormal retinogeniculate projection in Siamese cats. *Brain Res* 14:739–741.
Guillery RW (1971) An abnormal retinogeniculate projection in the albino ferret (Mustela furo). *Brain Res* 33:482–485.
Guillery RW, and Kaas, JH (1971) A study of normal and congenitally abnormal retinogeniculate projections in cats. *J Comp Neurol.* 143:73–100.
Guillery RW, Scott GL, Cattanach BM, and Deol MS (1973) Genetic mechanisms determining the central visual pathways of mice. *Science* 179:1014–6.
Guillery RW, LaMantia AS, Robson JA, and Huang K (1985a) The influence of retinal afferents upon the development of layers in the dorsal lateral geniculate nucleus of mustelids. *J Neurosci.* 5:1370–1379.
Guillery RW, Ombrellaro M, and LaMantia AL (1985b) The organization of the lateral geniculate nucleus and of the geniculocortical pathway that develops without retinal afferents. *Brain Res* 352:221–233.
Gunhan E, Choudary PV, Landerholm TE, and Chalupa LM (2002) Depletion of cholinergic amacrine cells by a novel immunotoxin does not perturb the formation of segregated on and off cone bipolar cell projections. *J Neurosci.* 22:2265–73.
Hahm JO, Cramer KS, and Sur M (1999) Pattern formation by retinal afferents in the ferret lateral geniculate nucleus: developmental segregation and the role of N-methyl-D aspartate receptors. *J Comp Neurol.* 411:327–345.
Hanson MG, and Landmesser LT (2004) Normal patterns of spontaneous activity are required for correct motor axon guidance and the expression of specific guidance molecules. *Neuron* 43:687–701.
Hebb DO (1949) Organization of Behavior: A Neuropsychological Theory (New York: John Wiley and sons).
Hererra E, Brown L, Aruga J, Rachel RA, Dolen G, Mikoshiba K, Brown S, and Mason CA (2003) Zic2 patterns binocular vision by specifying the uncrossed retinal projection. *Cell* 114:545–57.
Hill JA Jr, Zoli M, Bourgeois JP, and Changeux JP (1993) Immunocytochemical localization of a neuronal nicotinic receptor: the beta 2-subunit. *J Neurosci.* 13:1551–68.
Huberman AD, Stellwagen D, and Chapman B (2002) Decoupling eye-specific segregation from lamination in the lateral geniculate nucleus. *J Neurosci.* 22:9419–29.
Huberman AD, Wang GY, Liets LC, Collins OA, Chapman B, and Chalupa LM (2003) Eye-specific retinogeniculate segregation independent of normal neuronal activity. *Science* 300:994–8.
Huberman AD, Dehay C, Berland M, Chalupa LM, and Kennedy H (2005a) Early and rapid targeting of eye-specific axonal projections to the lateral geniculate nucleus in the fetal macaque. *J Neurosci.* 25:4014–4023.
Huberman AD, Murray KD, Warland DK, Feldheim DA, and Chapman B (2005b) Ephrin-As mediate targeting of eye-specific projections to the lateral geniculate nucleus. *Nature Neuroscience*, 8:1013–21.

Hutchins JB, and Casagrande VA (1990) Development of the lateral geniculate nucleus: interactions between retinal afferent, cytoarchitectonic, and glial cell process lamination in ferrets and tree shrews. *J Comp Neurol.* 298:113–128.

Jeffrey G (1984) Retinal ganglion cell death and terminal field retraction in the developing rodent visual system. *Brain Res.* 315:81–96.

Johnson PT, Williams RR, Cusato K, and Reese BE (2000) Rods and cones project to the inner plexiform layer during development. *J Comp Neurol.* 414:1–2.

Jones EG (1985) The Thalamus. (New York: Plenum Press).

Kawasaki H, Crowley JC, Livesey FJ, and Katz LC (2003) Molecular correlates of ocular dominance column formation in ferret thalamus and cortex. *J Neurosci.* 24:9962–70.

Linden DC, Guillery RW, and Cucchiaro J (1981) The dorsal lateral geniculate nucleus of the normal ferret and its postnatal development. *J Comp Neurol.* 203:189–211.

Lund RD, Cunningham TJ, and Lund JS (1973) Modified optic projections after unilateral eye removal in young rats. *Brain Behav Evol* 8:51–72.

Maffei L, and Galli-Resta L (1990) Correlation in the discharges of neighboring rat retinal ganglion cells during prenatal life. *Proc Natl Acad Sci USA* 87:2861–4.

McLaughlin T, Torborg CL, Feller MB, and O'Leary DD (2003) Retinotopic map refinement requires spontaneous retinal waves during a brief critical period of development. *Neuron* 40:1147–60.

Meister M, Wong RO, Baylor DA, and Shatz CJ (1991) Synchronous bursts of action potentials in ganglion cells of the developing mammalian retina. *Science* 252:939–43.

Morgan J, and Thompson ID (1993) The segregation of ON- and OFF-center responses in the ferret lateral geniculate nucleus of normal and monocularly enucleated ferrets. *Vis Neurosci.* 10:303–311.

Muir-Robinson G, Hwang BJ, and Feller MB (2002) Retinogeniculate axons undergo eye-specific segregation in the absence of eye-specific layers. *J Neurosci.* 22:5259–64.

Nakamoto M, Cheng HJ, Friedman GC, McLaughlin T, Hansen MJ, Yoon CH, O'Leary, DD, and Flanagan JG (1996) Topographically specific effects of ELF-1 on retinal axon guidance *in vitro* and retinal axon mapping *in vivo*. *Cell* 86:755–766.

Penn AA, Riquelme PA, Feller MB, Shatz CJ (1998) Competition in retinogeniculate patterning driven by spontaneous activity. *Science* 279:2108–2112.

Pfieffenberger C, Cutforth T, Woods G, Yamada J, Renteria RC, Copenhagen DR, Flanagan JG, Feldheim DA (2005) Ephrin-As and neural activity are required for eye-specific patterning during retinogeniculate mapping. *Nat Neurosci.* 8:1022–7.

Rakic P (1976) Prenatal genesis of connections subserving ocular dominance in the rhesus monkey. *Nature* 261:467–471.

Rakic P (1981) Development of visual centres in the primate brain depends on binocular competition before birth. *Science* 214:928–931.

Ramoa AS, and McCormick DA (1994) Enhanced activation of NMDA receptor responses at the immature retinogeniculate synapse. *J Neurosci.* 14:2098–105.

Rossi FM, Pizzorusso T, Porciatti V, Marubio LM, Maffei L, and Changuex JP (2001) Requirement of the nicotinic acetylcholine receptor β2 subunit for the anatomical and functional development of the visual system. *Proc Natl Acad USA* 98:6453–6458.

Shatz CJ (1983) The prenatal development of the cats retinogeniculate pathway. *J Neurosci.* 3:482–499.

Shatz CJ (1990). Competitive interactions between retinal ganglion cells during prenatal development. *J Neurobiol.* 21:197–211.

Shatz CJ (1996) Emergence of order in visual system development. *Proc Natl Acad Sci USA* 93:602–8.

Shatz CJ, and Stryker MP (1988) Prenatal tetrodotoxin infusion blocks segregation of retinogeniculate afferents. *Science* 242:87–89.

So KF, Schneider GE, and Frost DO (1978) Postnatal development of retinal projections to the lateral geniculate body in Syrian hamsters. *Brain Res.* 142:343–52.

Sretavan D, and Shatz CJ (1984) Prenatal development of individual retinogeniculate axons during the period of segregation. *Nature* 308:845–8.

Sretavan DW, and Shatz CJ (1986a) Prenatal development of retinal ganglion cell axons: segregation into eye-specific layers within the cat's lateral geniculate nucleus. *J Neurosci.* 6:234–51.

Sretavan DW, and Shatz CJ (1986b) Prenatal development of cat retinogeniculate axon arbors in the absence of binocular interactions. *J Neurosci.* 6:990–1003.

Sretavan DW, Shatz CJ, and Stryker MP (1988) Modification of retinal ganglion cell axon morphology by prenatal infusion of tetrodotoxin. *Nature* 336:468–71.

Stellwagen D, Shatz CJ, and Feller MB (1999) Dynamics of retinal waves are controlled by cyclic AMP. *Neuron* 24:673–685.

Stellwagen D, and Shatz CJ (2002) An instructive role for retinal waves in the development of retnogeniculate connectivity. *Neuron* 33:357–367.

Stryker MP, and Zahs KR (1983) On and off sublaminae in the lateral geniculate nucleus of the ferret. *J Neurosci.* 10:1943–1951.

Syed MM, Lee S, Zheng J, and Zhou ZJ (2004) Stage-dependent dynamics and modulation of spontaneous waves in the developing rabbit retina. *J Physiol.* 560:533–49.

Tavazoie SF, and Reid RC (2000) Diverse receptive fields in the lateral geniculate nucleus during thalamocortical development. *Nat Neurosci.* 3:608–16.

Thompson I, and Holt C (1989) Effects of intraocular tetrodotoxin on the development of the retinocollicular pathway in the Syrian hamster. *J Comp Neurol.* 282:371–388.

Torborg CL, Hanson KA, and Feller MB (2005) High frequency, synchronized bursting drives eye-specific segregation of retinogeniculate axons. *Nat Neurosci.* 8:72–8.

Uhlrich DJ, Tamamki N, Murphy PC, and Sherman SM (1994) Effects of brain stem parabrachial activation on receptive field properties of cells in the cat's lateral geniculate nucleus. *J Neurophysiol.* 73:2428–47.

Weliky M, and Katz LC (1999) Correlational structure of spontaneous neuronal activity in the developing lateral geniculate nucleus *in vivo*. *Science* 285:599–604.

Williams RW, Hogan D, and Garraghty PE (1994) Target recognition and visual maps in the thalamus of achiasmatic dogs. *Nature* 367:637–639.

Wong RO, Meister M, and Shatz CJ (1993) Transient period of correlated bursting activity during development of the mammalian retina. *Neuron* 11:923–938.

Wong RO (1999) Retinal waves and visual system development. *Annu Rev Neurosci.* 22: 29–47.

Wong RO (1999) Retinal waves: stirring up a storm. *Neuron* 24:493–5.

Wong WT, Myhr KL, Miller ED, and Wong RO (2000) Developmental changes in the neurotransmitter regulation of correlated spontaneous retinal activity. *J Neurosci.* 20:351–360.

Zahs KR, and Stryker MP (1985) The projection of the visual field onto the lateral geniculate nucleus of the ferret. *J Comp Neurol.* 241:210–224.

Zhou ZJ (1998) Direct participation of starburst amacrine cells in spontaneous rhythmic activities in the developing mammalian retina. *J Neurosci.* 18:4155–65.

Zoli M, Le Novere N, Hill JA Jr, and Changuex JP (1995) Developmental regulation of nicotinic ACh receptor subunit mRNAs in the rat central and peripheral nervous systems. *J Neurosci.* 15:1912–39.

15

LTD as a Mechanism for Map Plasticity in Rat Barrel Cortex

Kevin J. Bender, Suvarna Deshmukh, and Daniel E. Feldman

LTP and LTD as Cellular Mechanisms for Sensory Map Plasticity

The ability of neural circuits to adapt to new experiences and to store information about the environment is central to brain development and learning. An important paradigm for studying this adaptive ability is sensory map plasticity, in which sensory and motor maps are modified based on recent experience, including training on learning tasks. Map plasticity occurs with highly similar functional properties across many brain areas, including primary visual, auditory, somatosensory, and motor cortex (Wiesel and Hubel, 1963; Buonomano and Merzenich, 1998; Sanes and Donoghue, 2000). However, the cellular and synaptic mechanisms that mediate map plasticity are only beginning to be understood.

Long-term potentiation (LTP) and depression (LTD) of cortical synapses emerged as prominent candidate mechanisms for cortical map plasticity relatively soon after the discovery of ocular dominance plasticity in the visual cortex (Stent, 1973; Bear et al., 1987). These mechanisms instantiate Hebbian synaptic plasticity, which can explain many features of cortical map plasticity (Hebb, 1949; Bear et al., 1987; Buonomano and Merzenich, 1998). LTP and LTD are generally hypothesized to mediate rapid components of map plasticity, while anatomical changes that often occur during map plasticity may mediate slower components.

LTD has been hypothesized to play two major roles in map development and plasticity. First, during developmental refinement of topographic projections, LTD is thought to act to weaken aberrant synapses according to Hebbian learning rules, perhaps leading ultimately to synapse elimination (Stent, 1973; Buonomano and Merzenich, 1998). Second, even after maps have formed, patterns of sensory use and disuse powerfully regulate map topography. During this phase, LTD is thought to be involved in weakening excitatory synapses that are underused or behaviorally irrelevant, thus reducing the representation of these inputs in cortical maps (Bear et al., 1987; Singer, 1995; Buonomano and Merzenich, 1998; Ruthazer and Cline, 2004). Though the capacity for LTD may decline somewhat with age, recent studies have clearly demonstrated LTD in adults, indicating that it may contribute to both developmental and adult plasticity (Heynen et al., 1996; Manahan-Vaughan and Braunewell, 1999).

Though LTD has long been hypothesized to contribute to sensory cortical map plasticity, and despite strong evidence for LTD being involved in cerebellar learning (Boyden et al., 2004), direct evidence for LTD in cortical map plasticity was lacking until recently. In this chapter, we review recent evidence that LTD is involved in plasticity of the whisker map in rat primary somatosensory cortex (S1). In S1, both *in vivo* and *in vitro* techniques have been used to provide insight into the locus of LTD during plasticity and the induction mechanisms that drive LTD in response to altered experience. This evidence indicates that LTD is a major mechanism for a common feature of cortical map plasticity, the reduction in cortical responsiveness to deprived sensory inputs.

Map Plasticity in Barrel Cortex

In the rat primary somatosensory cortex, the ~30 large facial whiskers are represented by clusters of cells in cortical layer 4 (L4) called barrels. Barrels are arranged in a map isomorphic with the whiskers on the rat's snout (Woolsey and Van der Loos, 1970; Welker and Woolsey, 1974), and neurons in each barrel are driven best by deflection of a single whisker, termed the principal whisker, which corresponds to the identity of the barrel within the map. Excitatory cells in each L4 barrel make a dense, columnar projection onto layer 2/3 (L2/3) neurons in the cortical column surrounding that barrel, termed the barrel column (Petersen and Sakmann, 2001; Feldmeyer et al., 2002). The vast majority of neurons in each barrel column are driven most strongly by the anatomically appropriate principal whisker, and only weakly by neighboring, surround whiskers (Simons, 1978; Keller, 1995). Thus, an orderly map of whisker receptive fields is present across S1, and the barrels in L4 provide an anatomical reference for this functional whisker map.

The whisker receptive field map in S1 is modifiable by sensory experience. If a whisker is plucked or trimmed for several days or weeks in adolescent animals (7 to ~60 days of age), receptive fields of L2/3 cells within the corresponding column change in two ways. First, L2/3 neurons within the deprived column lose responses to the deprived principal whisker, a phenomenon called principal whisker response depression (PWRD). Second, neurons begin to respond more strongly to neighboring, spared whiskers, termed spared whisker response potentiation (SWRP). These two components of plasticity can be separated genetically and developmentally, indicating that they represent two independent mechanisms for plasticity in S1 (Glazewski and Fox, 1996; Glazewski et al., 2000). Together, PWRD and SWRP cause receptive fields in deprived columns to become dominated by neighboring,

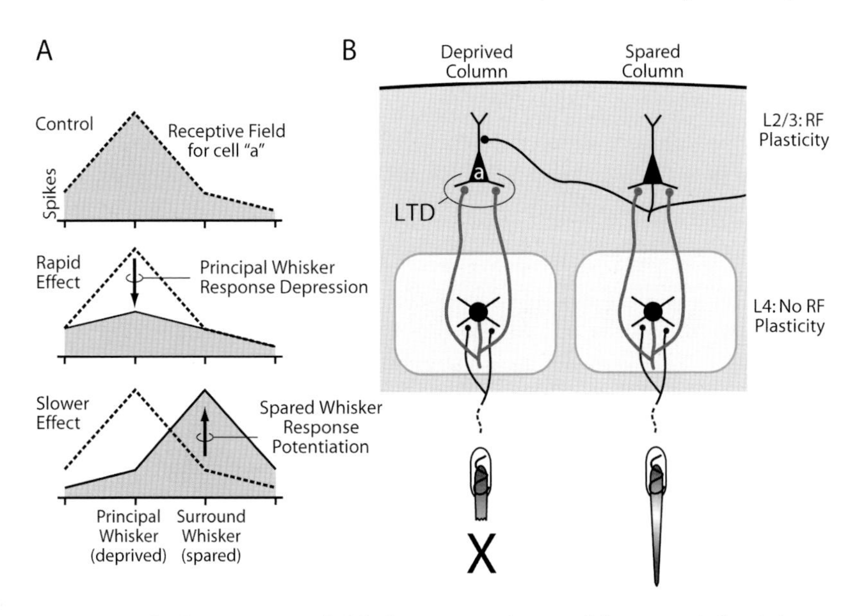

Figure 1 Whisker receptive field plasticity and a possible synaptic basis in rat S1 cortex. A, Receptive field of L2/3 neuron "a". Principal whisker deprivation causes a rapid (7 days) loss of responses to the deprived, principal whisker, and a slower (20 days) increase in responses to spared, surround whiskers. Dashed lines, control receptive field. Data schematized from Glazewski and Fox, 1996. B, Hypothesized site of LTD mediating principal whisker response depression in adolescent rats.

spared inputs, rather than deprived principal whisker inputs. This makes the representation of the spared whisker expand within the whisker map (Fox, 1992; Diamond et al., 1994) (Fig. 1A). Highly similar components of plasticity occur in visual cortex during monocular deprivation (Sawtell et al., 2003; Frenkel and Bear, 2004).

In animals older than the first postnatal week, PWRD and SWRP occur primarily and most rapidly in L2/3, with less or no receptive field plasticity in L4. This indicates that PWRD and SWRP are mediated by functional changes in intracortical, rather than subcortical, circuits. Substantial progress has been made in identifying the neural basis for PWRD. Fox's group originally hypothesized that PWRD is due to deprivation-induced weakening, perhaps LTD, of the excitatory L4 to L2/3 projection in deprived columns, which normally drives principal whisker responses in L2/3 (Glazewski and Fox, 1996; Fox, 2002) (Fig. 1B). Strong evidence now exists for this hypothesis (see below). In contrast, the mechanisms underlying SWRP are less clear. SWRP is likely to involve LTP, since transgenic mice with autophosphorylation-incompetent CaMKII, which lack cortical LTP, have substantially impaired SWRP (Glazewski et al., 2000; Hardingham and Fox, 2004). One possibility is that SWRP involves LTP of excitatory trans-columnar projections, which would increase surround whisker responses in L2/3 neurons However, the site(s) of LTP for SWRP are not yet known, and other mechanisms besides LTP and LTD are likely to contribute to this and other aspects of whisker map plasticity (eg., Lendvai et al., 2000; Knott et al., 2002; Shepherd et al., 2003).

Here we summarize recent work focusing on how LTD at the L4-L2/3 excitatory synapse might contribute to the first component of deprivation-induced plasticity, PWRD. This work shows that L4-L2/3 synapses are capable of LTD *in vitro*, and that whisker deprivation induces marked LTD-like depression of these synapses *in vivo*. Recordings of spiking patterns in L4 and L2/3 *in vivo* suggest that this LTD is induced by a reversal in the precise, millisecond-scale timing of L4 and L2/3 spikes during deprivation, which is known to induce spike timing-dependent LTD at this synapse. Finally, anatomical experiments suggest that large-scale changes in L4 neuron number or axonal anatomy do not occur during map plasticity. Therefore, at this synapse, experience primarily regulates synaptic efficacy, not large-scale axonal structure.

Deprivation Induces LTD-Like Weakening of L4-L2/3 Synapses *In Vivo*

To determine if deprivation weakens L4-L2/3 synapses, Allen *et al.* took advantage of the fact that synaptic and cellular plasticity induced by experience *in vivo* persists and can be measured in acute, *ex vivo* brain slices (McKernan and Shinnick-Gallagher, 1997; Finnerty et al., 1999; Rioult-Pedotti et al., 2000). Rats were raised with one or more rows of whiskers plucked starting at postnatal day (P) 12, and slices were prepared 10–20 days later, after whisker map plasticity had presumably occurred (Fig. 2A). Slices were cut in an "across-row" plane that contained one barrel column from each of the 5 rows (termed A–E), so that spared and deprived columns could be identified unambiguously in the slice (Fig. 2B). Bulk synaptic strength of the L4-L2/3 projection was assayed using input-output curves in field potential and whole cell recordings, and was found to be 30–40% weaker in deprived columns than either neighboring, spared columns (Fig. 2C, D) or control columns from sham-plucked littermates (not shown). Plucking did not affect measures of intrinsic postsynaptic excitability, suggesting that the measured depression was due to synaptic changes (Allen et al., 2003). In more recent experiments, deprivation was shown to increase paired pulse ratios, suggesting that deprivation may decrease release probability at L4-L2/3 synapses (Allen, 2004; K.J. Bender, C.B. Allen, and D.E. Feldman, unpublished data).

To determine whether this deprivation-induced synaptic depression represents a reduction in the strength of preexisting, strong synapses, versus a failure of initially weak synapses to strengthen with development, deprivation was begun at the older age of P20, when synapses are more developed. Four to six days of deprivation starting at P20 caused the same magnitude of synaptic depression as did deprivation from P12,

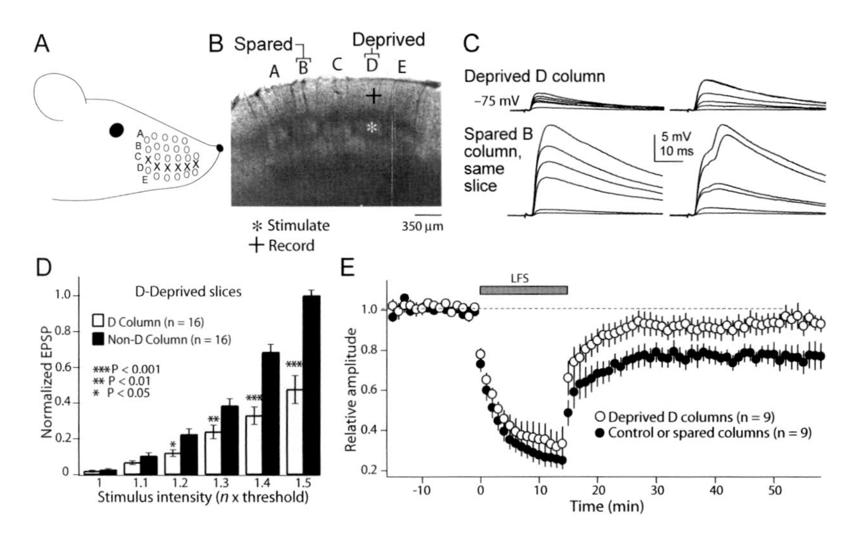

Figure 2 Whisker deprivation causes LTD-like weakening of L4 to L2/3 excitatory synapses. A, Deprivation of D row whiskers on the snout (X's). B, Living S1 slice containing five barrels corresponding to whisker rows A–E, visualized by transillumination. Stimulation and whole-cell recording sites for studying L4-L2/3 synapses are shown. C, Family of EPSPs in response to increasing stimulation intensity in L4, for two cells in a deprived D column, and 2 cells in the spared B column of the same slice. D, Comparison of mean EPSP amplitude between deprived and spared columns. All amplitudes are normalized to the mean maximal amplitude in the non-deprived column of each slice. E, Occlusion of LTD by whisker deprivation. LFS, 900 presynaptic stimuli at 1 Hz. Bars are S.E.M. Data from Allen et al., 2003.

suggesting that deprivation does not simply cause a failure of synaptic development, but actively weakens existing synapses. In addition, the time course of depression was determined by recording in slices made from animals deprived of whiskers for 3, 5, and 7 days, beginning at P12. Significant synaptic depression was observed after 5 days of deprivation, but not 3 days, suggesting that 4–5 days of deprivation are required to alter synaptic strength at these ages (Allen, 2004).

To determine whether deprivation-induced synaptic weakening represents LTD, Allen *et al.* tested for occlusion. Because LTD is typically a saturable phenomenon (Dudek and Bear, 1992; Mulkey and Malenka, 1992; Lebel et al., 2001), LTD induced by deprivation *in vivo* should occlude further LTD induction *in vitro*. Results showed that deprivation-induced synaptic weakening profoundly occluded LTD induction by low frequency stimulation (900 pulses at 1 Hz) (Fig. 2D). Consistent with the occlusion model, the capacity for LTP was enhanced by deprivation, indicating that deprived synapses were not merely deficient in plasticity. These findings were recently replicated (Hardingham and Fox, 2004). Thus, these experiments demonstrate that whisker deprivation reduces the physiological strength of L4-L2/3 synapses via LTD or an LTD-like mechanism. Similar results have been found for monocular deprivation, which causes both physiological and biochemical signatures of LTD at

L4-L2/3 synapses in visual cortex (Heynen et al., 2003). Together, these results suggest that LTD is likely to be an important mechanism for plasticity in S1 and V1. Whether deprivation also weakens circuits through reduction in synapse or neuron number is addressed by experiments below.

How is LTD Induced During Sensory Deprivation *In Vivo*?

At L4-L2/3 synapses *in vitro*, like at many excitatory synapses, LTP and LTD can be induced by multiple induction protocols. These include altering presynaptic firing rate (termed rate-dependent plasticity) (Madison et al., 1991; Linden and Connor, 1995), and modulating the relative timing of pre- and postsynaptic spikes on a millisecond timescale, largely independent of firing rate (spike-timing dependent plasticity, STDP) (Dan and Poo, 2004). Most models of experience-dependent cortical plasticity assume rate-dependent induction of LTP and LTD. However, Celikel *et al.* conducted experiments to determine which of these modes of LTP/LTD induction drives LTD at L4-L2/3 synapses in S1 in response to whisker deprivation and found strong evidence that STDP is the relevant mechanism (Celikel et al., 2004).

L4-L2/3 synapses in visual cortex exhibit a standard rate-dependent LTP/LTD learning rule in which presynaptic firing rates of a few Hz drive LTD, and rates >10 Hz drive LTP (Fig. 3A). Though the full learning rule is not known in S1, its basic form is similar, with a cross-over point between LTP and LTD at about 10 Hz (S. Bergquist and D.E. Feldman, unpublished data). To determine whether deprivation alters spike rate in a manner appropriate to drive rate-dependent LTD at L4-L2/3 synapses *in vivo*, Celikel *et al.* made extracellular recordings from L4 and L2/3 neurons in awake, behaving rats. When all whiskers were intact, L4 and L2/3 neurons fired at mean rates of 2.7 and 2.1 Hz, respectively, across several whisker-related behavioral states. Trimming of the

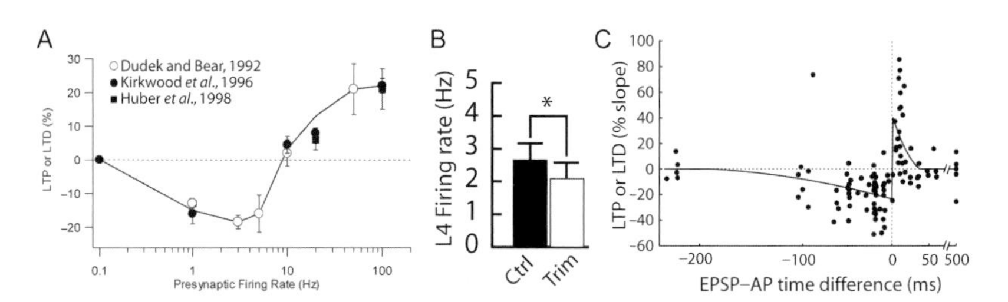

Figure 3 Rate-dependent and spike timing-dependent learning rules for LTP and LTD. A, Summary plot of the learning rule for firing rate-dependent LTP and LTD from Schaffer collateral-CA1 synapses (open symbols) and L4-L2/3 synapses in V1 (filled symbols). Data are from the indicated papers. B, Mean firing rate of L4 neurons in awake behaving rats recording when all whiskers were intact ("ctrl") and immediately after trimming the principal whisker to the level of the fur ("trim"). C, Learning rule for STDP at L4 to L2/3 synapses in S1. Data from Celikel et al., 2004.

principal whisker corresponding to the recorded column caused mean firing rates to reduce, but only modestly, to 2.1 and 1.7 Hz, respectively (Fig. 3B). Because 2–4 Hz firing elicits similar, near-maximal LTD at L4-L2/3 synapses in V1, as well as in CA1 hippocampus, it seems unlikely that these modest changes in spike rate could drive rate-dependent LTD *in vivo* (Dudek and Bear, 1992; Bear, 1996; Kirkwood et al., 1996; Huber et al., 1998) (Fig. 3A). Indeed, the low frequency of firing observed with all whiskers intact suggests that precise spike timing, rather than firing rate, may be most relevant for plasticity *in vivo*.

How spike timing may drive LTD *in vivo* can be inferred from the precise shape of the STDP learning rule measured *in vitro*. LTP is induced at L4-L2/3 synapses when presynaptic spikes lead postsynaptic spikes by 0–15 ms. In contrast, LTD results when postsynaptic spikes lead presynaptic spikes by a longer interval of 0–50 ms (Fig. 3C). The longer temporal window for LTD predicts that LTD can be induced *in vivo* by two means: either by reliable post-leading-pre firing within the LTD window, or by uncorrelated spiking at low rates, which drives net LTD because uncorrelated spike trains contain more interspike delays that fall within the long LTD window than delays that fall within the brief LTP window (Feldman, 2000).

To determine whether deprivation may drive spike timing-dependent LTD *in vivo*, Celikel *et al*. measured the spiking of L4 and L2/3 neurons simultaneously in the same barrel column in anesthetized rats. To mimic normal whisking, all whiskers were deflected together by inserting them into a piezoelectric-driven plastic mesh. Under this condition, L4 neurons faithfully spiked several milliseconds before neurons in L2/3, a pre-leading-post firing order that is appropriate to drive spike timing-dependent LTP (Fig. 4). To simulate whisker deprivation, the principal whisker was cut to narrowly escape the mesh, so that the mesh now deflected all whiskers but the principal whisker. This resulted in two immediate changes in L4 and L2/3 firing correlations in the deprived column. First, mean firing order reversed, with most L2/3 neurons now spiking before L4 neurons (Fig. 4). This reversal was most pronounced between L4 and L2 neurons. Second, overall firing correlations between pairs of L4 and L2/3 neurons significantly decreased (not shown). These changes recovered immediately when the principal whisker was reinserted into the mesh. Thus, whisker deprivation acutely altered spike timing at L4-L2/3 synapses in a manner that was exactly appropriate to drive spike timing-dependent LTD (Celikel et al., 2004).

These experiments suggest that spike timing, not spike rate, is the key feature of S1 spike trains that drives deprivation-induced weakening of L4-L2/3 synapses, and that STDP is the relevant mode of LTD induction. However, it will be critical to verify that these use-dependent changes in spike timing occur in awake-behaving, not just anesthetized, rats. In addition, the prevalence of STDP as a learning mechanism *in vivo* needs to be examined. Is it most relevant only in sparsely spiking brain regions, like S1, in which rate-dependent plasticity is unlikely, or is it utilized more generally?

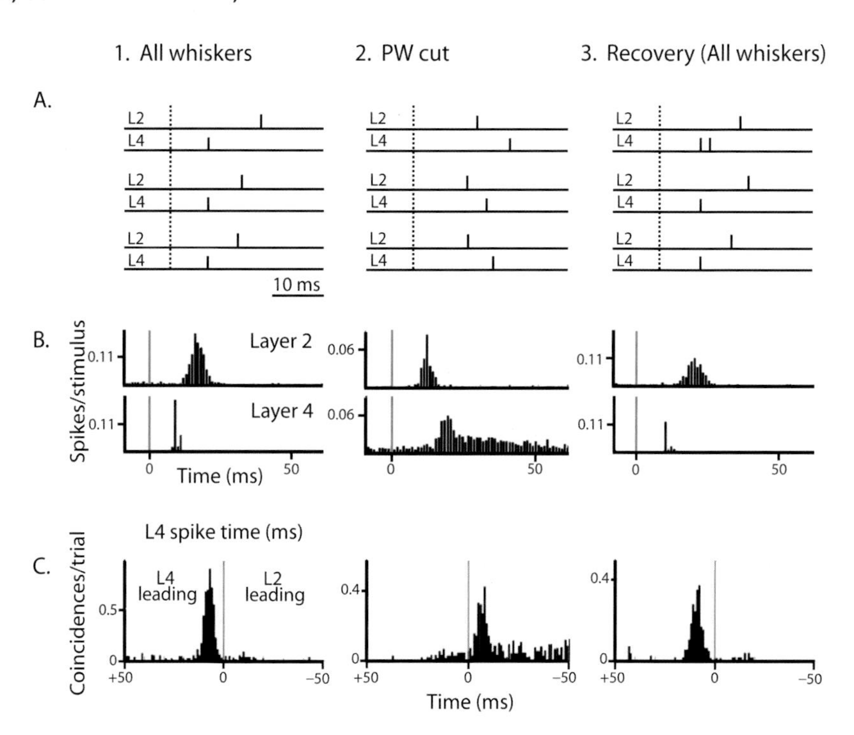

Figure 4 Acute deprivation of a single principal whisker causes a reversal in L4-L2/3 firing order appropriate to drive spike timing-dependent LTD *in vivo*. A, Spike trains of a pair of L4 and L2 neurons, recorded simultaneously in a single S1 column, under 3 sequential conditions: simultaneous deflection of all whiskers, deflection of all but the principal whisker (PW cut, to mimic acute deprivation of one whisker). and all-whisker deflection (recovery). Note reversal in L4-L2 firing order during PW cut. B, Peristimulus time histograms of L4 and L2 responses for each stimulus condition (900 stimulus repetitions). Stimulus onset, 0 ms. C, Cross-correlograms representing relative timing of L4 and L2 spikes during sensory responses in each condition. Data from Celikel et al., 2004.

Deprivation does not Alter the Anatomy of L4-L2/3 Projection

In mature S1, excitatory cells within L4 barrels extend dense, ascending axonal projections to L2/3. These projections are highly column-specific, preferentially targeting L2/3 within the home column (Petersen and Sakmann, 2001; Feldmeyer et al., 2002). This columnar precision is thought to be important for conferring appropriate principal whisker responses in L2/3 neurons. As for many projections, there is a debate over whether the L4-L2/3 axonal projection arises during development from an initially precise or imprecise projection. Axonal reconstructions from biocytin-filled L4 neurons in thick S1 slices (400 μm) showed that roughly one third of L4 spiny neurons extended non-column specific axonal arbors at P8-10, and that column specificity developed by P14 through selective addition of branches in the correct target column (Bender et al., 2003). However, another study that used axonal fills and functional mapping of projections in thinner slices (300 μm) found that initial axons showed adult-like columnar precision (Bureau et al., 2004). Whether this discrepancy arises from loss of longer, non-columnar axonal branches in thin slices is unclear. However, both studies do make clear that axons are still growing during the developmental period in which whisker deprivation induces weakening at L4-L2/3 synapses. Therefore, deprivation-induced changes in arbor size, arbor topography, and synapse number need to be considered as additional possible mechanisms for deprivation-induced weakening of the L4-L2/3 synaptic connection.

To test whether deprivation alters L4 axonal morphology or synapse number, axonal arbors of L4 excitatory cells projecting to L2/3 were examined using single-cell reconstructions (Bender et al., 2003). Animals were raised with all whiskers intact or the D-row deprived from P8 to P23-26, a manipulation known to drive synaptic depression at this projection (Allen et al., 2003). Slices were cut in the across-row plane to contain one barrel from each row. Excitatory spiny stellate and star pyramidal cells in the center of the D-barrel were filled with biocytin during whole-cell recording and visualized with a diaminobenzidine-based reaction (Fig. 5A). Axonal reconstructions were made relative to column boundaries, determined by counterstaining for L4 barrels with osmium tetroxide (Fig. 5B).

To determine whether deprivation reduced axonal length or distribution in L2/3, we quantified the length of axon in L2/3 as a function of location tangential to the pial surface. In control rats, the projection was largely columnar, with ~90% of axon in L2/3 contained in the home (D) column. D-row deprivation did not alter the tangential distribution of this projection, or the total length of each axon in L2/3 (Fig. 5C, D). Deprivation of all contralateral whiskers (A–E rows) also produced no detectable effect on axonal length or topography (Bender et al., 2003; Bureau et al., 2004).

As a first step in determining whether deprivation reduced synapse number, we calculated axonal bouton density for randomly selected axon segments in L2/3. Bouton density is relatively constant across axonal branches for cortical excitatory cells (Yabuta and Callaway, 1998; Bender et al., 2003). Deprivation did not detectably alter bouton density, suggesting that anatomical synapse number remains constant during sensory deprivation, despite the 40% reduction in bulk synaptic strength shown above (Fig. 5E). It is important to stress that these results show only that deprivation does not lead to massive synaptic withdrawal observable at the light level, and that ultrastructural changes including changes in the number of release sites per bouton may still occur.

Whisker deprivation may also reduce L4-L2/3 connection strength by reducing the number of L4 neurons in deprived barrels, rather than decreasing the number of L4-L2/3 synapses per L4 axon. Since barrel size does not change with deprivation (Fox, 1992), we estimated changes in cell number by calculating cell density within L4 barrels in control and whisker deprived animals. In these experiments, which have not

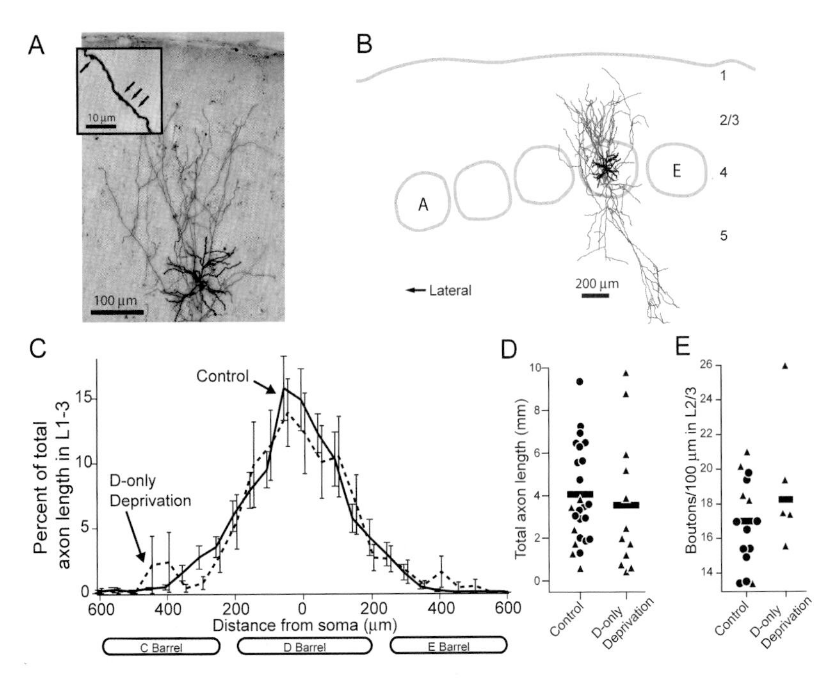

Figure 5 Whisker deprivation does not alter gross anatomy of L4 axons in L2/3. A, Single-section montage showing biocytin-labelled L4 spiny stellate neuron. Inset, High power view showing axonal boutons (arrows). B, Full reconstruction of neuron in (A). Barrel position was determined from neighboring osmium tetroxide-stained section. Black: dendrite. Grey: axon. Light grey: barrel outlines and pia. Numbers indicate layers. C, Distribution of axon tangential to the pial surface for control and D-row deprived rats. Bars are SEM. Ellipses show approximate barrel boundaries. D and E, Deprivation of D row whiskers did not alter total axon length or bouton density in layers 1–3. Bars are mean. Triangles, age-matched control and deprived rats. Data from Bender et al., 2003.

been previously reported, Long-Evans rats were D-row deprived from P12-P23, or had normal whisker experience. Rats were perfused at P23 with 4% paraformaldehyde, the contralateral hemisphere was sectioned at 40 µm in the across-row plane, and alternate sections were stained for NeuN, a neuron-specific nuclear protein, or cytochrome oxidase (CO) to visualize barrels. NeuN staining (mouse anti-NeuN, Chemicon, 1:1000 dilution, 18 hr at 4°C) was visualized using a fluorescent secondary antibody (Alexa-488 anti-mouse, Jackson ImmunoResearch, 1:1000, 1 hr at 25°C). NeuN-immunoreactive neurons were marked using Neurolucida software (Microbrightfield) with the experimenter blind to deprivation history and barrel boundaries. Barrel boundaries were then projected from neighboring CO sections (Fig. 6A, B). Neuronal density was calculated for B–E barrels in 3 separate slices per animal, and corrected for 2% tissue shrinkage, assessed by comparing the average distance between C, D, and E barrel centroids in fixed, CO-stained tissue, versus living, transilluminated acute brain slices. Shrinkage values matched previous measurements in our lab (Bender et al., 2003).

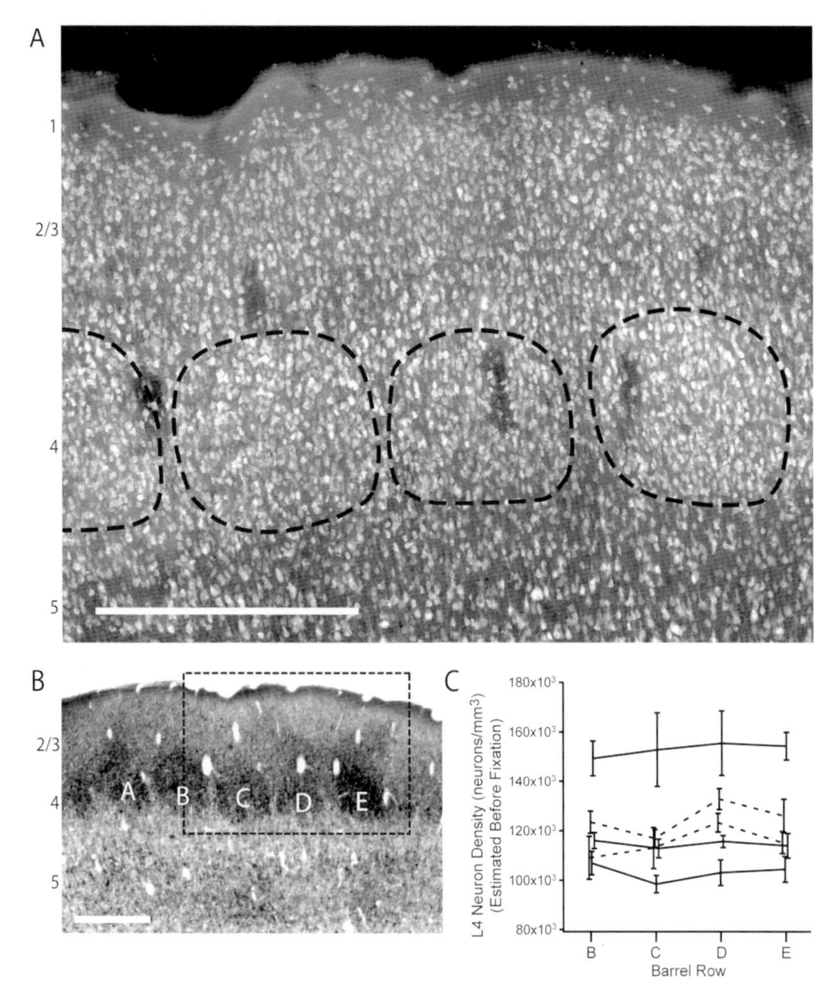

Figure 6 Whisker deprivation does not alter neuronal density in L4 barrels. A, Neu-N staining in single "across-row" section of S1. Barrel borders from neighboring CO-stained section (B) are shown. Dashed rectangle in B shows region corresponding to panel A. Scale bars in (A) and (B) are 500 μm. C, Absolute neuronal density in barrels B–E corrected for tissue shrinkage (see text). Solid lines: control animals. Dashed lines: D-row deprived. Error bars are S.E.M.

Results showed that neuronal density was constant across B–E barrels for both control rats ($n = 3$) and D-row deprived rats ($n = 2$) (Fig. 6C). Across all sections, the average neuronal density within barrels was $122{,}000 \pm 4{,}000$ neurons/mm^3 (mean $\pm$ S.E.M). At this density, an average barrel, approximated as a cube of across-row width 460 μm, within-row width 410 μm, and height 310 μm, would contain roughly 7000 neurons. Deprivation did not appear to alter neuronal density in the deprived D column (Fig. 6C).

Together, these anatomical results indicate that deprivation-induced map plasticity does not involve large scale loss of L4 neurons or axons.

Instead, deprivation seems to reduce the physiological strength of L4-L2/3 synapses while gross anatomical features of the projection remain intact. Similar results have been observed in the barn owl sound localization system. In the barn owl, auditory-visual misalignment induced by wearing prismatic spectacles causes visually guided learning of sound localization. This learning involves the loss of pre-existing, inappropriate auditory responses, and the growth of new axonal projections to mediate new auditory responses appropriate to the visual displacement. Axonal connections mediating normal auditory responses persist with extended prism experience, even though these responses are physiologically reduced or absent, suggesting that experience weakens functional synaptic efficacy but does not cause gross synaptic withdrawal (DeBello et al., 2001). Similarly, new axonal connections formed during learning remain anatomically intact even after they are functionally silenced by prism removal (Linkenhoker et al., 2005).

Why deprivation appears to affect synaptic efficacy but not axonal anatomy of L4 cells is unclear, especially because in visual cortex, axonal restructuring of thalamocortical and L2/3 horizontal axons does occur during deprivation-induced map plasticity (Antonini and Stryker, 1993; Darian-Smith and Gilbert, 1994). Similarly in S1, all-whisker unilateral deprivation does alter the branch structure of L2/3 pyramidal cells at these ages (Maravall et al., 2004), indicating that cortical neurons are capable of experience-dependent anatomical plasticity. Perhaps L4 axons could undergo anatomical plasticity given longer deprivation durations.

Conclusions

Work presented here suggests that LTD or LTD-like synaptic depression is an important component of developmental map plasticity in sensory cortex. A working hypothesis for how LTD is induced by deprivation is summarized in Fig. 7. In this hypothesis, feedforward connectivity from thalamus to L4 to L2/3 ensures that during normal sensory use, most L4 spikes occur before L2/3 spikes. Deprivation causes immediate reversal in firing order for L4 and L2/3 neurons (illustrated) and decreases overall firing correlations (not shown), with little change in spike rate. What cortical circuits mediate the firing order reversal are not yet clear, although trans-columnar excitatory inputs from surrounding columns with intact whiskers seem well-suited to mediate the residual L2/3 responses in deprived columns. We hypothesize that these acute changes in spike timing, over several days, drive spike timing-dependent LTD at L4-L2/3 synapses, and that this LTD is a primary mechanism for weakening of responses to the deprived principal whisker in L2/3. Anatomical measurements indicate that this reduction in L4-L2/3 synapse efficacy occurs without large-scale changes in L4 cell number, axonal anatomy, or bouton number, although changes in dendrites, spines, and synaptic ultrastructure may occur (Lendvai et al., 2000; Maravall et al., 2004).

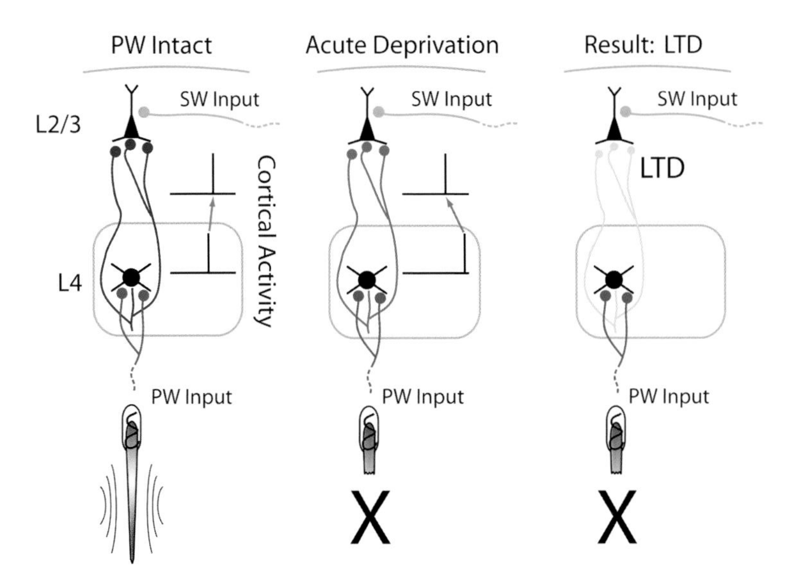

Figure 7 Model for deprivation-driven induction of LTD at L4-L2/3 synapses *in vivo*. Left, When all whiskers are intact, deflection of the principal whisker drives spikes from L4 neurons, which in turn activate L2/3 neurons via ascending, feedforward L4-L2/3 synapses. Thus, L4 neurons tend to spike before L2/3 neurons. Middle, whisker deprivation is known to acutely alter L4-L2/3 firing correlations in two ways: firing order reverses, so that L2/3 neurons tend to fire before L4 neurons (illustrated), and L4 and L2/3 spike trains become decorrelated (not illustrated). Together, these changes in spike timing are hypothesized to drive spike timing-dependent LTD at L4-L2/3 synapses (right panel).

If whisker deprivation causes immediate reversal of L4-L2/3 spike timing, and spike timing-dependent LTD occurs rapidly *in vitro* (within minutes), why is 5 days of deprivation required for measurable weakening of L4-L2/3 synapses *in vivo*? One likely factor is that in behaving animals, only a small fraction of total spikes in a given S1 column are driven by whisker deflections, with the rest being spontaneous or driven by whisker self-motion (Fee et al., 1997). Because deprivation only alters the timing of whisker-driven spikes, deprivation may produce only relatively small biases of overall spiking statistics, leading to relatively slow accrual of timing-dependent LTD. Another related factor is that ongoing, spontaneous network activity is known to powerfully reverse recently induced LTP and LTD, which could slow the accrual of these forms of plasticity *in vivo* (Xu et al., 1998; Zhou et al., 2003). Third, receptive field plasticity is known to be faster when whiskers are plucked singly or in a checkerboard pattern, so that each deprived column has many spared neighboring columns (Fox, 2002). In our studies of deprivation-induced synaptic weakening *in vivo*, we plucked whole rows of whiskers, thus leaving fewer spared whiskers around each deprived whisker. We are currently investigating whether this pattern of plucking may alter spike timing more modestly than single-whisker deprivation, which could explain the slower development of synaptic plasticity in this case.

Several key experiments are also necessary to test the working hypothesis in Fig. 7. First, it is unknown whether the changes in spike timing observed with acute whisker deprivation in anesthetized animals occur similarly in awake-behaving animals. Second, current evidence is only correlative that LTD contributes to receptive field plasticity; tests for causality are required. To test causality, it will be necessary to selectively block or manipulate LTD, either pharmacologically or genetically, and to determine if whisker map plasticity is altered or impaired. In visual cortex, this strategy has produced mixed results, so the causality of LTD in ocular dominance plasticity remains unknown (Hensch and Stryker, 1996; Renger et al., 2002; Fischer et al., 2004). To resolve this issue it will be critical to improve our understanding of the molecular basis of LTD at relevant cortical synapses, in order to develop more selective and effective reagents that interfere with LTD. These reagents could then be used to probe the role of LTD in cortical map plasticity.

References

Allen CB (2004) Synaptic depression induced by whisker deprivation in rat barrel cortex. PhD thesis University of California, San Diego.

Allen CB, Celikel T, Feldman DE (2003) Long-term depression induced by sensory deprivation during cortical map plasticity in vivo. Nat Neurosci 6:291–299.

Antonini A, Stryker MP (1993) Rapid remodeling of axonal arbors in the visual cortex. Science 260:1819–1821.

Bear MF (1996) A synaptic basis for memory storage in the cerebral cortex. Proc Natl Acad Sci U S A 93:13453–13459.

Bear MF, Cooper LN, Ebner FF (1987) A physiological basis for a theory of synapse modification. Science 237:42–48.

Bender KJ, Rangel J, Feldman DE (2003) Development of columnar topography in the excitatory layer 4 to layer 2/3 projection in rat barrel cortex. J Neurosci 23:8759–8770.

Buonomano DV, Merzenich MM (1998) Cortical plasticity: from synapses to maps. Annu Rev Neurosci 21:149–186.

Bureau I, Shepherd GM, Svoboda K (2004) Precise development of functional and anatomical columns in the neocortex. Neuron 42:789–801.

Celikel T, Szostak VA, Feldman DE (2004) Modulation of spike timing by sensory deprivation during induction of cortical map plasticity. Nat Neurosci 7:534–541.

Dan Y, Poo MM (2004) Spike timing-dependent plasticity of neural circuits. Neuron 44:23–30.

Darian-Smith C, Gilbert CD (1994) Axonal sprouting accompanies functional reorganization in adult cat striate cortex. Nature 368:737–740.

DeBello WM, Feldman DE, Knudsen EI (2001) Adaptive axonal remodeling in the midbrain auditory space map. J Neurosci 21:3161–3174.

Diamond ME, Huang W, Ebner FF (1994) Laminar comparison of somatosensory cortical plasticity. Science 265:1885–1888.

Dudek SM, Bear MF (1992) Homosynaptic long-term depression in area CA1 of hippocampus and effects of N-methyl-D-aspartate receptor blockade. Proc Natl Acad Sci U S A 89:4363–4367.

Fee MS, Mitra PP, Kleinfeld D (1997) Central versus peripheral determinants of patterned spike activity in rat vibrissa cortex during whisking. J Neurophysiol 78:1144–1149.

Feldman DE (2000) Timing-based LTP and LTD at vertical inputs to layer II/III pyramidal cells in rat barrel cortex. Neuron 27:45–56.

Feldmeyer D, Lubke J, Silver RA, Sakmann B (2002) Synaptic connections between layer 4 spiny neurone-layer 2/3 pyramidal cell pairs in juvenile rat barrel cortex: physiology and anatomy of interlaminar signalling within a cortical column. J Physiol 538:803–822.

Finnerty GT, Roberts LS, Connors BW (1999) Sensory experience modifies the short-term dynamics of neocortical synapses. Nature 400:367–371.

Fischer QS, Beaver CJ, Yang Y, Rao Y, Jakobsdottir KB, Storm DR, McKnight GS, Daw NW (2004) Requirement for the RIIbeta isoform of PKA, but not calcium-stimulated adenylyl cyclase, in visual cortical plasticity. J Neurosci 24:9049–9058.

Fox K (1992) A critical period for experience-dependent synaptic plasticity in rat barrel cortex. J Neurosci 12:1826–1838.

Fox K (2002) Anatomical pathways and molecular mechanisms for plasticity in the barrel cortex. Neuroscience 111:799–814.

Frenkel MY, Bear MF (2004) How monocular deprivation shifts ocular dominance in visual cortex of young mice. Neuron 44:917–923.

Glazewski S, Fox K (1996) Time course of experience-dependent synaptic potentiation and depression in barrel cortex of adolescent rats. J Neurophysiol 75:1714–1729.

Glazewski S, Giese KP, Silva A, Fox K (2000) The role of alpha-CaMKII autophosphorylation in neocortical experience-dependent plasticity. Nat Neurosci 3:911–918.

Hardingham NR, Fox K (2004) The relationship between spike timing plasticity and experience-dependent plasticity in αCAMKII-T286A Mutants. Soc Neurosci Abstr 857.21.

Hebb DO (1949) Organization of Behavior. John Wiley & Sons, New York.

Hensch TK, Stryker MP (1996) Ocular dominance plasticity under metabotropic glutamate receptor blockade. Science 272:554–557.

Heynen AJ, Abraham WC, Bear MF (1996) Bidirectional modification of CA1 synapses in the adult hippocampus in vivo. Nature 381:163–166.

Heynen AJ, Yoon BJ, Liu CH, Chung HJ, Huganir RL, Bear MF (2003) Molecular mechanism for loss of visual cortical responsiveness following brief monocular deprivation. Nat Neurosci 6:854–862.

Huber KM, Sawtell NB, Bear MF (1998) Brain-derived neurotrophic factor alters the synaptic modification threshold in visual cortex. Neuropharmacology 37:571–579.

Keller A (1995) Synaptic organization of the barrel cortex. In: The barrel cortex of rodents (Jones EG and Diamond IT, ed). New York: Plenum:221–262.

Kirkwood A, Rioult MC, Bear MF (1996) Experience-dependent modification of synaptic plasticity in visual cortex. Nature 381:526–528.

Knott GW, Quairiaux C, Genoud C, Welker E (2002) Formation of dendritic spines with GABAergic synapses induced by whisker stimulation in adult mice. Neuron 34:265–273.

Lebel D, Grossman Y, Barkai E (2001) Olfactory learning modifies predisposition for long-term potentiation and long-term depression induction in the rat piriform (olfactory) cortex. Cereb Cortex 11:485–489.

Lendvai B, Stern EA, Chen B, Svoboda K (2000) Experience-dependent plasticity of dendritic spines in the developing rat barrel cortex in vivo. Nature 404:876–881.

Linden DJ, Connor JA (1995) Long-term synaptic depression. Annu Rev Neurosci 18:319–357.

Linkenhoker BA, von der Ohe CG, Knudsen EI (2005) Anatomical traces of juvenile learning in the auditory system of adult barn owls. Nat Neurosci 8:93–98.

Madison DV, Malenka RC, Nicoll RA (1991) Mechanisms underlying long-term potentiation of synaptic transmission. Annu Rev Neurosci 14:379–397.

Manahan-Vaughan D, Braunewell KH (1999) Novelty acquisition is associated with induction of hippocampal long-term depression. Proc Natl Acad Sci U S A 96:8739–8744.

Maravall M, Koh IY, Lindquist WB, Svoboda K (2004) Experience-dependent changes in basal dendritic branching of layer 2/3 pyramidal neurons during a critical period for developmental plasticity in rat barrel cortex. Cereb Cortex 14:655–664.

McKernan MG, Shinnick-Gallagher P (1997) Fear conditioning induces a lasting potentiation of synaptic currents in vitro. Nature 390:607–611.

Mulkey RM, Malenka RC (1992) Mechanisms underlying induction of homosynaptic long-term depression in area CA1 of the hippocampus. Neuron 9:967–975.

Petersen CC, Sakmann B (2001) Functionally independent columns of rat somatosensory barrel cortex revealed with voltage-sensitive dye imaging. J Neurosci 21:8435–8446.

Renger JJ, Hartman KN, Tsuchimoto Y, Yokoi M, Nakanishi S, Hensch TK (2002) Experience-dependent plasticity without long-term depression by type 2 metabotropic glutamate receptors in developing visual cortex. Proc Natl Acad Sci U S A 99:1041–1046.

Rioult-Pedotti MS, Friedman D, Donoghue JP (2000) Learning-induced LTP in neocortex. Science 290:533–536.

Ruthazer ES, Cline HT (2004) Insights into activity-dependent map formation from the retinotectal system: a middle-of-the-brain perspective. J Neurobiol 59:134–146.

Sanes JN, Donoghue JP (2000) Plasticity and primary motor cortex. Annu Rev Neurosci 23:393–415.

Sawtell NB, Frenkel MY, Philpot BD, Nakazawa K, Tonegawa S, Bear MF (2003) NMDA receptor-dependent ocular dominance plasticity in adult visual cortex. Neuron 38:977–985.

Shepherd GM, Pologruto TA, Svoboda K (2003) Circuit analysis of experience-dependent plasticity in the developing rat barrel cortex. Neuron 38:277–289.

Simons DJ (1978) Response properties of vibrissa units in rat SI somatosensory neocortex. J Neurophysiol 41:798–820.

Singer W (1995) Development and plasticity of cortical processing architectures. Science 270:758–764.

Stent GS (1973) A physiological mechanism for Hebb's postulate of learning. Proc Natl Acad Sci U S A 70:997–1001.

Welker C, Woolsey TA (1974) Structure of layer IV in the somatosensory neocortex of the rat: description and comparison with the mouse. J Comp Neurol 158:437–453.

Wiesel TN, Hubel DH (1963) Single-Cell Responses in Striate Cortex of Kittens Deprived of Vision in One Eye. J Neurophysiol 26:1003–1017.

Woolsey TA, Van der Loos H (1970) The structural organization of layer IV in the somatosensory region (SI) of mouse cerebral cortex. The description of a cortical field composed of discrete cytoarchitectonic units. Brain Res 17:205–242.
Xu L, Anwyl R, Rowan MJ (1998) Spatial exploration induces a persistent reversal of long-term potentiation in rat hippocampus. Nature 394:891–894.
Yabuta NH, Callaway EM (1998) Cytochrome-oxidase blobs and intrinsic horizontal connections of layer 2/3 pyramidal neurons in primate V1. Vis Neurosci 15:1007–1027.
Zhou Q, Tao HW, Poo MM (2003) Reversal and stabilization of synaptic modifications in a developing visual system. Science 300:1953–1957.

16

High-Field (9.4T) Magnetic Resonance Imaging in Squirrel Monkey

Aimee J. Nelson, Cheryl A. Cheney, Yin-Ching Iris Chen, Guangping Dai, Robert P. Marini, Graham C. Grindlay, Yumiko Ishizawa, and Christopher I. Moore

Introduction

The recognition that adult sensory plasticity is both common and robust is one of the most important advances in neuroscience over the past 50 years. Many of the best cellular and systems level studies have explored the mechanisms and limitations of neural reorganization. Unique challenges exist, however, in the study of plasticity. In humans, it is typically impossible to obtain measures of brain organization prior to clinically relevant plasticity-inducing events (for example, stroke, spinal-cord injury or amputation: Cramer et al., 2000; Moore et al., 2000; Staines et al., 2002). Further, a classic paradox in the study of plasticity is that while intervention is necessary to induce reorganization (for example, training, deafferentation, or cortical stimulation) it is also typically required to assess reorganization (for example, craniotomies, injection of anatomical tracers or lesions). While the intent of these studies is to examine plasticity induced by a specific manipulation, the act of probing the system could in some non-linear way combine with the induction process to alter the findings.

The ideal technique for assessing plastic modification of the brain is, therefore, one that is minimally invasive and repeatable, permitting measurement before and after manipulation. Further, this approach should be applicable in non-human primates—respecting the cost and ethical issues that come into play in this regard—and in human subjects, so that parallel studies can be conducted. A technique that permits whole-brain coverage at a relatively fine scale resolution would be an additional benefit, especially for studies of sensory map organization where simultaneous estimation of multiple brain areas can be crucial to a proper understanding of the etiology of neural plasticity.

Non-invasive fMRI is ideally suited for studies of cortical plasticity. In animal model systems, neural organization can be repeatedly assessed following a variety of manipulations (some even performed during a single scanning session), allowing for short and long-term monitoring. Monkey imaging is a particularly powerful tool for studies of plasticity when it is performed using high-field magnets. High-fields deliver the spatial resolution necessary to reveal cortical column-level changes, while still providing whole-brain coverage. Human studies are also commonly conducted with fMRI, allowing direct comparisons between monkeys and humans.

This chapter describes methods we have developed for functional magnetic resonance imaging (fMRI) in squirrel monkeys using a 9.4 Tesla small diameter bore magnet. We begin with a discussion of the strengths and weaknesses of this approach relative to traditional electrophysiology. We then review the methods developed, with an emphasis on the logic driving experimental decisions that have proven critical to the success of imaging experiments. Examples of the anatomical and functional resolution obtained using this paradigm are provided.

Our current discussion is targeted to neuroscientists who may be considering this approach. As such, we attempt to provide a specific and practical discussion of the methods developed, and an introductory background to the factors driving different MR imaging choices.

Classical Sensory Neurophysiology and fMRI: Complimentary Strengths and Weaknesses

A central goal of neurophysiology is to understand how neural activity contributes to sensory perception. To this end, single-neuron electrophysiological techniques, including single electrode, multi-electrode (stereotrode, tetrode) and intracellular recording approaches are invaluable. These approaches are, however, limited in several ways. Each electrode has a small field of view (FOV) and can only sample the activity of, at most, several neurons. Also, despite many recent advances in implant technology, only $\leq$200 electrodes can be maintained in a single animal. Given the presence of millions of neurons in a single cortical area, this sampling coverage is restrictive. Another problem faced by single neuron recording techniques is that they are invasive. They require surgical intervention, leading to the termination of the animal at the cessation of acute experiments, or to implantation of a systemic alteration in the anatomy and physiology of the animal (e.g., the addition of a novel head apparatus). Further, these techniques are inappropriate for studying human subjects except in the case of brain surgery.

Perhaps the most profound problem with sensory electrophysiology is that an investigator must choose the optimal brain site for electrode placement prior to investigation. In some cases, this assumption can be guided by prior research using alternate methods: In other cases, substantial effort can be exerted to explicate the function of a brain area that is only tangentially related to the execution of a given sensory processing function.

Functional magnetic resonance imaging (fMRI) provides a complimentary set of strengths and weaknesses to those of classic electrophysiology. The weaknesses of fMRI derive from the uncertainties surrounding the origin of the blood flow signal and the resolution of fMRI measurements. The most common method used to measure fMRI activity is the blood-oxygen level dependent signal (BOLD) (Ogawa et al., 1990). Several studies suggest that the BOLD signal is correlated with electrophysiological activity, and that this hemodynamic measure may reflect more subtle, subthreshold potential changes (Logothetis et al., 2001; Heeger et al., 2000; Rees et al., 2000; Backes et al., 2000; Arthurs et al., 2000; Ances et al., 2000). Despite these several correlative findings, the mechanisms linking neural activity and BOLD signals are not fully understood. Further, even if a perfect correlation between net changes in activity level and BOLD can be assumed, this measure is likely 'blind' to changes in temporal patterning that do not require increased neural activity (but see Thompson et al., 2004).

A second weakness is that, relative to electrophysiology measures, the resolution of fMRI is limited in spatial and temporal specificity. The spatial resolution of human fMRI studies is typically on the order of millimeters. The temporal resolution of fMRI is limited due to the slow onset of the BOLD signal, and because sampling intervals (TR) are typically $\geq$1 sec.

Despite these inherent limitations, the complimentary strengths of fMRI are unique, and the technique is ideally positioned for integration with electrophysiological approaches. First, fMRI provides a remarkable field of view: In many experiments, functional activation can be measured across the entire brain. Second, it is not surgically invasive, and does not require the injection of radioisotopes as used in PET imaging. As such, fMRI is safe for repeated use in humans and animals. Third, fMRI offers the best spatial resolution of the non-invasive neuroimaging techniques available, and advances in MR tools (e.g., scanners, coils) and techniques (e.g., scan sequence design, analysis approaches) continue to enhance the precision of this approach.

Monkey fMRI

For several reasons, conducting fMRI in monkeys has the potential to allow for the 'best of both measurement worlds.' Repeated non-invasive imaging can be combined with subsequent invasive approaches such as electrophysiology and optical imaging, permitting whole brain coverage or, alternatively, high spatial resolution in a region of interest.

Historically, electrophysiology has been conducted in monkey subjects under the assumption that this model provides the most accurate parallel to neural mechanisms employed in the human. This assumption, while logical, has seldom been directly tested using identical measurement techniques in the two species. The use of fMRI in humans and monkeys provides the opportunity for parallel investigation. Using this approach, the areas activated by identical stimuli can be compared, as can aspects of the dynamics of activation in these regions (e.g., adaptation patterns). If parallel activations are observed, then extensive and detailed electrophysiological studies conducted in non-human primates provide the best inferential link currently possible between normal human brain function and the activity of single neurons. Interestingly, using fMRI, differences between species have already been noted in what are believed to be homologous brain areas (Vanduffel et al., 2002).

Monkey fMRI also provides an ideal system for longitudinal monitoring of plastic changes in brain organization. Because fMRI can be used repeatedly within subjects, baseline data can be acquired by imaging for months prior to the introduction of a manipulation. The plasticity induced by this change (or, the lack thereof) can then be tracked systematically and without further damage to the research subject (e.g., Smirnakis et al., 2005).

Importantly, monkey fMRI can also provide a guide for electrophysiological studies. This prior estimation allows a researcher to screen for the brain areas of greatest potential interest, and allows for targeting recordings to precise sub-regions of a given a brain area. As fMRI resolution increases, for example, specific column(s) of interest within a cortical area can be identified (Kim & Duong, 2002; Duong et al., 2001; Duong et al., 2000; Kim et al., 2000; Menon et al., 1997) and measured using electrophysiology.

High-Field Imaging

One advance in technology that has helped overcome some of the concerns regarding fMRI resolution is the increasing use of high magnetic field strengths (e.g., 9.4T). The signal to noise ratio (SNR) of anatomical and functional MRI increases with the static magnetic field strength: Thus, higher spatial sampling is achievable while still maintaining high SNR within each sample volume, improving both anatomical and functional resolution.

Despite these benefits, machines with high-field strengths such as 9.4 T are currently not available for human testing. One reason is that the small diameter of the magnet bore on most high-field machines precludes the use of humans or large primates. Additionally, there are safety issues that arise when imaging at high-fields (Bottomley & Andrew, 1978). For example, human subjects may experience dizziness or other balance-related symptoms (e.g., vertigo, nausea) upon entering or exiting high-field magnets (Schenck et al., 2000), and there is also a risk of peripheral nerve stimulation (Schaefer et al., 2000; Ham et al., 1997). The full scope of these physiological effects has not been determined, therefore restricting high-field imaging to non-human subjects.

Squirrel Monkey Scanning at 9.4 T

To take advantage of the several benefits of monkey MR for sensory neurophysiology, we have developed techniques for imaging squirrel monkeys (SM) at 9.4 T. The squirrel monkey (SM) was selected as our model system for several reasons. First, SM are semi-lissencephalic (Emmers & Alkert, 1963; Benjamin & Welker, 1957; Welker et al., 1957). Their relatively flat cortical surface is ideal for mapping studies using techniques such as optical imaging (Tommerdahl et al., 2002; Chen et al., 2003) and electrophysiology (Sur, 1984), and facilitate the transition from voxel localization in fMRI to these other approaches. Second, a great deal is already known about the cortical organization of SM. Specifically, the representations and receptive field properties in many sensory modalities have been characterized, including tactile (Jain et al., 2001; Merzenich et al., 1987; Sur et al., 1984), visual (Livingstone, 1996), auditory (Cheung et al., 2001), and vestibular (Akbarian et al., 1992; Guldin et al., 1992) cortices. This species has also been used extensively as a model for studies of cortical plasticity (Plautz et al., 2003; Frost et al., 2003; Nudo et al, 2003; Churchill et al., 2001; Xerri et al., 1996; Merzenich et al., 1993; Garraghty & Kaas, 1991), basal ganglia (Flaherty and Graybiel, 1991;1993;1994;1995), and disease (e.g., dystonia and Parkinson's disease: Blake et al., 2002; Rupniak et al., 1992; Boyce et al., 1990).

Several practical considerations also recommend SM use. They habituate well to handling (Abee, 2000) and, relative to macaque monkeys, there is a reduced risk of zoonotic disease transmission to experimenters. Also, because they are bred successfully in captivity, there is greater ease of acquisition. Last, and perhaps most important to our logic in selecting this model system, SM are relatively small. They have a body weight of $\sim$717 $\pm$ 170.4 g (Gergen & MacLean, 1962), and have a slender maximal body width: As such, SM can fit within the tight space limitations of higher field scanners (e.g., the 11.7 cm gradient-insert diameter of the 9.4 T we currently use). The primary drawback to using SM is that they are not favored for behavioral studies in primates: However, as discussed below, limitations on the tolerance for subject motion at high-fields likely preclude the use of a monkey behavioral preparation with our current methods.

Methods

Overview of an Experiment

In a typical experiment, monkeys are obtained from our vivarium and transported to the imaging center. In a surgical preparation room on site, anesthesia is induced, the monkey is intubated and catheterized, and subsequently maintained on isoflurane anesthesia for the duration of the experiment. The animal is positioned in an MR-compatible holding device that reduces head movement and secures one of several surface coils on the head. Sensory presentation equipment is positioned (e.g., a tactile stimulator or a visual screen), and the animal placed in the scanner. A series of anatomical and functional images are then acquired over a 3–6 hour period. Following scanning, anesthesia is terminated, and the animal is transferred back to our home facility.

Non-MR Aspects of the Experimental Approach

Transport

The monkeys are kept in a temperature (20 – 23°C), humidity (30 – 70%), and photoperiod (12 h dark/12 h light) controlled environment where they are single-housed in standard cages with a variety of perches and enrichment devices. The evening prior to an experiment, animals are placed on overnight food restriction. Animals are transferred to off-site experiments in a customized transport box (Primacarrier, by Primate Products) in a climate-controlled vehicle. To minimize the stress incurred by direct handling, animals are trained to enter the box, and the majority of our SM (3 of 4 tested) will enter without further interaction.

Anesthetic Induction

The transport box was customized for these studies. The front opening of the box has a sliding plastic door, through which the animal climbs when being collected for transport, and the two longer side panels have an array of openings for ventilation and observation. Two modifications were made. First, a clear, acrylic panel was inserted and secured against one interior side of the box. A pair of removable handles can be attached to receptacles built onto this panel. These handles pass through the holes in the outer wall of the box and enable one to push the panel to the opposite interior side of the box, creating a squeeze apparatus for anesthetic injection. Second, additional openings were added to the side of the box opposite the acrylic panel, to improve access to the lower limb for the initial intramuscular (IM) anesthetic injection.

Using the squeeze-box apparatus, initial sedation is achieved with Telazol delivered IM to a thigh muscle. Active components of Telazol include tiletamine, a dissociative anesthetic that blocks NMDA receptors (Fish, 1997), and zolazepam, a benzodiazepine tranquilizer that potentiates GABA receptors (Reves & Glass, 1990). The dosages used (6.7–8.8 mg/kg) are similar to those recommended for use in dogs: For a ~700 g squirrel monkey, 7.1 mg/kg (0.05 ml) of Telazol provides sufficient sedation (~30 min) to intubate and prepare the intravenous (IV) catheter. In several experiments, we have observed that higher concentrations of Telazol, or repeated injections of low doses, compromise the measurement of stimulus-evoked BOLD signals and delay anesthetic recovery after isoflurane is terminated. Though we have not tested other drugs or procedures for induction, a similar drug, ketamine, is often used in SM (Greenstein, 1975) and, if used at low-levels in macaques (1–2 mg/kg), preserves the fMRI BOLD signal (Leopold et al., 2002).

Atropine sulfate (0.04 mg/kg IM) is administered with the initial Telazol injection (same syringe). The anti-cholinergic action of atropine reduces respiratory secretions, keeping airways clear. Once the animal is positioned in the magnet, atropine is delivered through the IV catheter every 45–60 minutes. When access to the IV line is not an option, (e.g., during long anatomical sequences), glycopyrrolate is administered IM (0.01 mg/kg) prior to the scan. The anti-cholinergic action of glycopyrrolate has a longer acting duration (2–3 hours) than atropine.

Intubation and Catheterization
Endotracheal tubes (ET) used with SM must have an appropriately small diameter, smaller than those used for pediatric purposes. We employ customized, re-usable silicone cuffed 2.5 mm (inner diameter) ETs without wire reinforcement (Med-Caire, Vernon, CT). The length of each tube is customized to fit each monkey such that it spans the distance from the mouth to the manubrium sternum (6.7 cm and 8.0 cm for two monkeys whose data is presented in this Chapter). The patency of the ET cuff is tested prior to intubation by inflating with a syringe. Two to three minutes prior to intubation, a single spray of the topical anesthetic, Cetacaine, is delivered to the glottis to reduce the incidence of laryngospasm. The ET is coated with Lidocaine Hydrochloride Oral Topical Solution and inserted into the trachea using a stylet and laryngoscope (size 1 Macintosh blade). Successful ET placement is determined by observing motion of hairs held at the opening of the ET, condensation on a mirror, and/or expansion/contraction of a latex covering placed at the end of the ET tube. The cuff is then inflated with air (~2–2.5 ml). The ET is secured by way of a velcro strap customized with an opening that fits around the ET connector (15 mm) and wraps around the head.

An IV catheter is then placed in the lateral or medial tarsal, metatarsal or saphenous vein, or alternatively in the lateral tail vein (Brady, 2000) using a 24G × 3/4″ Surflo catheter with a 27G needle and Surflo injection plug (Terumo Medical Corporation). A small splint is typically used to prevent the ankle from rotating and corrupting the catheter. To mitigate venous injury, catheter placement alternates between the left or right lower limb, and occasionally the tail, across experiments. Currently, this IV line is used to deliver Lactated Ringers solution throughout scanning (7.5 ml/kg/hr), atropine at 45–60 minute intervals, and slow injections of dextrose following scanning (100–250 mg/kg IV; 5% dextrose in water as a single ~3.0 ml slow dosage). In experiments now under way, the IV permits the infusion of contrast agents (e.g., MION see discussion) prior to or during imaging.

Anesthesia, Respiration Rate and CO_2-Level Maintenance
Following ET and catheter placement, the monkey is transported to the shielded imaging room and placed on mechanical ventilation (SAR-830 Series Small Animal Ventilator, CWE, Inc.). Anesthesia is maintained via isoflurane in balance oxygen (0.5–0.6L O_2/min). During animal positioning (ear bar insertion, head restraint, placement in cradle), isoflurane is maintained at 1.5% and is subsequently reduced in three incremental steps over ~30 minutes to achieve ~0.5–0.6% expired isoflurane (measured with a V9004 Capnograph Series with inspired/expired anesthetic gas, Surgivet). Ventilation is maintained at a rate between 34–39 breaths

per minute with an inspiration time of 0.5 s and expiration duration of ~1.1 s. The inspired/expired duration ratio of the SAR-830 may be adjusted, an important feature to counteract the effects of positive pressure ventilation. Such ventilation may impede venous return via changes in pleural pressure during the inspiration phase; reducing the inspiration duration relative to expiration time blunts the negative effects of positive pressure ventilation.

Isoflurane is a recommended anesthetic for use in SM (Brady, 2000), and for maintaining stimulus-evoked fMRI signals (Nair & Duong, 2004; Sicard & Duong, 2005; Liu et al., 2004). Isoflurane is a volatile anesthetic that causes a dose-dependent decrease in blood pressure through vasodilation, though the effect of increasing cerebral blood flow is less than that caused by halothane (Reinstrup et al., 1995). The molecular mechanisms of isoflurane action are not completely understood, but it is considered to act on multiple neural membrane proteins, including GABA A chloride channels. Isoflurane has a wide safety margin, analgesic properties, and is associated with a relatively rapid recovery. Importantly, this anesthetic is sufficient to induce muscle relaxation in SM, and muscle relaxants such as mivacurium, often used when imaging larger monkeys (Logothetis et al., 1999), are not necessary for our research.

At the end of the experiment, SM awaken quickly after low level isoflurane is discontinued (~5–15 minutes). Therefore, it is important to maintain anesthesia while removing items that could potentially damage an awakening monkey (ear bars, head restraints, surface coils). This transient consciousness is followed by ~2 hours of recovery and the monkey is kept warm using a hot water blanket or rechargeable hot packs during transport. Pulse oximetry is monitored during recovery and fluids (Lactated Ringer's) are delivered IV.

The use of anesthesia is essential for collecting high-resolution primate data in our paradigm. In our preparation, and at the resolution employed, the motion generated in the anesthetized animal is already in some cases in excess of required tolerances (e.g., see Figure 2). The motion observed by an awake, head restrained monkey will almost certainly exceed the spatial resolution at 9.4T and create signal artifacts that are further enhanced because of the greater susceptibility at higher fields. There are nevertheless clear drawbacks to the use of anesthesia. First, anesthesia may depress the BOLD signal, requiring averaging across several fMRI runs for a sufficient contrast to noise ratio. Second, fMRI BOLD signals obtained under anesthetic can be difficult to interpret, as isoflurane has a direct impact on central neural processing and substantial vasodilatory properties (Warltier & Pagel, 1992). Further studies directed to examine the impact of anesthesia on BOLD signals and central nervous system function are essential (Disbrow et al., 2000; Ishizawa et al., 2005). However, preliminary data suggest that the increased SNR afforded by higher field strength may recoup some of the anesthetic induced suppression of the BOLD signal (see Figure 13B).

Table 1 Physiological measures (5 scan sessions, 3–6 hours each).

Physiological measures	Mean (n = 5 sessions)
[Inspired isoflurane]	0.65% ± 0.061
[Expired isoflurane]	0.65% ± 0.056
End-tidal CO_2	32.6 ± 9.92 mm Hg
Systolic blood pressure	127.01 ± 15.21 mm Hg
Diastolic blood pressure	79.54 ± 14.35 mm Hg
Mean arterial blood pressure	96.84 ± 13.76 mm Hg
SpO_2	97.9 ± 1.54
Heart rate (beats per minute)	240 ± 24 BPM
Rectal Temperature	36.0 ± 1.5 C

Physiological Monitoring

Proper physiological monitoring is particularly important where narrow, long bores prevent visual inspection of the animal. Variables we typically measure include: end-tidal CO_2, expired/inspired isoflurane concentration (V9004 Capnograph Series, Surgivet), non-invasive blood pressure from the femoral artery measured between EPI scans (V6004 Series Non-Invasive Blood Pressure, Surgivet), heart rate and arterial oxygen saturation via a pulse oximetry sensor secured to the palm (Nonin 8600V), and rectal temperature. Published normative physiological data for anesthetized SM is scarce, and for this reason Table 1 lists the physiological data (mean, standard deviation) recorded over five typical scan sessions. In the awake restrained SM, body temperature is 37–39°C (Brady, 2000; Pinneo, 1968), mean arterial blood pressure is 140 ± 4 mmHg (Byrd & Gonzalez, 1981), and the heart rate may exceed 300 beats per minute (Pinneo, 1968). As Table 1 indicates, isoflurane anesthesia decreases body temperature, blood pressure and heart rate. Stimulus-evoked activation is robust in sessions when end-tidal CO_2 levels range between 35–40 mm Hg; sessions with lower capnic levels require greater signal averaging to reveal significant activation.

Monitoring animal physiology in small bore magnets is challenging. One challenge is to locate devices that are MR compatible at high-field strengths: In our experience, devices advertised as 'MR compatible' can have components that are readily corrupted at 9.4 T. Also, as mentioned above, there is virtually no visual information available to indicate the health condition of the monkey, making physiological monitoring across several variables a necessity. Another practical consideration is that input/output connections (e.g. the pulse oximeter cable) should avoid coursing beneath the surface coil (Figure 1).

One benefit of fMRI is that it can be repeatedly conducted in the same subject (animal or human). Nevertheless, the anesthetized preparation described above involves repeated procedures that could impact the health of the animal. These include the increased risk of infection associated with repeated intubation, anesthesia and IV, a risk of sensory damage to repeated placement of the animal in the noise environment

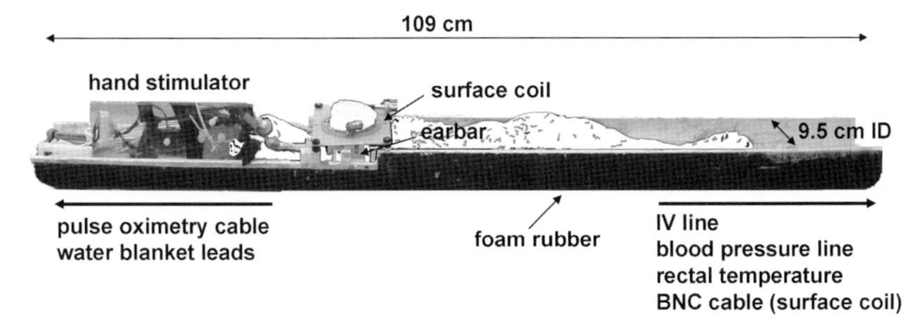

Figure 1 Animal positioning in cradle The scanning cradle is shown with a cartoon image of the basic monkey position. The arms are extended forward, and one hand is secured to the tactile stimulator hand mold, while the other is used for pulse oximetry measures. The cradle with foam rubber application on the base fits precisely in the diameter of the 11.7 cm 9.4 T bore. Arrows at each end of the cradle indicate the direction that specified cables exit the cradle.

of the scanner, and potential adaptation to anesthetic state. Thus far, we have observed that the above procedures promote a healthy recovery and allow longevity of the animals in experiments: The fMRI data and most anatomical images reported here are from two monkeys scanned repeatedly (~twice per month) over the course of 1 year. In two cases where significant complications were observed, monkeys had not been sufficiently acclimated to the anesthetic treatment and/or to travel to the scanning facility: These steps are now standard procedures in our design.

Body Positioning and Head Stabilization During Scanning
Following intubation, catheter placement, and ~10 minutes of isoflurane, muscle tone is sufficiently low and the monkey is placed in the custom-made cradle shown in Figure 1. The length of the cradle encases the entire elongated outstretched body, including the tail. The body of the cradle is made from plastic piping (ID 9.5 cm, OD 11 cm), and the outer surface is covered in rubber foam pipe insulation tape (~1 mm thick) to dampen the transfer of magnet vibrations to the preparation, and to provide frictional resistance to micro-motions of the apparatus. Placement of SM in the cradle is in the prone position, atop a heated water blanket (Gaymar Therma Pump, Harvard Apparatus). The body is extended with the arms outstretched in front of the animal for presentation of tactile stimuli.

One advantage of using an anesthetized preparation is that subject motion is minimized. However, the resolution of EPI images (e.g., 625 μm in-plane) and typical anatomical images (e.g., 195 μm) demands minimal motion, and motion-related signal artifacts are amplified at high-fields. Therefore, proper head restraint is mandatory for successful experiments. We have found that a triangulation of restraint—chin rest, ear bars and head piece—is necessary to reduce subject motion to an acceptable level. Examples of subject motion in a functional scan taken with and without this triangulation are shown in Figure 2. With each

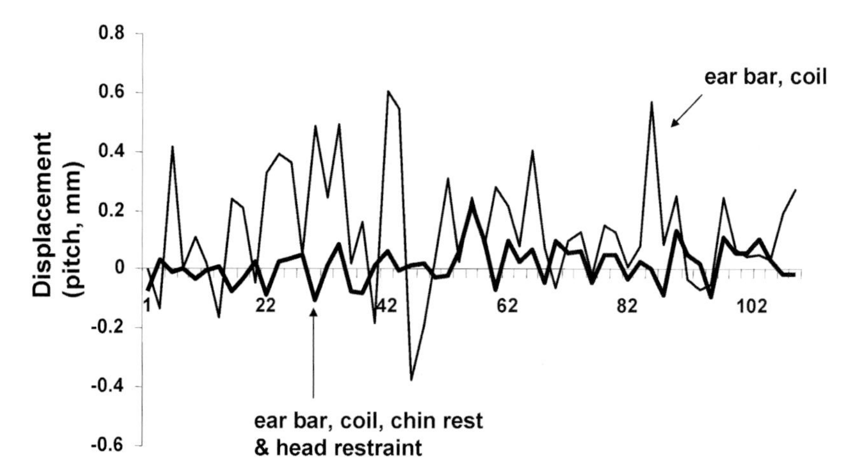

Figure 2 Head motion Head motion plots (mm of displacement in the pitch direction) are shown for two functional scans. A clear reduction in head motion was observed following the use of the chin rest and head restraint elements (see Figure 8C). Motion of the amplitude observed in the 'ear bar, coil' condition, where ear bars and mild padding between the fixed coil and the head were employed, leads to non-usable functional data.

restraint in place, subject motions of less than ~190 μm are routinely observed across a ≥3 hour scanning period. Because subject motion poses a substantial challenge at the resolution we employ, each functional scan is subjected to motion assessment immediately following its acquisition (*AFNI*: Analysis of Functional Neuroimages). The on-line motion estimation provides a rapid assessment of the effectiveness of the head restraint and anesthesia depth, either of which can be subsequently adjusted.

Using three points of restraint is key to the precise repetition of the head position across experiments and to the reduction of head motion within experiments. These points are the *chin rest*, *ear bars* and *head piece*. The chin rest consists of a hard rubber stopper (2.4 cm height) secured on the bottom of the cradle. This piece prevents downward motions of the head, ensures accuracy of the height and angle of head placement, and helps prevent the ET from being inadvertently dislodged or compressed (see Figure 8A). The ear bars are cylinders (3.7 cm length × 0.4 mm diameter) tapered at their insertion tip to be non-rupturing. They are positioned at a height of 4.2 cm from the floor of the cradle. The ear bars are held in a 0.5 mm slot with a tapped opening on the posterior side for a delrin plastic thumb screw. Prior to insertion into the ears, the bars are coated with topical anesthetic (Lidocaine HCL Jelly, 2%, Teva). The head piece, as shown in Figure 8C, consists of a horizontal rubber slab joined at 2 points to a 'Y' support. The anterior-posterior position of the Y-piece is adjusted to compensate for the head dimensions of each monkey. To achieve identical cradle placement across scan sessions, a peg is inserted through an opening in the posterior base of the cradle into a hole in the gradient coil insert.

Tactile Sensory Stimulation

We have conducted successful tactile and visual BOLD fMRI of SM in the 9.4 T scanner: Because the focus of our studies thus far has been on the tactile, we describe our approach in that modality here. In tactile experiments, accurate between-session hand placement and consistent site of stimulus delivery within a session are essential. To provide stability, the left hand of the monkey is fitted into a custom rubber mold made from a double casting of the monkey's hand (Mix-a-Mold, AMACO, Indianapolis, IN), the positive is then cast into a rubber mold (PMC-121, Smooth-on, Inc., Easton, PA). This mold regularizes hand placement, separates the fingers and damps non-specific vibration transmission. The hand mold is mounted to an L-shaped acetyl plastic housing (2.0 cm thickness) that also secures the vibrotactile elements. The finger position is maintained via plastic cable ties, and two velcro strips maintain the wrist. Figure 3 shows a schematic of the stimulator. Piezoelectric (PZ) elements (Noliac, Denmark: $3.2 \times 0.78 \times 0.18$ cm) are used to deliver mechanical vibrotactile stimulation to the glabrous surface of the hand. These elements are favored because their multi-layered PZ synthesis provides a high relative force generation and a high fundamental resonance (typically >700 Hz in a fixed-free condition). Stimuli are usually applied to the distal and middle segments of the second digit, though fMRI-compatible stimulators have been made for human and monkey with a greater number of elements (e.g., 9 independent stimulators). Each PZ is equipped with a 3 mm diameter delrin post that vibrates perpendicular to the skin surface through an opening in the mold. A third PZ element secured at a 2.0 cm distance from the hand (not in contact

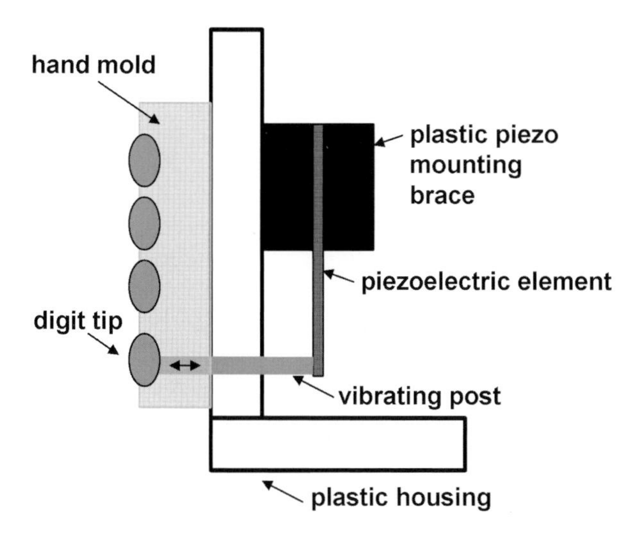

Figure 3 Vibrotactile stimulator A side profile schematic of the tactile stimulator in contact with the digit tip. The PZ element is mounted in a plastic brace, and a small plastic post that contacts the skin is slotted into a base affixed to the PZ element.

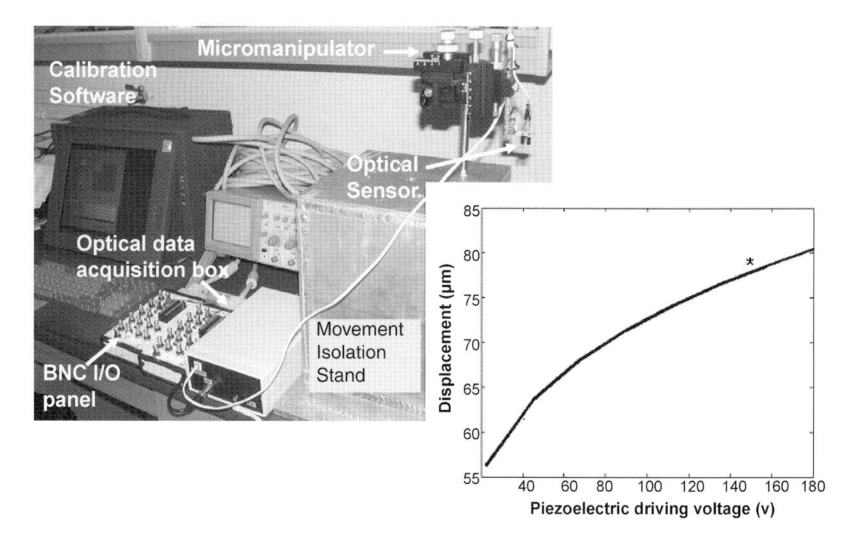

Figure 4 Stimulus calibrator *Left* A picture of the calibrator hardware set-up. Waveform signals are sent from the calibration software (portable computer, BSI) specifying the desired frequency and amplitude of PZ vibration. Actual PZ excursion is measured via an optical sensor mounted on a micromanipulator and captured by an analogue input PCI card (IOTech). An immovable steel platform is used to eliminate vibration that arises from non-PZ sources. *Right* A plot of PZ displacement as a function of driving voltage. These data were obtained while the monkey's finger was in contact with the vibrating probe. The asterisk indicates the voltage (150 V) and displacement (~80 μm) typically used in our tactile studies.

with the skin) has been used to deliver vibration to the device but not directly to the finger, to emulate non-specific effects of PZ activation during control, non-stimulation runs.

Tactile presentation is controlled via custom software developed in MATLAB. Using a portable computer (BSI) with slots for four full sized PCI cards, signals are sent through an array of National Instruments digital output cards connected to a BNC panel. Currently, the system controls up to 16 tactile and 2 audio independent channels, though the software is designed to accommodate additional output. For a typical experiment, a ≤10V signal is sent through a 15X amplifier (Sensor Technologies) to the PZ. During imaging, timing of stimulus presentation is yoked to data sampling to prevent errors due to drift in scanner timing: The scanner sends a TTL output at the beginning of the scan that is routed through a digital port on the BNC panel. The MATLAB program registers this pulse and triggers the program to output a signal that contains the programmed on/off durations, waveform type, amplitude and frequency.

To calibrate the amplitude of PZ movement, we built an optical sensor system with custom software. This calibration is important for reducing between-session variability in the output of the PZ, as these elements can degrade steadily or suddenly over time. As shown in Figure 4 (left), the optical sensor (Fairchild Semiconductors, QVE11233) is mounted on a micromanipulator, and recordings are made on an immovable steel platform. While the plastic post extension of the PZ is in contact with the monkey skin, the optical sensor registers changes in the displacement of the vibrating probe by detecting motion of a side attachment that breaks the light beam, and changes an input driving voltage. An example of a calibration experiment is shown in Figure 4 (right). The input voltage used in typical tactile experiments, indicated by an asterisk, evokes ~80 μm of indentation to the skin surface.

MR Aspects of the Experimental Approach

The MRI system we currently use is a Magnex Scientific 9.4 T 20 cm inner diameter horizontal bore magnet, with a gradient strength of 200 mT/m with fast gradient switching (100 μs rise time). The system is equipped with Bruker Avance console, and has an effective ID of 11.7 cm with the gradient inset. The advantage of higher static field strength (B_0) is increased SNR, due to the greater proportion of proton magnetization, with the net gain of SNR increasing as the square root of the static magnetic field (Gati et al., 1997). One benefit of greater SNR is the gain in anatomical resolution, which permits the identification of subtle features (e.g., cortical laminae). Similarly, for functional imaging purposes, the contrast to noise ratio of magnetization differences between oxy- and deoxy-hemoglobin is much greater at higher field (Yacoub et al., 2003; Yacoub et al., 2005), enhancing the BOLD signal.

The BOLD signal is the most common contrast agent used to measure functional activity, and there are many excellent reviews that describe what is known of its neural origins (Arthurs & Boniface, 2002; Logothetis & Pfeuffer, 2004; Logothetis & Wandell, 2004). In brief, the BOLD signal depends on blood flow, blood volume and the ratio of deoxygenated to oxygenated hemoglobin. An increase in neural activity evokes a concomitant increase in the BOLD signal due to an increase in the relative concentration of oxygenated hemoglobin that exceeds the local requirement for oxygen.

The BOLD signal can be obtained using *gradient echo* (GRE) or *spin echo* (SE) imaging. In GRE imaging, the BOLD effect is derived from both microvascular (e.g., those that perfuse brain tissue) and macrovascular (e.g., large draining veins) sources, and generally provides greater SNR than SE. In SE imaging, a refocusing pulse reduces the contribution of large blood vessels to the BOLD signal, thereby improving the spatial localization of fMRI activity to the activated neural tissue (Lee et al., 1999; Yacoub et al., 2003; Yacoub et al., 2005). For example, visually evoked signal changes obtained with SE have been localized to the approximate position of layer IV in cat visual cortex (the input layer), while under the same paradigm GRE BOLD signals were observed at the cortical surface (Zhao et al., 2004; see also Yacoub et al., 2005). Because of the increased SNR at high-field, which helps compensate for potential loss of signal

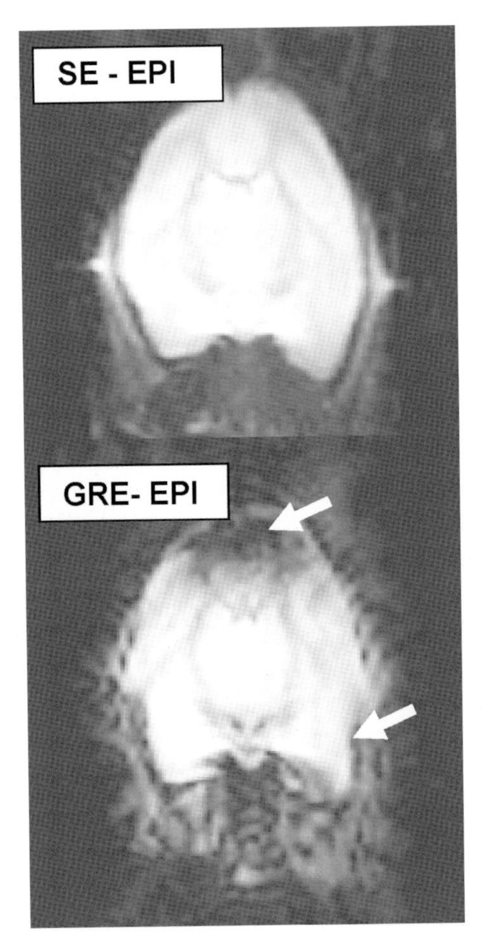

Figure 5 Spin echo versus gradient echo imaging Coronal images taken at identical slice positions using GRE (TR 2.0 s, TE = 13.0 ms) and SE (TR 2.0 s, TE = 25.4 ms) sequences. Signal loss at regions with high magnetic susceptibility is observed in the GRE images (arrow). FOV =5.0 cm, 80 × 80 matrix, 625 μm, 1 mm slice thickness.

when not using GRE, SE imaging provides an excellent opportunity to reveal the spatial specificity of the BOLD signal.

The second reason for choosing SE imaging relates to signal loss at tissue interfaces. Each tissue type (e.g., bone, dura, brain) exhibits its own magnetic properties when subjected to a static magnetic field. The interface at mismatched tissue types creates a local magnetic gradient that results in an inhomogeneous magnetic field, an effect called magnetic susceptibility. This effect is amplified at high-field, and poses a greater challenge to imaging across a large, inhomogeneous sample. The GRE sequence is specifically sensitive to the susceptibility effects, and signal dropout is often seen at locations where mismatched tissue types meet (e.g., near the sinuses and ear canal). Effects of the susceptibility induced signal inhomogeneity on SE and GRE images are shown in Figure 5.

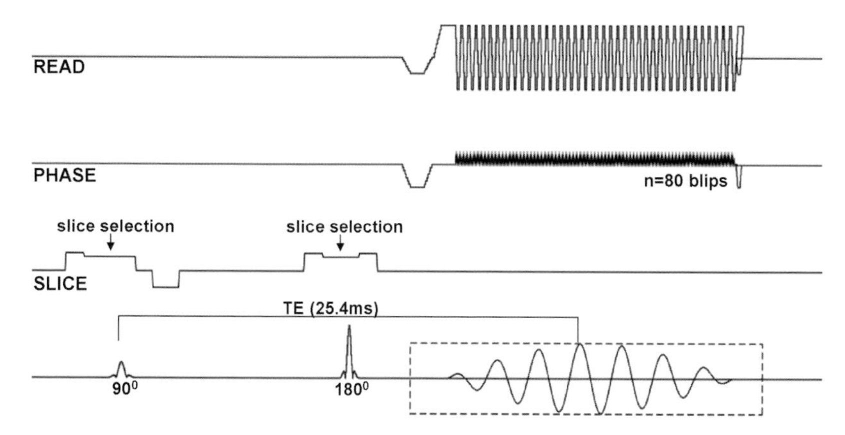

Figure 6 Single shot spin-echo pulse sequence The pulse sequence schematic displays the frequency encoding performed along the x axis ('Read') with 80 points and 100 μs rise time, the phase encode along the y axis ('Phase') with 80 phase encode lines, and the slice excitation ('Slice') surrounded by crusher gradients. The bottom line displays the occurrence of the first radiofrequency pulse (flip angle = 90°) and the second refocusing pulse (flip angle = 180°) that creates the spin echo with maximum amplitude at the echo time (TE). Data were sampled at TE = 25.4 ms. Typical EPI parameters used for fMRI data include a TR of 2.0 s; 17 coronal slices, 1 mm thick; FOV is 5.0 cm; 80 × 80 acquisition matrix; reconstructed using 128 × 128 matrix with zero filling.

For the reasons highlighted above, our BOLD fMRI sequence is a single-shot SE sequence and is depicted in Figure 6. Using this sequence, the entire spatial frequency domain (k-space) is acquired with a single repetition (90° RF, 180° RF pair), requiring only ~40 ms per 2-D image set. However, as is common in echo-planar imaging, in exchange for rapid data acquisition, images suffer from geometric image distortions. The fMRI acquisition can stretch or compress images when compared to the non-distorted anatomical images. These geometric distortions may be caused by magnetic susceptibility and are particularly severe at high-field strengths. Distortion reduction requires improvement in the magnetic homogeneity over the sample volume by optimizing electric currents in shim coils. Although we had some success with automated shimming routines (e.g., FASTMAP; Greuetter & Tkac, 2000), the improvements were modest, and we found that manual shimming improved image quality substantially. Using this technique, the sample volume was determined to be a cuboidal region of interest that encompassed the entire monkey brain: Manual shimming was then performed using linear, second and higher-order polynomials. The results of such a shim are shown in Figure 7, where EPI slices and analogous anatomical images are comparable in global brain shape and local features. Each monkey subject has a unique shim parameter, and these provide a good initial shimming basis for each fMRI session.

The surface coil plays an important role in optimizing SNR. Features to be considered in designing or purchasing a coil include its size,

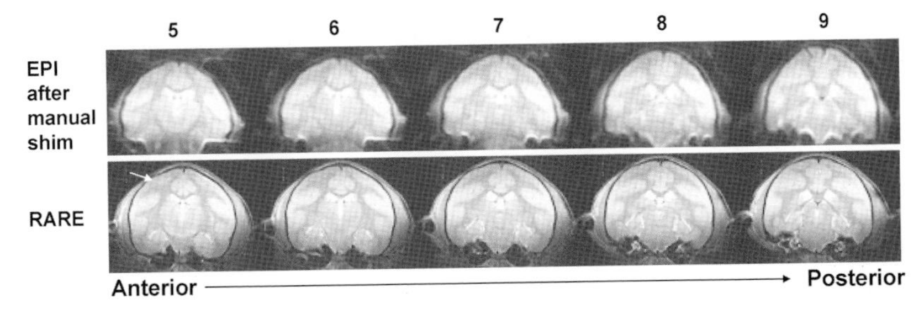

Figure 7 EPI and corresponding anatomical images The fMRI and corresponding anatomical images taken during a single imaging session with coil C in Figure 8. Numbers indicate slices moving from the anterior to posterior direction. Anatomical images were collected with a RARE sequence. To achieve the grey-white matter contrast shown in RARE images, the following parameters were used; TE = 12.447 ms, TR = 10000 ms, RARE factor = 8, 256 × 256 matrix, 17 × 1 mm thick slices, 195 μm in-plane, 1 mm thick; acquired in 340 s. The EPI images do not exhibit gross distortions in geometry and mirror the anatomical data in the right to left and superior to inferior dimensions. The arrow (slice 5) points to layer IV in primary somatosensory cortex.

position and shape. In choosing the *size* of the coil, a compromise must be made between greater SNR with smaller coils or greater depth of coverage with larger coils. Irrespective of size, the SNR will decrease with increasing distance from the center of the coil, thereby limiting the sensitivity of the coil to roughly its radius. Our custom-made coils are shown in Figure 8. The coils in Figure 8A and 8C are similar in size, and when positioned 1.0 cm above the ear bars, provide excellent full brain coverage. The oval coil in 8C is smaller in length and width than the circular coil (8A) and fits snuggly around the circumference of the SM head, maximizing SNR for our preparation. The images shown in Figure 7, and fMRI data reported here, were taken with the coil shown in Figure 8C. The small coil in Figure 8B provides increased SNR over a small circumference and depth, and is used in applications requiring high resolution over small spatial volumes. All of the coils shown in Figure 8 are 'receive-transmit'—they transmit RF pulses and receive the subsequent signal. Adjustable capacitors were incorporated into those shown in Figure 8B and 8C, to compensate for different loads. The coil in Figure 8A was tuned outside of the magnet to 400 MHz (Larmor frequency for the 9.4 T magnet).

Anatomical MRI Paradigm

Anatomical imaging is most frequently performed with RARE (Rapid Acquisition Relaxation Enhancement) and MSME (multi-slice multi-echo) sequences. Using either of these SE sequences, high resolution and high gray-white matter contrast images are acquired. For our purposes, the RARE sequence is ideal for fast 2-D anatomy to align with fMRI EPI data. Examples of RARE images are shown in Figures 7 and 10A (bottom). The data shown in Figure 7 was acquired in 340 seconds and reveals the laminar structure of primary somatosensory cortex (SI:

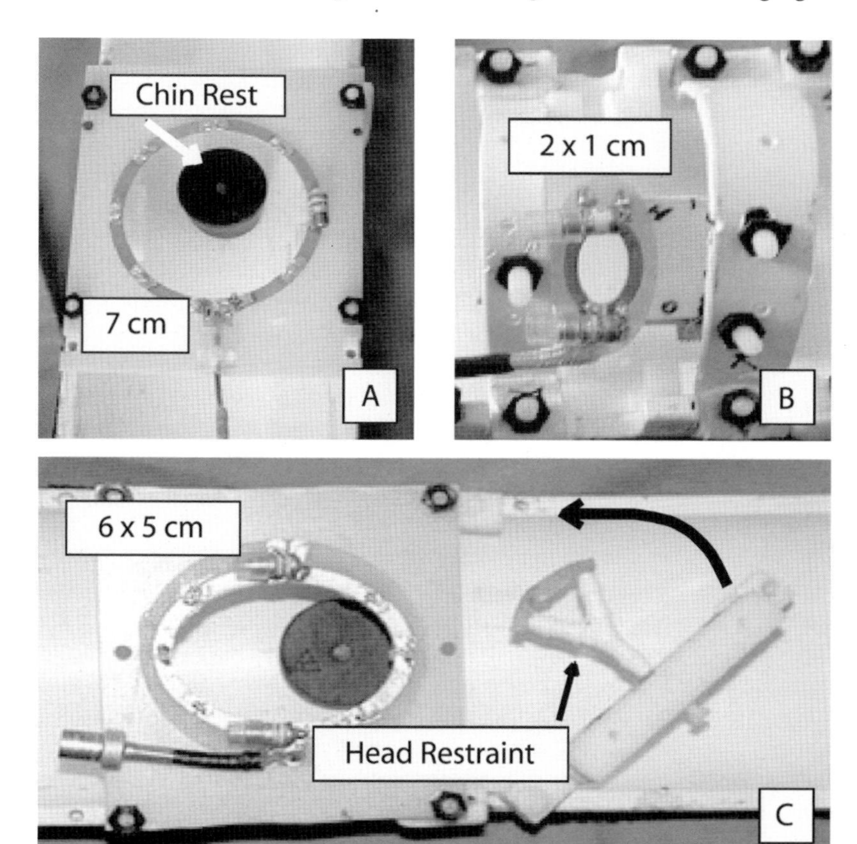

Figure 8 Receive-transmit custom-made surface coils A. Top view of 7 cm circular coil mounted in plastic and secured to the cradle. This coil is used for 3-D anatomical imaging with a 6.4 cm FOV. B. A 2 × 1 cm oval coil. This coil is used for high-resolution anatomy and fMRI over a small region (FOV = 3.0 cm). The matching and tuning capacitors for coils A and B are accessed remotely by a tuning rod. C. A 6 × 5 cm oval coil (FOV = 5.0 cm). Also shown are the chin rest and Y-piece head restraint. Once a coil is positioned, the head piece is swiveled forward and secured by a second screw. The rubber end of the Y-piece sits at the level of the brow on the forehead and is manipulated in the anterior-posterior direction by a screw.

slice 5, arrow). The MSME sequence is a conventional SE anatomical acquisition, and requires a longer time scale (TR × 256, if a 256 × 256 matrix). Examples of MSME are shown in Figure 10A (top) and 10B, and also are used in later figures for overlaying fMRI statistical maps. Either sequence is used to collect data over the entire brain or a region of interest using slice thicknesses ranging from 80–500 μm.

Functional MRI Paradigm and Data Analysis

In our current studies, we have had success applying vibrotactile stimuli in a blocked design with alternate periods of stimulation (8 seconds) and no stimulation (12 seconds). The off-on pattern is repeated eight times

(8 epochs) during a single functional scan for a total run length of 160 s. Due to the known decreases in BOLD signal with isoflurane anesthesia (Disbrow et al., 2000), averages across several runs are required to detect stimulus-evoked activity. All runs deemed acceptable (motion of less than 200 μm) are averaged for each stimulus condition (~10–15 runs of 160 s each) to create a grand average for each stimulus condition. Using an orthogonalized boxcar correlation and the *AFNI* software, the grand averaged time series is correlated with the hypothesized hemodynamic response function. The resulting statistical maps are typically smoothed at 625 μm (one pixel, in-plane) or not spatially smoothed.

Stimulus conditions include vibration of one focus on the distal digit tip, and of two foci placed on the distal and middle segment of the same digit, vibrated simultaneously or with inter-stimulus onset asynchrony of 100 ms offset (Nelson et al., 2005). The latter stimulus condition elicits the percept of tactile apparent motion in humans (Kirman, 1974; Aparicio and Moore, 2005).

A consistent challenge in functional imaging is to find an appropriate 'significance' level for the determination of functional activation. While the risk of false positives is high due to the enormous sample space (often thousands of voxels), many corrections are overly conservative (Locascio et al., 1997). To determine criteria for deeming activation statistically significant, we constructed statistical maps during 'no stimulus' presentation, in which a PZ embedded in an identical holder just distal to the hand was activated, but without direct skin contact. This stimulus condition is used to estimate the non-physiological noise potentially induced by the PZ elements. The runs were acquired in the same session in alternation with vibrotactile stimulation, and an equivalent number of 'no' stimulus runs were acquired (10–15). Using data from several sessions across 2 monkeys, we empirically defined the probability of aberrant activation in our scanning conditions using the coil shown in Figure 8C. Specifically, the correlation threshold level at which $p = 0.005$ in the 'no' stimulation data—5 aberrant voxels are activated in 1000 voxels—is typically set as our threshold in the vibrotactile scans. Figure 9 shows an example of non-overlapping distributions of correlation values for a stimulus run versus a 'no' stimulus run. As noted by the black line and asterisk in this example, significance was determined at $r = .12$.

Results

Anatomical Images at 9.4 T

The ability to obtain detailed anatomy is a clear advantage of imaging at high-field strengths. A hallmark of high-resolution brain MRI is the ability to detect layer IV in primary visual cortex. This signature feature is known as the stria of Genari (Gross, 1998), and reflects the dense cell body and thalamocortical axonal termination layer in primary visual cortex. The stria can be seen with the naked eye in unstained tissue, and is easily observed in images at high-field strengths. Compared with the

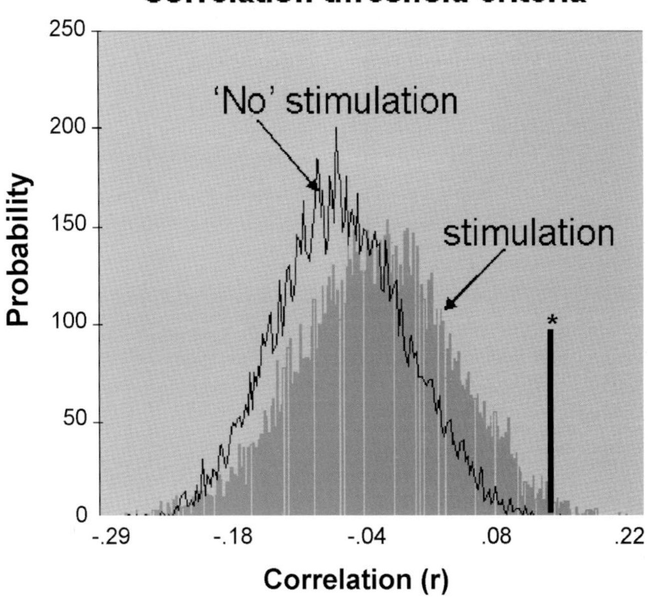

Figure 9 Correlation threshold criteria Plotted are the frequency distributions of correlation values for a vibrotactile condition and a 'no' stimulus condition, during which a PZ element attached to the hand holder was driven using the on/off paradigm but was not in direct contact with the skin surface. Thresholds in functional imaging studies were empirically determined using 'no' stimulus false positive distributions to define the correlation threshold cutoff for $p < 0.005$. In this example (one slice with ~1000 voxels), only responses in the stimulus condition with correlation values greater than $r = .12$ would be considered significant (black bar and asterisk).

prominent layer IV in visual cortex, layer IV in primary somatosensory cortex (SI) is more subtle. However, as shown in the coronal slices in Figure 10A (bottom), at 9.4 T with an in-plane resolution of 195 μm, layer IV in SI is clearly defined. These coronal slices were taken at a slice traversing the central sulcus (CS: shown in a 3-D rendered brain in the top image in 10A). A clear macro-anatomical marker for the 'Rolandic' cortex of the SM can also be appreciated in this coronal image, the thickening and bending of the gray matter at approximately mid-way through its medio-lateral course.

Another preparation that we have found useful for high-resolution imaging is the post-mortem SM brain. While there are obvious issues in making inferences from this non-living brain tissue preparation, high-resolution anatomical scans can be run for several hours. An example of such an image is shown in Figure 10B (40 hour scan). In the post-mortem SM brain, layer IV in SI is readily observed (thin arrow), as are electrode track penetrations (thick arrow). This approach allows relatively precise localization of the track orientation and depth. Using high-field imaging, electrode track information could also be obtained from living primates

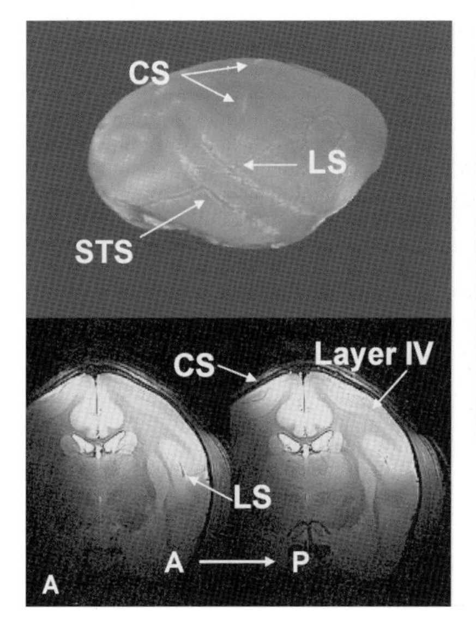

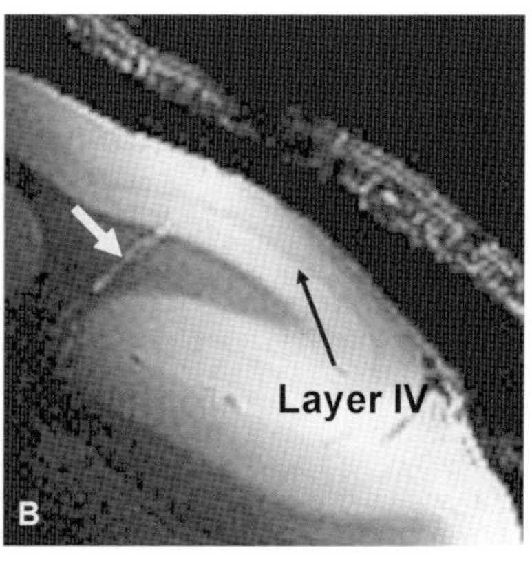

Figure 10 Anatomical imaging, primary somatosensory cortex *Top* A three dimensional volume rendered image of a SM brain to display locations of the central sulcus (CS), lateral sulcus (LS) and superior temporal sulcus (STS). The 3-D image was acquired with coil A in Figure 8 using an MSME sequence. *Bottom* Coronal images taken through the central sulcus. Data were acquired using a RARE sequence, with an oval coil (3 × 2 cm, not shown) positioned unilaterally over the central sulcus (FOV 5.0 cm isotropic, 195 μm in-plane, 1 mm slice thickness). Layer IV appears darker than the surrounding gray matter in these T2 images, indicating a higher density of white matter. B. An image from a post-mortem monkey brain imaged with small oval coil (coil B, Figure 8) taken through the central sulcus region. Note the prominent layer IV and the microelectrode track penetrating through all layers and white matter (arrows). Data were acquired using MSME (FOV 3.0 × 2.5 × 2.5 cm; matrix 300 × 256 × 256; isotropic resolution 100 μm). Four echoes were collected (15 ms, 30 ms, 45 ms, 60 ms), data shown are from TE = 15 ms. The scan lasted 40 hours.

and could ultimately reduce the need for euthanasia and increase the lifespan of research monkey subjects.

Functional MRI at 9.4 T

Electrophysiological maps in SM reveal discrete representations of the distal fingertips in areas 3b and 1, separated by ~2–3 mm in the anterior-posterior axis (Sur et al., 1982). Vibrotactile stimulation to the second digit tip was, therefore, predicted to evoke BOLD signal increases in areas 3b and 1, paralleling these maps. In the example shown in Figure 11, activation occurred in an anterior focus, putative area 3b, and at a region located ~2 mm posterior, putative area 1. Anatomical and electrophysiological considerations from other species suggest that the more posterior activation may also encompass part of area 2 (Pons et al., 1985). The distance between the fMRI activation foci is similar to the area 3b to 1 distance obtained from electrophysiology maps (Figure 11A, top). The

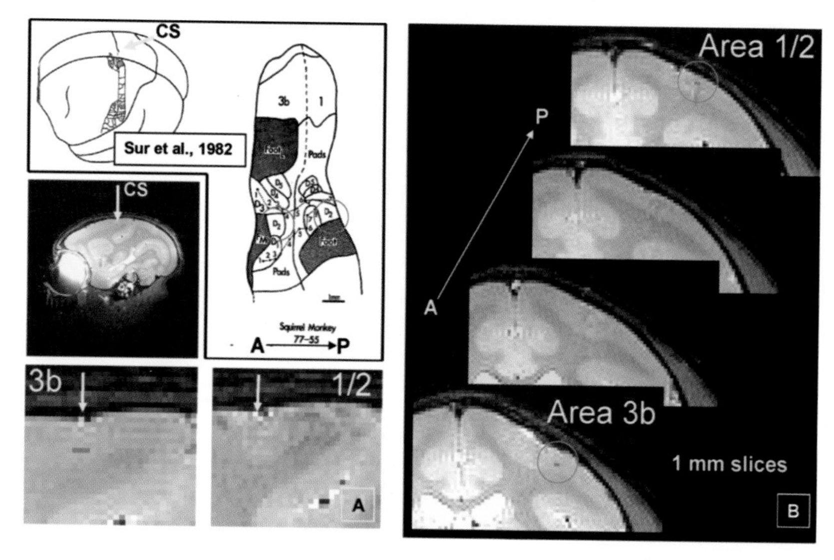

Figure 11 Functional imaging, primary somatosensory cortex A. *Top diagrams* Schematic figures from Sur et al., (1982) depicting areas 3b and 1 in the SM. *Top inset* A sagittal slice through a 3-D rendering showing the position of the central sulcus (yellow arrow). *Below* Functional activity overlaid on sagittal images showing activation of putative areas 3b and 1/2 (p3b and p1/2). B. p3b and p1/2 activation superimposed on coronal images. A characteristic thickening of the cortical mantle is localized to the p3b activation (see also Figure 10). The distance between the p3b and p1/2 activation regions is between 2 and 3 mm, as predicted by electrophysiology maps (A, top).

ability to distinguish between these regions in the SM requires higher spatial resolution than is typically employed in human imaging at lower field strengths (e.g., 3 mm voxels: Moore et al., 2000; Nelson et al., 2004). In the medio-lateral direction, the location of each BOLD focus shows good correspondence to the location of the 2nd digit receptive field maps in areas 3b and 1, and the thickening of the central sulcus in Figure 11B (bottom) again provides an anatomical correlate of the Rolandic region.

Time courses for the putative area 3b and 1/2 activation clusters are shown in Figure 12. The figure on the left reveals the full time course over the eight stimulation periods (shaded in gray) for 3b and 1/2. The response patterns are similar for the two different clusters, including within-epoch parallels (e.g., see the 4th stimulation epoch). A decline of stimulus-evoked response towards the end of the EPI run was also observed, an effect commonly observed in human fMRI (CIM and AJN, unpublished observations). The 'on-off' cycle averaged across all 8 epochs is shown on the right and reveals a similar response in both regions.

The lateral sulcus of New World monkeys has several distinct somatosensory regions (Krubitzer et al., 1995). An example of SM activation in this region is shown in Figure 13 with recent data from the New World Titi monkey (top right: Coq et al., 2004). From the fMRI statistical map, three distinct activation foci were observed. In accordance with the

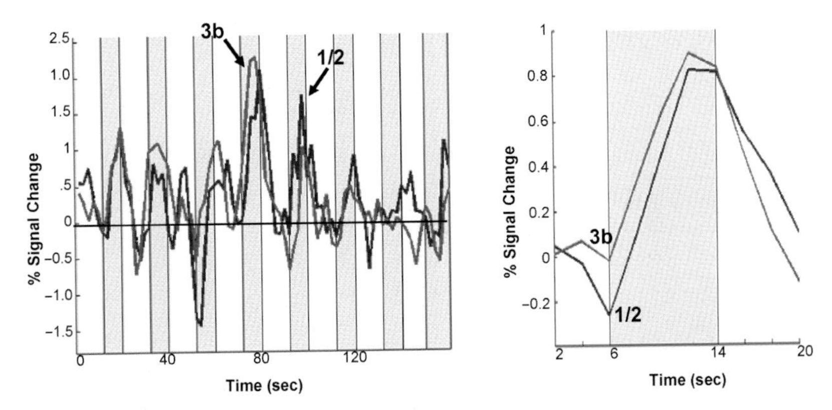

Figure 12 BOLD signal time courses for activation in putative areas 3b and 1/2 *Left* Time series from p3b and p1/2 averaged over 10 EPI runs. *Right* The average 'on/off' stimulus cycle (averaged from the full time series on the left). The vibrotactile stimulus 'on' period is indicated with gray background.

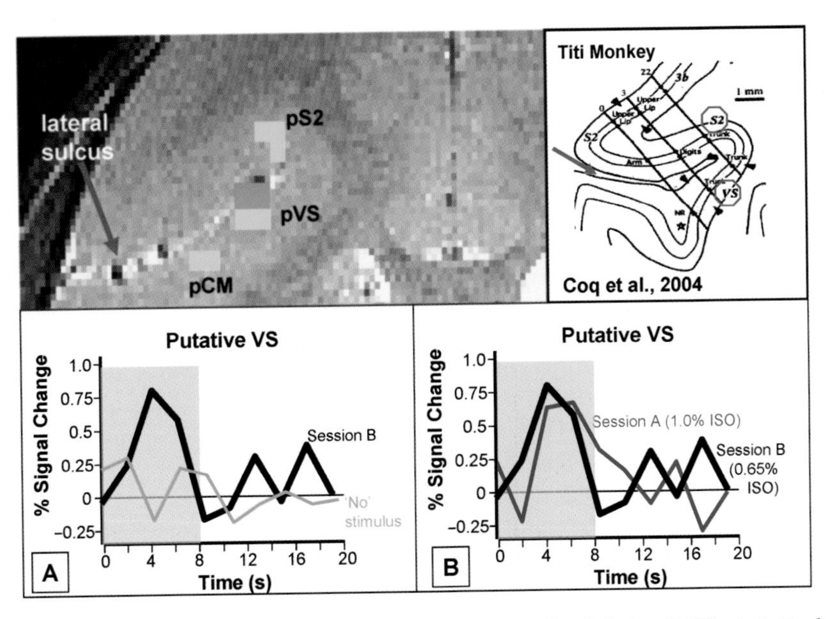

Figure 13 Functional imaging, lateral sulcus A. *Top left* An fMRI statistical map of three distinct activation foci in the lateral sulcus, the putative ventral somatosensory (pVS), second somatosensory (pS2), caudo-medial (pCM) areas. *Top right* A schematic from Coq et al., (2004) showing the position of tactile receptive fields and cortical areas in the lateral sulcus of New World Titi monkeys. *Bottom left* Average BOLD time courses for pVS voxels during tactile stimulation and also 'no' stimulation. B. An average BOLD time course for two sessions with different levels of isoflurane anesthesia. Session A (1.0% isoflurane) reveals a slower onset and an initial negativity in response to the vibrotactile stimulus than Session B (0.65%), differences that may reflect anesthetic concentration differences. The vibrotactile stimulus 'on' period is indicated with gray background.

electrophysiology map, two loci have been labeled the putative ventral somatosensory (pVS) and secondary somatosensory (pS2) regions. A third activation focus, located on the inferior bank of the lateral sulcus is labeled the putative caudo-medial region (pCM). This area may be the homologue of macaque area CM, a region that is responsive to both auditory and tactile stimulation (Schroeder et al., 2001).

While preliminary, our data also suggest that higher field imaging allows observation of activation at deeper anesthetic levels than at lower fields. Previous studies of lateral sulcus somatosensory regions in the human and macaque at 1.5 T reported that 0.8% isoflurane anesthesia suppressed all significant BOLD activation, even when using a GRE sequence (Disbrow et al., 2000). We have, however, consistently observed activation at higher isoflurane levels using SE imaging and the paradigms described above. The percent signal change at 0.65% and 1% were approximately equivalent, though the lower anesthetic level appeared to have a faster onset time.

Conclusion

Squirrel monkey imaging at 9.4 T is a promising technique for noninvasive studies of the primate brain, and the anatomical and functional resolution obtained with this approach is complimentary to electrophysiological and optical techniques. We emphasize in closing that the SM model presented is applicable to studies beyond the tactile-related examples described. This model is also ideal for performing longitudinal studies in lesioned or pathological states. Future directions in our research include parallel sensory mapping in other modalities—we are currently conducting visual fMRI studies with a projection beam focused within the bore, and have obtained preliminary data demonstrating significant activation of multiple visual cortical areas. Another advantage of using a smaller animal in a higher field is the relatively greater resolution obtained in subcortical structures, making this preparation potentially ideal for studies of subcortical anatomy and functional organization of structures such as the thalamus and basal ganglia.

Another important future direction is the enlistment of contrast agents. While there is a clear advantage of using BOLD—because an identical measure can be obtained in humans—animal models permit the use of contrast agents that enhance functional signals and, potentially, provide a closer match to the electrophysiological signals of interest. To this end, we are beginning to scan with Dextran-coated Monocrystalline iron oxide nanoparticle (MION). This contrast agent has been used in repeat monkey and rat imaging studies where enhanced contrast to noise has been observed (Leite et al., 2002; Vanduffel et al., 2001), even at the high field strengths employed here (Mandeville et al., 2004).

Acknowledgments

The authors would like to thank Limin Chen, Randolph Nudo, Christopher Wiggins, Graham Wiggins, Lawrence Wald, Vitaly Napadow, Jason Ritt, Mark Andermann, Christina Triantafyllou and Joseph Mandeville.

References

Abee CR (2000) Squirrel monkey (Saimiri spp) research and resources. ILAR J 41:2–9.
Akbarian S, Grusser OJ, Guldin WO (1992) Thalamic connections of the vestibular cortical fields in the squirrel monkey (Saimiri sciureus). J Comp Neurol 326:423–41.
Ances BM, Zarahn E, Greenberg JH, Detre JA (2000) Coupling of neural activation to blood flow in the somatosensory cortex of rats is time-intensity separable, but not linear. J Cereb Blood Flow Metab 20:921–30.
Aparicio P, Moore CI (2005) Neural processing streams in tactile apparent motion. Society for Neuroscience abstracts, 2005.
Arthurs OJ, Boniface SJ (2002) How well do we understand the neural origins of the fMRI BOLD signal? Trends Neurosci 25(1):27–31.
Arthurs OJ, Williams EJ, Carpenter TA, Pickard JD, Boniface SJ (2000) Linear coupling between functional magnetic resonance imaging and evoked potential amplitude in human somatosensory cortex. Neuroscience 101:803–6.
Backes WH, Mess WH, van Kranen-Mastenbroek V, Reulen JP (2000) Somatosensory cortex responses to median nerve stimulation: fMRI effects of current amplitude and selective attention. Clin Neurophysiol 111:1738–44.
Benjamin RM, Welker WI (1957) Somatic receiving areas of cerebral cortex of squirrel monkey (Saimiri sciureus). J Neurophysiol 20:286–299.
Blake DT, Byl NN, Cheung S, Bedenbaugh P, Nagarajan S, Lamb M, Merzenich M (2002) Sensory representation abnormalities that parallel focal hand dystonia in a primate model. Somatosens Mot Res 19:347–57.
Bottomley PA, Andrew ER (1978) RF magnetic field penetration, phase shift and power dissipation in biological tissue: implications for NMR imaging. Phys Med Biol 23:630–43.
Boyce S, Rupniak NM, Steventon MJ, Iversen SD (1990) Differential effects of D1 and D2 agonists in MPTP-treated primates: functional implications for Parkinson's disease. Neurology 40:927–33.
Brady AG (2000) Research techniques for the squirrel monkey (Saimiri sp). ILAR J 41:10–8.
Byrd LD, Gonzalez FA (1981) Time-course effects of adrenergic and cholinergic antagonists on systemic arterial blood pressure, heart rate and temperature in conscious squirrel monkeys. J Med Primatol 10:81–92.
Chen LM, Friedman RM, Roe AW (2003) Optical imaging of a tactile illusion in area 3b of the primary somatosensory cortex. Science 302:881–5.
Cheung SW, Bedenbaugh PH, Nagarajan SS, Schreiner CE (2001) Functional organization of squirrel monkey primary auditory cortex: responses to pure tones. J Neurophysiol 85(4):1732–49.
Churchill JD, Arnold LL, Garraghty PE (2001) Somatotopic reorganization in the brainstem and thalamus following peripheral nerve injury in adult primates. Brain Res 910:142–52.

Coq JO, Qi H, Collins CE, Kaas JH (2004) Anatomical and functional organization of somatosensory areas of the lateral fissure of the New World titi monkey (Callicebus moloch). J Comp Neurol 476:363–87.

Cramer SC, Moore CI, Finklestein SP, Rosen BR (2000) A pilot study of somatotopic mapping after cortical infarct. Stroke 31(3):668–71.

Disbrow EA, Slutsky DA, Roberts TP, Krubitzer LA (2000) Functional MRI at 1.5 tesla: a comparison of the blood oxygenation level-dependent signal and electrophysiology. Proc Natl Acad Sci 97:9718–23.

Duong TQ, Kim DS, Ugurbil K, Kim SG (2001) Localized cerebral blood flow response at submillimeter columnar resolution. Proc Natl Acad Sci 98:10904–9.

Duong TQ, Kim DS, Ugurbil K, Kim SG (2000) Spatiotemporal dynamics of the BOLD fMRI signals: toward mapping submillimeter cortical columns using the early negative response. Magn Reson Med 44:231–42.

Emmers R, Akert K (1963) A Stereotaxic Atlas of the Brain of the Squirrel Monkey (Saimiri Sciureus). Madison, WI: The University of Wisconsin Press.

Fish RE (1997) Pharmacology of Injectable Anesthetics. In: Anesthesia and Analgesia of laboratory Animals. (Kohn DF, Wixson SK, White WJ, Benson GJ Ed), pp. 2–28. New York: Academic press.

Flaherty AW, Graybiel AM (1991) Corticostriatal transformations in the primate somatosensory system. Projections from physiologically mapped body-part representations. J Neurophysiol 66(4):1249–63.

Flaherty AW, Graybiel AM (1993) Two input systems for body representations in the primate striatal matrix: experimental evidence in the squirrel monkey. J Neurosci 13(3):1120–37.

Flaherty AW, Graybiel AM (1994) Input-output organization of the sensorimotor striatum in the squirrel monkey. J Neurosci 14(2):599–610.

Flaherty AW, Graybiel AM (1995) Motor and somatosensory corticostriatal projection magnifications in the squirrel monkey. J Neurophysiol 74(6):2638–48.

Frost SB, Barbay S, Friel KM, Plautz EJ, Nudo RJ (2003) Reorganization of remote cortical regions after ischemic brain injury: a potential substrate for stroke recovery. J Neurophysiol 89:3205–14.

Gati JS, Menon RS, Ugurbil K, Rutt BK (1997) Experimental determination of the BOLD field strength dependence in vessels and tissue. Magn Reson Med 38(2):296–302.

Garraghty PE, Kaas JH (1991) Large-scale functional reorganization in adult monkey cortex after peripheral nerve injury. Proc Natl Acad Sci 88:6976–80.

Gergen JA, MacLean PD (1962) A Stereotaxic Atlas of the Squirrel Monkey's Brain: Public Health Service Publication No 933. Bethesda, MD: NIH.

Greenstein ET (1975) Ketamine HCl, a dissociative anesthetic for squirrel monkeys (Saimiri sciureus). Lab Anim Sci 25:774–7.

Gross CG (1998) Brain, Vision, Memory: Tales in the History of Neuroscience. Cambridge, MA: The MIT Press.

Gruetter R, Tkac I (2000) Field mapping without reference scan using asymmetric echo-planar techniques. Magn Reson Med 43:319–23.

Guldin WO, Akbarian S, Grusser OJ (1992) Cortico-cortical connections and cytoarchitectonics of the primate vestibular cortex: a study in squirrel monkeys (Saimiri sciureus). J Comp Neurol 326:375–401.

Ham CL, Engels JM, van de Wiel GT, Machielsen A (1997) Peripheral nerve stimulation during MRI: effects of high gradient amplitudes and switching rates. J Magn Reson Imaging 7:933–7.

Heeger DJ, Huk AC, Geisler WS, Albrecht DG (2000) Spikes versus BOLD: what does neuroimaging tell us about neuronal activity? Nat Neurosci 3: 631–3.

Ishizawa Y, Nelson AJ, Cheney C, Brown E, Moore CI (2005) Effects of isoflurane on somatosensory stimuli-induced brain activation in monkeys: a fMRI study. Annual meeting of American Society of Anesthesiologists, New Orleans.

Jain N, Qi HX, Catania KC, Kaas JH (2001) Anatomic correlates of the face and oral cavity representations in the somatosensory cortical area 3b of monkeys. J Comp Neurol 429:455–68.

Kim DS, Duong TQ, Kim SG (2000) High-resolution mapping of iso-orientation columns by fMRI. Nat Neurosci 3:164–9.

Kim SG, Duong TQ (2002) Mapping cortical columnar structures using fMRI. Physiol Behav 77:641–4.

Kirman JH (1974) Tactile apparent movement: the effects of number of stimulators. J Exp Psychol 103:1175–80.

Krubitzer L, Clarey J, Tweedale R, Elston G, Calford M (1995) A redefinition of somatosensory areas in the lateral sulcus of macaque monkeys. J Neurosci 15:3821–39.

Lee SP, Silva AC, Ugurbil K, Kim SG (1999) Diffusion-weighted spin-echo fMRI at 9.4 T: microvascular/tissue contribution to BOLD signal changes. Magn Reson Med 42:919–28.

Leite FP, Tsao D, Vanduffel W, Fize D, Sasaki Y, Wald LL, Dale AM, Kwong KK, Orban GA, Rosen BR, Tootell RB, Mandeville JB (2002) Repeated fMRI using iron oxide contrast agent in awake, behaving macaques at 3 Tesla. Neuroimage 16:283–94.

Leopold DA, Plettenberg HK, Logothetis NK (2002) Visual processing in the ketamine-anesthetized monkey. Optokinetic and blood oxygenation level-dependent responses. Exp Brain Res 143:359–72.

Liu ZM, Schmidt KF, Sicard KM, Duong TQ (2004) Imaging oxygen consumption in forepaw somatosensory stimulation in rats under isoflurane anesthesia. Magn Reson Med 52:277–85.

Livingstone MS (1996) Oscillatory firing and interneuronal correlations in squirrel monkey striate cortex. J Neurophysiol 75:2467–85.

Locascio JJ, Jennings PJ, Moore CI, Corkin S (1997) Time series analysis in the time domain and resampling methods for studies of functional magnetic resonance brain imaging. Hum Br Map 5:168–193.

Logothetis NK, Guggenberger H, Peled S, Pauls J (1999) Functional imaging of the monkey brain. Nat Neurosci 2:491–2.

Logothetis NK, Pauls J, Augath M, Trinath T, Oeltermann A (2001) Neurophysiological investigation of the basis of the fMRI signal. Nature 412:150–7.

Logothetis NK, Wandell BA (2004) Interpreting the BOLD signal. Annu Rev Physiol. 66:735–69.

Logothetis NK, Pfeuffer J (2004) On the nature of the BOLD fMRI contrast mechanism. Magn Reson Imaging 22(10):1517–31.

Mandeville JB, Jenkins BG, Chen YC, Choi JK, Kim YR, Belen D, Liu C, Kosofsky BE, Marota JJ (2004) Exogenous contrast agent improves sensitivity of gradient-echo functional magnetic resonance imaging at 9.4 T. Magn Reson Med 52:1272–81.

Menon RS, Ogawa S, Strupp JP, Ugurbil K (1997) Ocular dominance in human V1 demonstrated by functional magnetic resonance imaging. J Neurophysiol 77:2780–7.

Merzenich MM, Nelson RJ, Kaas JH, Stryker MP, Jenkins WM, Zook JM, Cynader MS, Schoppmann A (1987) Variability in hand surface representations in areas 3b and 1 in adult owl and squirrel monkeys. J Comp Neurol 258:281–96.

Merzenich MM, Jenkins WM (1993) Reorganization of cortical representations of the hand following alterations of skin inputs induced by nerve injury, skin island transfers, and experience. J Hand Ther 6:89–104.

Moore CI, Stern CE, Corkin S, Fischl B, Gray AC, Rosen BR, Dale AM (2000) Segregation of somatosensory activation in the human rolandic cortex using fMRI. J Neurophysiol 84(1):558–69.

Moore CI, Stern CE, Dunbar C, Kostyk SK, Gehi A, Corkin S (2000) Referred phantom sensations and cortical reorganization after spinal cord injury in humans. Proc Natl Acad Sci USA 19, 97(26):14703–8.

Nair G, Duong TQ (2004) Echo-planar BOLD fMRI of mice on a narrow-bore 9.4 T magnet. Magn Reson Med 52:430–4.

Nelson AJ, Staines WR, Graham SJ, McIlroy WE (2004) Activation in SI and SII: the influence of vibrotactile amplitude during passive and task-relevant stimulation. Brain Res Cogn Brain Res 19(2):174–84.

Nelson AJ, Cheney CA, Chen I, Dai G, Grindlay G, Kempadoo K, Ramanathan A, Moore CI (2005) Tactile activation of putative area MT+ in the squirrel monkey reveal with 9.4 T fMRI. Society for Neuroscience abstracts, 2005.

Nudo RJ, Larson D, Plautz EJ, Friel KM, Barbay S, Frost SB (2003) A squirrel monkey model of poststroke motor recovery. ILAR J 44:161–74.

Ogawa S, Lee TM, Kay AR, Tank DW (1990) Brain magnetic resonance imaging with contrast dependent on blood oxygenation. Proc Natl Acad Sci 87:9868–72.

Pinneo LR (1968) Brain Mechanisms in the Behavior of the Squirrel Monkey. In: The Squirrel Monkey (Rosenblum LA, Cooper RW, eds.), pp 319–346. New York: Academic Press.

Plautz EJ, Barbay S, Frost SB, Friel KM, Dancause N, Zoubina EV, Stowe AM, Quaney BM, Nudo RJ (2003) Post-infarct cortical plasticity and behavioral recovery using concurrent cortical stimulation and rehabilitative training: a feasibility study in primates, Neurol Res 25:801–10.

Pons TP, Garraghty PE, Cusick CG, Kaas JH (1985) The somatotopic organization of area 2 in macaque monkeys. J Comp Neurol 241:445–66.

Rees G, Friston K, Koch C (2000) A direct quantitative relationship between the functional properties of human and macaque V5. Nat Neurosci 3:716–23.

Reinstrup P, Ryding E, Algotsson L, Messeter K, Asgeirsson B, Uski T (1995) Distribution of cerebral blood flow during anesthesia with isoflurane or halothane in humans. Anesthesiology 82:359–66.

Reves JG, Glass PSA (1990) Nonbarbiturate Intravenous Anesthetics. In: Anesthesia 3rd Ed. (Ronald Miller, Ed), pp 243–280. New York, Churchill Livingstone.

Rupniak NM, Boyce S, Steventon MJ, Iversen SD, Marsden CD (1992) Dystonia induced by combined treatment with L-dopa and MK-801 in parkinsonian monkeys. Ann Neurol 32:103–5.

Schaefer DJ, Bourland JD, Nyenhuis JA (2000) Review of patient safety in time-varying gradient fields. J Magn Reson Imaging 12:20–9.

Schenck JF (2000) Safety of strong, static magnetic fields. J Magn Reson Imaging 12:2–19.

Schroeder CE, Lindsley RW, Specht C, Marcovici A, Smiley JF, Javitt DC (2001) Somatosensory input to auditory association cortex in the macaque monkey. J Neurophysiol 85:1322–7.

Sicard KM , Duong TQ (2005) Effects of hypoxia, hyperoxia, and hypercapnia on baseline and stimulus-evoked BOLD, CBF, and CMRO2 in spontaneously breathing animals. Neuroimage 25:850–8.

Smirnakis SM, Brewer AA, Schmid MC, Tolias AS, Schuz A, Augath M, Inhoffen W, Wandell BA, Logothetis NK (2005) Lack of long-term cortical reorganization after macaque retinal lesions, Nature 435:300–7.

Staines WR, Black SE, Graham SJ, McIlroy WE (2002) Somatosensory gating and recovery from stroke involving the thalamus, Stroke 33:2642–51.

Sur M, Nelson RJ, Kaas JH (1982) Representations of the body surface in cortical areas 3b and 1 of squirrel monkeys: comparisons with other primates. J Comp Neurol 211:177–92.

Sur M, Wall JT, Kaas JH (1984) Modular distribution of neurons with slowly adapting and rapidly adapting responses in area 3b of somatosensory cortex in monkeys. J Neurophysiol 51:724–44.

Thomsen K, Offenhauser N, Lauritzen M (2004) Principal neuron spiking: neither necessary nor sufficient for cerebral blood flow in rat cerebellum, J Physiol 560:181–9.

Tommerdahl M, Favorov O, Whitsel BL (2002) Optical imaging of intrinsic signals in somatosensory cortex. Behav Brain Res 135:83–91.

Vanduffel W, Fize D, Peuskens H, Denys K, Sunaert S, Todd JT, Orban GA (2002) Extracting 3D from motion: differences in human and monkey intraparietal cortex. Science 298:413–5.

Vanduffel W, Fize D, Mandeville JB, Nelissen K, Van Hecke P, Rosen BR, Tootell RB, Orban GA (2001) Visual motion processing investigated using contrast agent-enhanced fMRI in awake behaving monkeys. Neuron 32:565–77.

Warltier DC, Pagel PS (1992) Cardiovascular and respiratory actions of desflurane: is desflurane different from isoflurane? Anesth Analg 75(4 Suppl):S17–29; S29–31.

Welker WI, Benjamin RM, Miles RC, Woolsey CN (1957) Motor effects of stimulation of cerebral cortex of squirrel monkey (Saimiri sciureus). J Neurophysiol 20:347–364.

Xerri C, Coq JO, Merzenich MM, Jenkins WM (1996) Experience-induced plasticity of cutaneous maps in the primary somatosensory cortex of adult monkeys and rats. J Physiol Paris 90:277–87.

Yacoub E, Van De Moortele PF, Shmuel A, Ugurbil K (2005) Signal and noise characteristics of Hahn SE and GE BOLD fMRI at 7 T in humans. Neuroimage 24:738–50.

Yacoub E, Duong TQ, Van De Moortele PF, Lindquist M, Adriany G, Kim SG, Ugurbil K, Hu X (2003) Spin-echo fMRI in humans using high spatial resolutions and high magnetic fields. Magn Reson Med 49:655–64.

Zhao F Wang P, Kim SG (2004) Cortical depth-dependent gradient-echo and spin-echo BOLD fMRI at 9.4 T. Magn Reson Med 51:518–24.

Index